Comparative Karyology of Primates

World Anthropology

General Editor

SOL TAX

Patrons

CLAUDE LÉVI-STRAUSS
MARGARET MEAD†
LAILA SHUKRY EL HAMAMSY
M. N. SRINIVAS

MOUTON PUBLISHERS · THE HAGUE · PARIS · NEW YORK

Comparative Karyology of Primates

Editors

B. CHIARELLI
ANN L. KOEN
G. ARDITO

MOUTON PUBLISHERS · THE HAGUE · PARIS · NEW YORK

ISBN 90-279-7850-6 (Mouton)
0-202-90065-7 (AVC Inc.)
Indexes by Society of Indexers, Great Britain
Jacket photo by permission of ABC Press
Cover and jacket design by Jurriaan Schrofer
Phototypeset in V.I.P. Times by
Western Printing Services Ltd, Bristol
Printed in Great Britain

General Editor's Preface

The excellent bibliography included in this book shows that there were no published studies of the chromosomes of nonhuman primates before 1922, only five before 1955, and fifteen from 1955 through 1959. The field burgeoned in the 1960's; but karyology is still a term as strange to most anthropologists as the information in this book is to medical research, where attention remains exclusively devoted to our own species. The present volume is in fact, therefore, a pioneer in a specialty which hardly existed ten years before the congress for which it was organized, a congress which encouraged anthropologists to bring together scientists of neighboring disciplines at every margin of anthropology. It was part of a movement to give recognition to new problems and new points of view accompanying the increasing heterogeniety of the international community itself.

Like most contemporary sciences, anthropology is a product of the European tradition. Some argue that it is a product of colonialism, with one small and self-interested part of the species dominating the study of the whole. If we are to understand the species, our science needs substantial input from scholars who represent a variety of the world's cultures. It was a deliberate purpose of the IXth International Congress of Anthropological and Ethnological Sciences to provide impetus in this direction. The *World Anthropology* volumes, therefore, offer a first glimpse of a human science in which members from all societies have played an active role. Each of the books is designed to be self-contained; each is an attempt to update its particular sector of scientific knowlege and is written by specialists from all parts of the world. Each volume should be read and reviewed individually as a separate volume on its own given subject. The set as a whole will indicate what changes are in store for anthropology as scholars from the developing countries join in studying the species of which we are all a part.

The IXth Congress was planned from the beginning not only to include as many of the scholars from every part of the world as possible, but also with a view toward the eventual publication of the papers in high-quality volumes. At previous Congresses scholars were invited to bring papers which were then read out loud. They were necessarily limited in length; many were only summarized; there was little time for discussion; and the sparse discussion could only be in one language. The IXth Congress was an experiment aimed at changing this. Papers were written with the intention of exchanging them before the Congress, particularly in extensive pre-Congress sessions; they were not intended to be read aloud at the Congress, that time being devoted to discussions—discussions which were simultaneously and professionally translated into five languages. The method for eliciting the papers was structured to make as representative a sample as was allowable when scholarly creativity—hence self-selection—was critically important. Scholars were asked both to propose papers of their own to suggest topics for sessions of the Congress which they might edit into volumes. All were then informed of the suggestions and encouraged to re-think their own papers and the topics. The process, therefore, was a continuous one of feedback and exchange and it has continued to be so even after the Congress. The some two thousand papers comprising *World Anthropology* certainly then offer a substantial sample of world anthropology. It has been said that anthropology is at a turning point; if this is so, these volumes will be the historical direction-markers.

As might have been foreseen in the first post-colonial generation, the large majority of the Congress papers (82 percent) are the work of scholars identified with the industrialized world which fathered our traditional discipline and the institution of the Congress itself: Eastern Europe (15 percent); Western Europe (16 percent); North America (47 percent); Japan, South Africa, Australia, and New Zealand (4 percent). Only 18 percent of the papers are from developing areas: Africa (4 percent); Asia-Oceania (9 percent); Latin America (5 percent). Aside from the substantial representation from the U.S.S.R. and the nations of Eastern Europe, a significant difference between this corpus of written material and that of other Congresses is the addition of the large proportion of contributions from Africa, Asia, and Latin America. "Only 18 percent" is two to four times as great a proportion as that of other Congresses; moreover, 18 percent of 2,000 papers is 360 papers, 10 times the number of "Third World" papers presented at previous Congresses. In fact, these 360 papers are more than the total of *all* papers published after the last International Congress of Anthropological and Ethnological Sciences which was held in the United States (Philadelphia, 1956).

The significance of the increase is not simply quantitative. The input of scholars from areas which have until recently been no more than subject matter for anthropology represents both feedback and also long-awaited theoretical contributions from the perspectives of very different cultural, social, and historical traditions. Many who attended the IXth Congress were convinced that anthropology would not be the same in the future. The fact that the Xth Congress (India, 1978) was our first in the "Third World" may be symbolic of the change. Meanwhile, sober consideration of the present set of books will show how much, and just where and how, our discipline is being revolutionized.

Readers of this book will be interested in the others in this series dealing with primates and adaptation, human biology and evolution, and physical anthropology as it is developing in every part of the world.

Chicago, Illinois Sol Tax
April 18, 1979

Preface

Mapping of the progress of primate chromosomal evolution, from the lowest up to and including man, has been brought closer to a definitive science by the introduction of chromosome-banding techniques. These have been used extensively in the studies reported here. Delineation of the course of evolution has very direct applications to the interpretation of human chromosome patterns, both normal and abnormal, as some of the papers describe. Humans are primates, and as such they take their place on the evolutionary tree. It remains to be seen whether they are indeed the topmost branch, or if one or more of the chromosomal aberrations which appear in the human line may signal the first growth of a new and higher twig on the tree. Perhaps patterns discerned from past evolutionary events will help us to forecast what the future holds in store.

This book is the result of a Symposium and Workshop on Primate Karyology held at the C. S. Mott Center for Human Growth and Development, Wayne State University School of Medicine, on August 29–31, 1973. It was supported in part by the National Foundation for the March of Dimes and by the C. S. Mott Center for Human Growth and Development. To them we extend our deep appreciation for making the meetings possible.

We also thank Mrs. Shirley Leach and Mr. Kyu Bahng for their untiring help in making a success of the Symposium–Workshop, Dr. G. Ardito and Dr. E. Rabino Massa for their tutorial assistance in the workshop, and Mrs. S. Albright and Miss C. Bullo for their secretarial help. The enthusiastic interest and support of Dr. Gabriel W. Lasker, Department of Anatomy, Dr. Tommy N. Evans, chairman of the Department of Gynecology–Obstetrics, Wayne State University School of Medicine, is also gratefully acknowledged.

In an attempt to facilitate comparison and exact references, a tentative Atlas of conventionally stained karyotypes of the different species has been prepared by G. Ardito and included as section six of this volume.

December, 1973

B. Chiarelli
Ann L. Koen
G. Ardito

Table of Contents

SECTION TWO: RECENT STUDIES ON PRIMATE CHROMOSOMES

SECTION THREE: CHROMOSOME STUDIES IN HUMANS

APPENDICES

Introduction

B. CHIARELLI

Dear colleagues!

With great pleasure and interest I am opening this conference and workshop which I do not take credit for having organized.

All credit for your being here is owed to Ann Koen and to the staff of this laboratory.

I feel responsible for all the things which will not materialize properly during the workshop. In fact I took the responsibility of cooperating with her in organizing this workshop almost one year ago; however I started working on this project only yesterday when all was already properly organized.

To Ann Koen and to the School of Medicine of Wayne State University, therefore, we must convey our warmest thanks for the organization of this primate chromosome conference. Over the last fifteen years there has been an outburst of perspectives for the study of human chromosomes and those of the vertebrates in general.

The importance of chromosomes in taxonomic and evolutionary study was no doubt the stimulus for most of the investigations undertaken on primates, in order to assess phylogenetic relationships among the various species of this group and between them and our own species.

However, technological progress leaves but little time for meditation on the obtained results. Many workers, who have been taken up, like me, for instance, by teaching and administrative duties during the last couple of years, are aware of the ineluctable rush of events in the field of karyology and we are having difficulties in keeping up to date with the everyday increase in information.

These remarks were delivered at the opening of the Symposium and Workshop on Primate Karyology at the C. S. Mott Center for Human Growth and Development, Detroit, Michigan, on August 27, 1973.

We are in fact the beholders of a new era of karyology: the band or chromosome-banding era.

No need to envision it historically: we do live in it. New techniques are published every day and now the competition has shifted from those who are the first to "invent" a new "zebratin" method to those who succeed in securing the priority of publication of their discovery in some wide-ranging journals or even daily newspapers. Generally, the most successful in this race are the English-speaking people. The story of the publication priority for chromosome banding by trypsin is enlightening in this respect.

I do not want to go into detail on the difficulties we had in securing in an English journal the publication of our results of the treatment of the chromosomes with trypsin already obtained in the spring of 1971!

I want, however, to take this occasion to explain why I delayed publishing results in the use of banding techniques in the comparison of the chromosomes of the different primates.

The main reason is simply a matter of caution. In fact, before jumping to phylogenetic conclusions on the basis of the results yielded by the banding technique, I feel we must gain a clear insight into their actual meaning. Mine is chronic ignorance, however; a really logical explanation of the banding phenomenon is still beyond my reach.

The hypothesis of DNA denaturation and renaturation seems no longer tenable, neither is the hypothesis that these bands are sites particularly rich in G-C or A-T nucleotides.

The results obtained in our laboratory by means of enzymatic digestion suggest rather the denaturation of the structural protein frame of the chromosome. Actually, trypsin acts first by denaturing the protein substrate and second by digesting it. This denaturation involves changes within the secondary and the tertiary structure of the protein molecule which result in the uncoiling of histones, which are the major chromosome protein components.

The bands, in our opinion, are therefore the result of the alteration occurring in the protein component rather than in the DNA.

It is conceivable that in the same species homologous chromosomes have an identical structure, being composed of the same proteins, thus showing the same banding pattern under the same experimental conditions, possibly showing their chromomeric structure, as we assume. However, no definite conclusion regarding their absolute genetic homology, that is, identical DNA sequences, can be reached with this method even for the chromosomes of the same species. Even less safely can we speculate on any genetical identity between chromosomes from different species.

This does not mean at all that I do not believe in bands, in their repeatability, and in their weight in cytotaxonomic and evolutionary studies. As witnessed by the papers just published in the last issue of the *Journal of Human Evolution* (volume 2, number 4), I do. But I wish to stress the need for greater caution and for a clear insight into the logical process of any method before one resorts to it.

Attempts at interpreting homologies between chromosomes from different species systematically distant based on similarities of bands, recently reported by different authors, are perplexing indeed. Obviously it is conceivable that some chromosomes remained constant during evolution (as is the case for chromosome X), but the concern of those who deal with mechanics of karyological evolution is centered on the differences between karyotypes rather than on homologies. However, differences based on the presence or absence of a single bright spot on a chromosome on the basis of only two or three individuals examined, as has recently been proposed by Pearson for the chimpanzee–gorilla–man Y chromosome, or the expulsion of the gibbons from the primate stock for their lack of bright spots on the chromosomes as has been recently published by Lejeune, are not sufficient! Homologies between chromosomes of related species moreover are most obvious, or at any rate an easy result of *a priori* assumptions based on their phylogenetic relationship.

As regards the comparison between the chromosomes of the apes and those of man, likely to appeal more directly to the human cytogeneticists, a really huge bulk of data is now available. We might even affirm that chromosome banding makes it possible to put Eve's chromosomes together again, reconstructing all the steps that lead up to mankind living today. Dr. Turleau is going to give us a detailed account of this chain of events. However, are we absolutely certain, as we say, that the bands indicate genetic homology? or identical DNA sequences? and on the other hand, are we absolutely certain of the unicity of origin of mankind? and what about the possible variability existing in the karyotype of the different apes? Could not cellular hybridization help us further in our understanding of these homologies? No doubt identification of identical genetical markers between chromosomes from different species is a more reliable test for genetic homologies. Practically two thirds of human chromosomes are nowadays identifiable by at least one genetic marker. Could not this technique be applied to the chromosomes of the different primate species? Dr. Creagan will provide us with some positive information in this respect.

Moreover could it not be possible that chromosomes in the interphase nucleus of the eukaryote are arranged in a circular sequence so that the identification of this sequence could be determinant to identify the possible chromosomal mutation which occurred during evolution.

As you see, a lot of open questions are facing us in these two days of work together! I hope that to me and all of you this workshop will be of help in grasping such insight.

Thanks to the efficiency of Dr. Emma Rabino Massa and Dr. Giuseppe Ardito, a workshop could be added to this conference. I hope all of you will enjoy your stay here and will find the next few days profitable from a scientific point of view.

SECTION ONE

Theory of Chromosome Banding

Thermal and Chemical Denaturation of Nucleohistone Compared with the Reactions of Chromatin

BODIL E. LOMHOLT

Differentiating staining techniques of human chromosomes include a large number of procedures (Nilsson 1973) leading to five major groups of banding patterns, the Q-, G-, R-, C-, and T-bands. When slides are colored by a fluorescent dye without any pretreatment, the Q pattern is revealed, while the remaining patterns arise by various pretreatments of preparations before they are stained. While the Q- and G-bands are similar, the R-bands complement these patterns. Pronouncedly stained centromeric regions characterize the C-bands. The T-bands display staining of the telomeres. In contrast with T- and C-bands, the Q-, G-, and R-bands allow a reliable identification of the 46 human chromosomes.

The heat–Giemsa technique (Lomholt and Mohr 1971) has been worked out as a modification of the C-band method described by Arrighi and Hsu (1971), and it belongs to the G-banding procedures.

Banding patterns obtained from human lymphocytes, fibroblasts, and meiotic cells are identical, indicating that they neither reflect the genetic activity of the cells of a particular tissue nor result from stimulation by phytohemagglutinine. It could be inferred that banding patterns obtained by the differentiating techniques reflect structural changes that are not well understood at present.

In an attempt to bridge the gap between cytological and molecular investigations it appears reasonable to correlate specific reactions of native or salt-extracted nucleohistone with cytological procedures for conventional and differentiating treatments of chromosomes.

Subirana (1973) has compared thermal denaturation profiles of nucleohistones of calf thymus and two echinoderm species under various conditions. The melting profile of either native or salt-extracted nucleohistone approaches that of a bimodal curve. This could indicate that the two parts of the curve represent distinct physical phenomena which take place during denaturation. Subirana anticipates that the first step in the melting profile measures the local denaturation of less-stable regións in the nucleohistone complex, and that the second step measures the stability against denaturation of the most protected parts of the complex. These parts appear to be saturated with histone. The melting temperature of free native double-stranded DNA is much lower than the melting temperature of native DNA in the presence of heat-denatured nucleohistone. DNA is then protected by the histones released from the nucleohistone. Partly denatured nucleohistone does not renature upon cooling; the conformation of the complex is permanently modified. Most of the DNA, however, does renature upon cooling. The conformational modification of nucleohistone obtained after thermal denaturation is consistent with the general behavior of proteins subjected to heat treatment. The basic proteins of the complex are mainly arginine-rich and lysine-rich histones. They both contribute to the stability of nucleohistone conformation, even though the effects of their individual participation differ distinctly. With reference to Subirana, the lysine-rich histones appear to be associated with the coiled or folded nucleohistone complex and act as a zipper, giving additional stability to nucleohistone, but without influence on the tertiary structure of the DNA in the complex. Consistent with other investigators, Mirsky and Silverman (1973) indicate that saline–citrate salt solutions selectively extract lysine-rich histones from nuclei, while mixtures of alcohol and acid remove arginine-rich histones. With the extraction of 25 percent lysine-rich histones, the morphology of the nucleohistone complex changes to a network of diffuse fibrils, whereas more than 50 percent of the arginine-rich histones may be extracted without any change in the morphology of the complex.

If we assume that banding patterns observed by light microscopy reflect, at least in part, the physiochemical reactions at the molecular level, I suggest the following hypothesis.

During harvest, cells are submitted to hypotonic treatment and to mixtures of alcohol and acid. The complex now looses histones, and this does not induce morphological changes. The combined effects of added hydrogen ions and organic solvent, however, accompanied by a decrease in temperature, induce denaturation of DNA. As the transition of DNA involves both reversible and irreversible reactions, it seems reasonable to expect a modification of the nucleohistone complex.

A comparison between scanning curves of heat–Giemsa treated and conventionally stained chromosomes reveals a fair correspondence of profiles. The similarity of curves suggests that the nucleohistone complex, already at the stage of fixation, displays conformational changes. The close resemblance between the scanning results of these patterns and those of Q-banding patterns further suggests that areas of dense fluorescence reflect stable regions.

After heat treatment in saline–citrate solutions, preparations develop G-patterns revealed by Giemsa staining. At these steps, selective loss of lysine-rich histones causes local changes of the complex toward a diffuse network of fibrils. Less stable regions, modified by extraction and morphological changes, are then likely to appear faintly stained, as opposed to intensely colored stable regions.

R-banding patterns that complement both G- and Q-bands may be produced by treatment of slides in basic solutions brought close to the boiling point. Under these conditions less-stable regions may well denature completely. Abundant hydrolysis could then leave these regions with a large surface area and a greatly increased binding capacity which would facilitate an uptake of acid-basic dyes. One could explain the faint staining of these areas by the assumption that stable regions denature to a state comparable to less-stable regions subjected to milder thermal and chemical conditions.

At present the banding patterns are open to a large number of interpretations, so this hypothesis may encourage further research to correlate molecular and cytological investigations.

REFERENCES

ARRIGHI, F. E., T. C. HSU
1971 Localization of heterochromatin in human chromosomes. *Cytogenetics* 10:81–86.

LOMHOLT, BODIL, J. MOHR
1971 Human karyotyping by heat-Giemsa staining and comparison with fluorochrome techniques. *Nature* 234 (47):109–110.

MIRSKY, A. E., B. SILVERMAN
1973 Effects of selective extraction of histones on template activities of chromatin by use of exogeneous DNA and RNA polymerases. *Proceedings of the National Academy of Science* 70(7):1,973–975.

NILSSON, BIRGIT
1973 A bibliography of literature concerning chromosome identification, with special reference to fluorescence Giemsa staining techniques. *Hereditas* 73:259–270.

SUBIRANA, J. A.
1973 Studies on the thermal denaturation of nucleohistones. *Journal of Molecular Biology* 74:363–386.

*Trypsin and Trypsin/Versene Banding on Chromosomes**

G. ARDITO and B. CHIARELLI

After Caspersson's experiments with quinacrine (Caspersson, Farber, et al. 1968; Caspersson, Zech, Modest, et al. 1969; Caspersson, Zech Johansson, et al. 1970) and, shortly thereafter, the first description of the effect of heat denaturation reported by Dutrillaux and Lejeune (1971), the study of chromosome banding was undertaken by testing proteolytic enzymes such as pronase, trypsin, and α-chymotrypsin in order to ascertain the ability, if any, of these enzymes to produce banding.

During the 1971 Paris Conference, the bands obtained by the use of these enzymes were defined as G-bands. Moreover, it was shown that these bands correspond not only to those obtained by other Giemsa techniques but also to those obtained by quinacrine treatment. The correspondence with the last compound consists of the fact that each bright Q-band corresponds to a dark G-band. Conversely, to each weak or negative-fluorescent Q-band, a weakly stained G-band was seen to correspond.

In our laboratory, both trypsin alone and a trypsin solution with versene addition were used. The results yielded by trypsin alone are similar to those reported by other workers (Seabright 1971; Dutrillaux, de Grouchy, et al. 1971; Chiarelli and Sarti-Chiarelli 1972; Chiarelli, Sarti-Chiarelli, et al. 1972). Using this treatment banding is associated with chromosome swelling, so that except in the initial stages, shown in Plate 1, the chromatids become indistinguishable, resulting in the pattern visible in Plate 2.

To explain these facts, a hypothesis on the mechanism of the action of trypsin was tentatively formulated. Two stages were proposed: denaturation followed by digestion (Chiarelli 1973). The postulated steps are (1) the banding process itself and (2) chromosome breakdown.

*A similar paper has in the meantime been published in *Bollettino di Zoologia* 41(2):127–128 (1973).

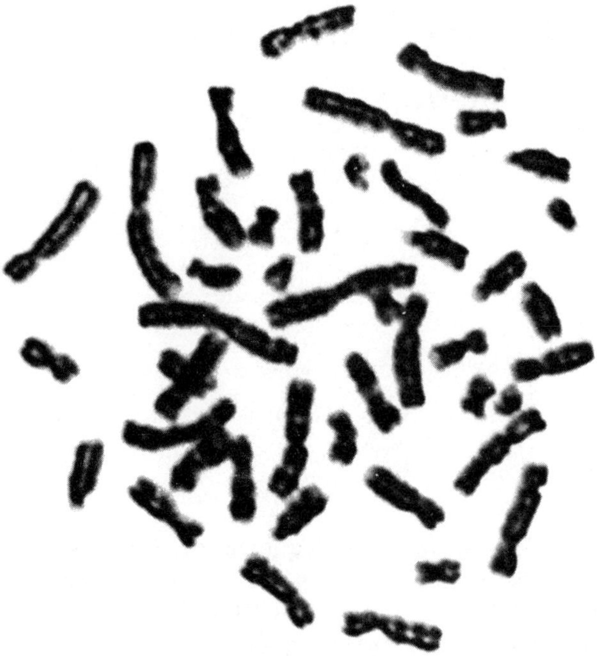

Plate 1. Initial stage of banding of human chromosomes treated with trypsin 0.25 percent

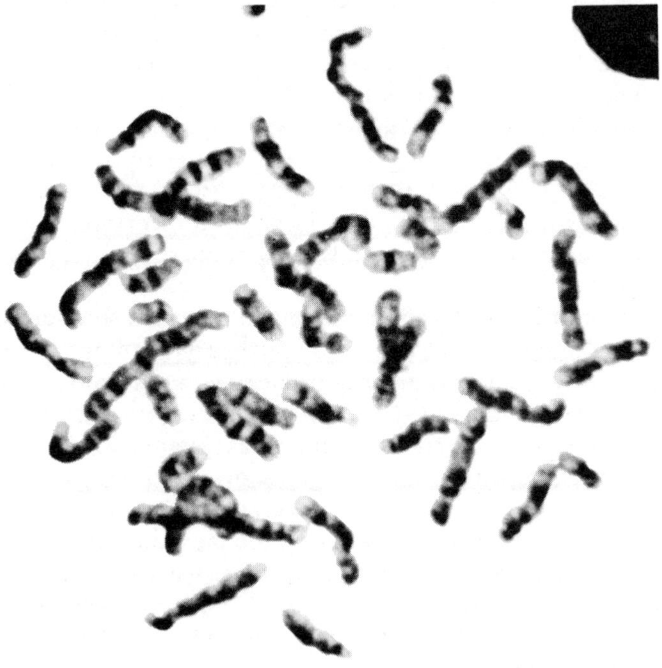

Plate 2. Chromosome swelling: the chromatids are not distinguishable (human metaphase plate treated with trypsin 0.25 percent)

More complex results were produced by the addition of versene to the trypsin solution (Wang and Fedoroff 1972). Versene delays trypsin action, thereby permitting better banding. Unfortunately, some unexpected banding patterns were frequently observed. Very often metaphase plates, instead of actual bands, displayed "foldings" of the chromosome arms, which we called "false bands"; an example is shown in Plate 3. We believe that these images simply indicate the uncoiling of proteins, resulting in "overlaps." In fact, a homology between these folds was not observed in homologous chromosomes, as could be expected. On the contrary, homology is evident in bands obtained by the other methods.

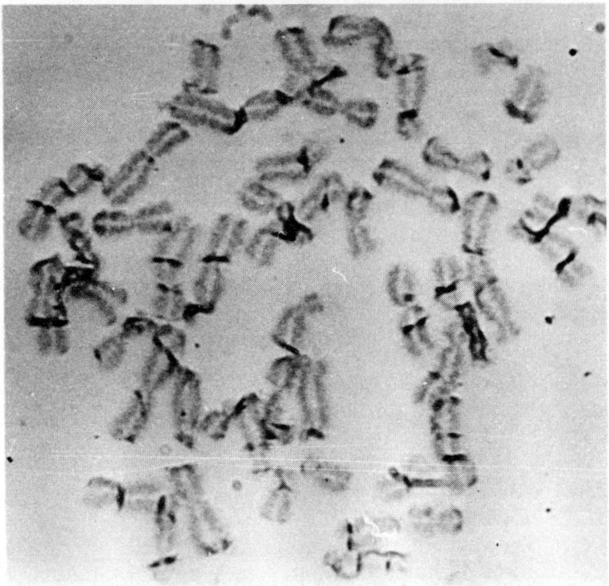

Plate 3. "False bands" (foldings) of chimpanzee chromosomes treated with a trypsin/versene solution (trypsin 0.03 percent; versene 0.02 percent)

A still different banding pattern was obtained when the concentration of trypsin versus versene was increased. Instead of "foldings," we observed very thick bands, mostly telomeric, as shown in Plate 4. The meaning of these is not yet clear. It may be that they represent the result of movement of the chromatic material toward regions more stably attached to the glass support.

For this reason, before studying possible homologies, it is advisable to clarify what these bands actually represent and why the same bands are consistently obtained when substances specific for proteins (e.g. trypsin in fixed preparations or actinomycin D *in vivo*: Hsu et al. 1973; Shafer 1973) and/or for DNA are used.

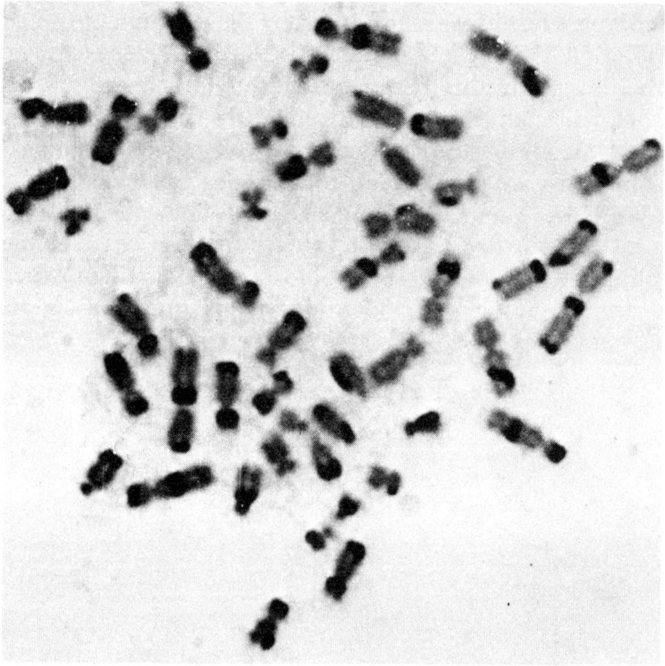

Plate 4. Accumulation of chromatic material toward the telomeres (human chromosomes treated with a solution of trypsin 0.05 percent; versene 0.02 percent)

REFERENCES

CASPERSSON, T., S. FARBER, G. E. FOLEY, J. KUDINOWSKI, E. J. MODEST, *et al.*
1968 Chemical differentiation along metaphase chromosomes. *Experimental Cell Research* 49:219–222.

CASPERSSON, T., L. ZECH, C. JOHANSSON, E. J. MODEST
1970 Identification of human chromosomes by DNA-binding fluorescent agents. *Chromosoma* 30:215–227.

CASPERSSON, T., L. ZECH, E. J. MODEST, G. E. FOLEY, V. WAGH, E. SIMONSSON
1969 Chemical differentiation with fluorescent alkylating agents in *Vicia faba* metaphase chromosomes. *Experimental Cell Research* 58:128–140.

CHIARELLI, B.
1973 Hypothetical mechanism of action of trypsin and other substances producing bands in chromosomes. *Genen en Phaenen* 16(1):3–6.

CHIARELLI, B., M. SARTI-CHIARELLI
1972 Bandeggiamento dei cromosomi con tripsina. *Rivista di Antropologia* 57(1970–1971):269–272.

CHIARELLI, B., M. SARTI-CHIARELLI, D. SHAFER
1972 Chromosome banding with trypsin. *Genetica* 43:190–194.

DUTRILLAUX, B., J. DE GROUCHY, C. FINAZ, J. LEJEUNE
1971 Mise en évidence de la structure fine des chromosomes humains par digestion enzymologique (pronase en particulier). *Comptes Rendus Hebdomadaires des Sciences de l'Académie des Sciences* 273:587–588.

DUTRILLAUX, B., J. LEJEUNE
1971 Sur une nouvelle technique d'analyse du karyotype humain. *Comptes Rendus Hebdomadaires des Sciences de l'Académie des Sciences* 272:2,638–2,640.

HSU, T. C., S. PATHAK, D. SHAFER
1973 Induction of chromosome crossbanding by treating cells with chemical agents before fixation. *Experimental Cell Research* 79:484–487.

SEABRIGHT, M.
1971 A rapid banding technique for human chromosomes. *Lancet* 2:971–972.

SHAFER, D.
1973 Banding human chromosomes in cultures with actinomycin D. *Lancet* 14:828.

WANG, H. C., F. FEDOROFF
1972 Banding in human chromosomes treated with trypsin. *Nature New Biology* 235:52–53.

ABSTRACT:

Photodensitometric Techniques in Evaluating Chromosome Banding

T. CASPERSSON

The basic prerequisite for the use of any chromosome-banding technique in biology and medicine is that it can be shown that the resultant patterns differ for all the various chromosome types and the main regions within the chromosomes, and, furthermore, that they are constant and reproducible.

The proper technique for the recognition of the fluorescence patterns is photoelectric evaluation, and this technique is applicable to practically any O.M. preparation.

A photograph is taken of the metaphase in the fluorescence microscope, and either the photographic negative or a positive print thereof is evaluated in the densitometer–reflectometer, in which the photograph is automatically moved through a measuring system and the patterns recorded on an XY recorder. The primary aim is to determine the location of fluorescence maxima and minima, as well as inflection points on the intensity curves.

The print is mounted on the stage of a microscope system, evenly illuminated with two lamps. The image is projected onto a photocell system, in front of which there is an aperture that is either a slit or a spot. Under continuous visual control from a side viewer, the chromosome image is maneuvered in such a way that the split aperture sweeps broadside down the length of the chromosome (slit measurement). Alternatively, the small circular aperture can be swept, also under visual control, down the midline of a chromatid (chromatid measurement). The movement of the stage is connected mechanically to a potentiometer that serves as the input for the X-movement in an XY recorder, the intensity of the reflected light being recorded along the Y-axis. The graph then gives the distribution of fluorescence intensity along the chromosome.

An Outline of the Application of the Method of Nucleic Acid Hybridization to Chromosomes in situ

K. W. JONES

Native DNA usually occurs in the form of helical double-stranded molecules in which the strands are antiparallel with their component bases hydrogen bonded, guanine with cytosine and adenine with thymine. On heating to boiling in solutions of dilute salts (less than 0.015M Na^+) or being subjected to high or low pH or organic denaturants, the paired strands fall apart into random coils and the DNA is said to be denatured or single stranded. Restoration of higher salt concentrations (above 0.15M Na^+) in neutral solution and moderate heat (about 65°C) promotes the restoration of the base-paired structure. Such incubation is termed *annealing* and the process of restoration, *renaturation*.

Renaturation depends on the existence of single DNA strands which are complementary in base sequence. Its rate depends upon the concentration of such complementary sequences. Renaturation therefore implies base sequence complementarity and is an important measure of species relatedness in evolution (see Kohne 1970).

The DNA of eukaryotes contains major proportions of nucleotide sequences which are repeated or reiterated a very large number of times (Britten and Kohne 1968). Such parts of the genomal DNA renature more or less rapidly depending upon their representation in the genome. Relatively large proportions of the genome which react quickly may be mapped directly on chromosomes by annealing purified radioactive copies of the particular fraction of interest with cytological preparations of chromosomes which have previously had their DNA denatured. The radioactive molecules compete for and "hybridize" with complementary sites within the chromosomal DNA and are detected by autoradiography. In practice the radioactive copies used are RNA molecules which are synthesized enzymatically upon a template of pure DNA *in vitro*. Such RNA is called complementary RNA (cRNA); unreacted RNA is removed from the specimen by pancreatic RNase to which hybridized RNA is resistant.

The method of investigating chromosome structure by *in situ* nucleic acid hybridization has a twofold advantage for evolutionary studies because evolution ultimately involves changes in DNA as well as alterations in its distribution in chromosomes. In a following article (Jones, this volume) its use in investigations of nucleotide homologies among the chromosomes of high primates is described. The present account is concerned with the purely technical aspects of the method.

THE METHOD OF *IN SITU* HYBRIDIZATION

In the following account, the schedule of the method is set out first, followed by explanatory notes relating to specific steps. The schedule shown is that presently used in the author's laboratory and it contains steps original to this laboratory as well as procedures adopted from other sources, where these have been shown to constitute improvements in, or useful variations of, the method. General references to original papers are John, Birnstiel, and Jones (1969) and Gall and Pardue (1969). Reviews of the technique have appeared in Gall and Pardue (1971); Steffenson and Wimber (1972); Jones (1973a); and Hennig (1973). These reviews should be consulted for a fuller account of the technique.

Schedule

1. Prepare and fix cytological material.
2. Air dry onto glass slides, postfix.
3. Pretreat with RNase.
4. Wash thoroughly to remove RNase.
5. Denature chromosomal DNA.
6. Dehydrate through alcohol, air dry.
7. Apply radioactive RNA solution.
8. Cover and seal preparation.
9. Anneal to allow hybridization.
10. Remove coverslip and wash off excess RNA in saline.
11. Incubate in RNase to remove nonspecific contamination.
12. Wash off RNase.
13. Wash for a further two to four hours at 4°C in a stirred bath.
14. Dehydrate through alcohols and air dry.
15. Apply autoradiographic emulsion, air dry.
16. Store for exposure in the dark at 4°C.
17. Develop, rinse, fix, and wash autoradiographs.
18. Stain and mount for microscopical examination.

Notes on Schedule

Step 1. Chromosomes prepared according to the commonly used phytohaemagglutinin-treated peripheral blood method and fixed in acetic ethanol, followed by air drying onto slides, are ideal. The chromosomes need not be freshly prepared and will be satisfactory after prolonged storage.

Step 2. Post fixation in acetic ethanol or 45 percent acetic acid promotes adhesion to the glass and serves to remove additional proteinaceous material which may impede hybridization.

Step 3. RNase pretreatment is sometimes used to remove from the specimen cellular RNA which may conceivably compete with the radioactive input RNA for sites on DNA. 20μg/ml in 2 × SSC, 20 minutes at 35°C is suitable.

Step 4. Three washes in 2 × SSC (standard saline citrate; 0.3M NaCl, 0.03M trisodium citrate, pH 7.0–7.2) usually will suffice.

Step 5. Specimens may be denatured by any of four different methods: immersion in boiling DSC (0.1 SSC) for 15 to 30 seconds followed by immediate immersion in ice-cold DSC; or incubation in 0.2N HCl for 20 minutes at room temperature; or similar incubation for 1 minute in 0.07N NaOH; or incubation for 2.5 hours in 90 percent formamide in DSC at 65°C. Each procedure produces some morphological alteration in chromosomes, but HCl is least damaging, followed by formamide with heat, NaOH being most disruptive. Formamide appears to produce the most reliable denaturation and gives rather more efficient hybridization.

Stept 6. Brief immersion in 70 percent, 90 percent, and absolute ethanol.

Step 7. Usually about 3 to 5 μl RNA is sufficient to occupy completely the space between slide and coverslip. RNA is usually dissolved in 2–6 × SSC, and several thousand CPM are usually applied to one preparation.

Step 8. The edges of the coverslip used to cover the RNA solution over the preparation are sealed with a thin rubber solution in petroleum ether. Sometimes nail polish has been used.

Step 9. The choice of annealing temperature depends critically on the fraction of nucleic acids being renatured or hybridized. In general, nucleic acids with low GC contents will require lower temperatures than will those with high GC contents. In critical work it is necessary first to determine the optimum hybridization temperature (T. opt) by filter methods (see for examples Birnstiel 1972). Annealing time has varied from one to sixteen hours in successful experiments, and the effects of prolonged annealing are not as critically obvious for hybridization *in situ* as they are *in vitro*, where the product of time and nucleic acid concentration (C_0t) (Britten and Kohne 1968) are importantly interrelated variables. This is because the renaturation behavior of DNA in chromosomes under the conditions of *in situ* hybridization is different from that of pure DNA in solution and is imperfectly understood. In practice, temperatures of 65° to 67°C have been used to anneal a wide variety of different nucleic acid fractions. Slides are floated in a steel tray in a covered water bath.

Step 10. Coverslips are detached with a finely pointed forceps and slides rinsed twice in cold 2 × SSC.

Step 11. Pancreatic RNase at 20μg/ml in 2 × SSC at 37°C for 20 minutes.

Steps 12 and 13. 2 × SSC at 4°C.

Steps 15 through 18. Usual autoradiographic methods. The procedure may be interrupted for convenience after steps 2, 6, and 14. Slides should be stored in airtight or vacuum desiccators to avoid contamination. All glassware and apparatus must be absolutely clean and handled with gloves to avoid possible nuclease contamination from the skin.

Preparation of cRNA

This is prepared using radioactive nucleoside triphosphates and RNA polymerase which is commercially available or which can be prepared according to the method of Chamberlain and Berg (1961). Details of a preparative method for cRNA are fully set out in Jones (1973a).

Preparation of Satellite DNA

Satellite DNAs which are important in the investigation of karyotypic evolution are prepared from nuclear DNA by centrifugation in gradients of heavy metal salts (Cs_2SO_4 and CsCl) in the ultracentrifuge. The methods of preparation are detailed in Flamm et al. (1968); Corneo, Ginelli, and Polli (1968); Corneo, Ginelli, Soave, et al. (1968); Prosser et al. (1973). A general account of human satellite DNA is to be found in Jones (1973b).

EVALUATION OF THE METHOD

The value of the *in situ* hybridization method for anthropological studies lies in its capacity to detect the specific chromosomal locations of highly repetitive DNA. This enables the mapping of such fractions in human and primate chromosomes and allows the detection of cross-hybridizing, and therefore related, fractions within the primate group as a whole. The chromosomal distribution of related DNA fractions in closely related species reveals details of the manner in which chromosome structures have evolved and diverged within the group and unequivocally testifies to the prior existence of common ancestors. Given this information it is then possible to extend the study in depth by examining the relative extent of nucleotide sequence divergence in the related fractions in each of the species concerned. This suggests the likely order of species divergence and thus the sequence of evolution within the group.

REFERENCES

BIRNSTIEL, M., B. H. SELLS, I. F. PURDOM

1972 Kinetic complexity of RNA molecules. *Journal of Molecular Biology* 63:21–39.

BRITTEN, R. J., D. E. KOHNE

1968 Repeated sequences in DNA. *Science* 161:529–540.

CHAMBERLAIN, M., P. BERG

1961 Deoxyribonucleic acid-directed synthesis of ribonucleic acid by an enzyme from *Escheria coli*. *Proceedings of the National Academy of Sciences* 48:81–94.

CORNEO, G., E. GINELLI, E. POLLI

1968 Isolation of the complementary strands of human satellite DNA. *Journal of Molecular Biology* 33:331–335.

CORNEO, G., E. GINELLI, C. SOAVE, G. BERNARDI

1968 Isolation and characterization of mouse and guinea-pig satellite deoxyribonucleic acids. *Biochemistry* 7(3):4,373–4,379.

FLAMM, W. G., M. L. BIRNSTIEL, P. M. B. WALKER

1968 "Isopynic centrifugation of DNA methods and application," in *Subcellular components*. Edited by G. Birnie, 279–310. London: Butterworth Science Publications.

GALL J. G., M. L. PARDUE

1969 Formation and detection of RNA–DNA hybrid molecules in cytological preparations. *Proceedings of the National Academy of Sciences* 63:378–383.

1971 "Nucleic acid hybridisation in cytological preparations," in *Methods in enzymology*, volume twenty-one: 470–480. New York: Academic.

HENNIG, W.

1973 "Molecular hybridization of DNA and RNA *in situ*," in *International review of cytology*, volume thirty-six. Edited by G. H. Bourne and J. F. Danielli, 1–44. London: Academic.

JOHN, H. A., M. L. BIRNSTIEL, K. W. JONES

1969 RNA–DNA hybridisation at the cytological level. *Nature* 223:582–587.

JONES, K. W.

1973a "The method of *in situ* hybridisation," in *New techniques in biophysics and cell biology*. Edited by B. J. Smith and R. H. Pain, 29–66. London: Wiley.

1973b Satellite DNA. *Journal of Medical Genetics* 10:273–281.

KOHNE, D. E.

1970 Evolution of higher organism DNA. *Quarterly Reviews of Biophysics* 3:327–375.

PROSSER, J., M. MOAR, M. BOBROW, K. W. JONES
1973 Satellite sequences in chimpanzee (*Pan troglodytes*). *Biochimica et Biophysica Acta* 319:122–134.
STEFFENSON, D. M., D. E. WIMBER
1972 "Hybridization of nucleic acids to chromosomes", in *Nucleic acid hybridisation in the study of cell differentiation*. Edited by H. Ursprung, 47–63. Berlin: Springer.

SECTION TWO

Recent Studies on Primate Chromosomes

Problems of Primate Karyology

B. CHIARELLI

A general account of the literature available on primate chromosomes up to 1972 has been collected in the last part of this book. In the second part a bibliography of those papers has been published which are considered basic references in this field.

A collection of recent papers on primate karyology was also published in the *Journal of Human Evolution* (volume 2, number 4, 1973). The literature on the subject is therefore covered up to the beginning of 1973. These papers give us some idea of the potentialities of the new techniques, suggesting, however, some caution in the generalization of the results.

My task is now to bring into focus some of the most important and debated aspects of the study of chromosomes in primate taxonomy and phylogeny, which we hope will receive a solution or at least some clarification by means of the new techniques.

Among prosimians, although some fairly clear notions on karyological relationships between different species from the same family have been gained by means of the morphological analysis of chromosomes, the relationship between these species at the level of the various suborders (Tupaioidea, Lorisoidea, Lemuroidea, and Tarsioidea) is, however, rather speculative from a karyological point of view. They are indeed phylogenetically very distant groups. Nevertheless, identification of some type of chromosomal homology and its quantification would be determinant in outlining the history of their separation.

Recent reevaluation of Wegener's theory of the continental drift throws a new and stimulating light on the systematic position of the South American monkeys. In this group, too, traditional karyological investigations have clarified many controversial aspects of their systematics at a subfamily level. It fails, however, to solve the problem of the relationship between Callithricidae and Cebidae, or to provide some plausible explanation about the relationships existing between them and the Old World monkeys and/or the prosimians.

The same is true, in general, with regard to the Old World monkeys. Although some sufficiently clear ideas have been reached about the relationships between allied species and genera, mere assumptions, hypotheses, or speculations have been extrapolated to explain the relationships between more distant species groups. Maybe, rather than clarifying it, karyological analysis has complicated the problem of the relationships between the various groups of the Old World primates! To date it is clear enough that the genera *Macaca, Papio, Theropithecus*, and *Cercocebus* comprise a homogeneous group (called Papinae by us since 1966) which is absolutely foreign to the various species of the genus *Cercopithecus*. The homology of the chromosomes in this group of species is proved not only by the similarity of their karyotype but by the great frequency of hybrids, according to the available data.

It was clear from our earliest studies of chromosomes in 1960 and 1961 that African apes (gorilla and chimpanzee) and man make up a homogeneous group, and that the difference in number was due to a centric fusion mechanism which probably produced chromosome 2 in man, while other rearrangements produced differences in the chromosome morphology. These events seem to be confirmed by recent findings on chromosome banding. It was also clear, however, that the gibbon has nothing in common with the Pongidae; since 1962, in fact, we have allocated the gibbons close to the Colobinae from a karyological point of view.

Indeed, these two conflicting aspects of traditional systematics (the *Cercopithecus*–Papinae and the Hylobatinae–Colobinae relationships) require clarification by the new techniques.

The variability in the chromosome number in the various species from the *Cercopithecus* genus is interesting in itself in view of an interpretation of the mechanism of karyological variation, but it would be of even more use in the determination of phylogenetic relationships between this species group and the other Old World monkeys. Perhaps it can provide information on possible links between these monkeys and the platyrrhines from South America!

An utterly open question concerns the Hylobatinae. How did the chromosome differences between *Hylobates*, *Symphalangus*, and *Nomascus* come about? Some hypotheses, at least for the variations in the small chromosomes and in the marker chromosome, are available; but what about the other chromosomes?

These are some of the problems which are facing us today. Any attempt to provide a solution is most welcome.

Cytogenetic Study of the Indridae Bennet, 1832, an Endemic Malagasy Lemur Family

Y. RUMPLER and R. ALBIGNAC

ABSTRACT
A cytogenetic study has been made from leukocyte cultures of the peripheral blood of the different species of Indridae. The results permit us to give a schema of the chromosomal evolution of these animals, ranging in order from the most primitive to the most evolute: *Avahi, Propithecus verreauxi, Propithecus diadema*, and *Indri*. Only centric fusion and pericentric inversion seem to have occurred in the chromosomal evolution of this family.

The Indridae comprise three different genera: *Indri* E. Geoffroy and Cuvier, 1795; *Propithecus* Bennet, 1832; and *Avahi* (Gmelin 1828). The genus *Propithecus* is composed of two species: *Propithecus verreauxi* Grandidier, 1867, and *Propithecus diadema* Bennet, 1832. According to their morphological and behavioral characteristics, the animals of this family are the most evolved of the lemurs. To date the chromosomal complements of only two subspecies of *Propithecus verreauxi* have been published, and we felt that making cytogenetical studies of other members of this family would be interesting.

MATERIAL AND METHODS

We karyotyped two male and two female *Propithecus v. verreauxi* Grandidier, 1867; one male and one female *Propithecus v. deckeni* Peters, 1870; one male and one female *Propithecus v. coronatus* Milne-Edwards, 1871; one male *Propithecus v. coquereli* Milne-Edwards, 1867; three male *Propithecus d. diadema* Bennet, 1832; one male *Propithecus d. perrieri* Lavauden, 1931; one male *Avahi laniger* Gmelin, 1788; and one male *Indri indri* (Gmelin 1788).

Chromosome preparations were made using a modified leukocyte cultivation method of Lejeune, reported in an earlier paper (Rumpler

1970). For the karyotype analysis of each animal, about three to ten mitotic metaphase plates were examined, except in the case of the *Avahi laniger*, for which only one mitotic of rather mediocre quality was available, because it died shortly after capture.

RESULTS

The only *Avahi laniger* metaphase plate has a dipliod number of 66. Its karyotype comprises two pairs of submetacentric chromosomes, thirty pairs of acrocentric chromosomes, and one pair of sex chromosomes in which the X is a medium-sized metacentric and the Y a very small acrocentric. The fundamental number is 68.

All the *Propithecus verreauxi* have the same karyotype. They have a diploid number of 48. Their karyotype is composed of six pairs of metacentric chromosomes, eight pairs of submetacentric chromosomes, nine pairs of acrocentric chromosomes, and two sex chromosomes, shown in Plate 1.

A small pair of chromosomes is difficult to identify: it appears in many preparations as acrocentric and in others as metacentric.

The X chromosome is a submetacentric and the Y chromosome a small acrocentric. The fundamental number is 76.

All the *Propithecus diadema* studied have the same karyotype. Their diploid number is 42. Their karyotype comprises seven pairs of metacentric chromosomes, eight pairs of submetacentric chromosomes, five pairs of small acrocentric chromosomes, and one pair of sex chromosomes, as shown in Plate 2. One of the smallest pairs of chromosomes is difficult to interpret, as in the case of *Propithecus verreauxi*. The X chromosome is a metacentric and the Y a small acrocentric. The fundamental number is 72.

The *Indri* has a diploid number of 40. His karyotype is composed of five pairs of metacentric, ten pairs of submetacentric, and four pairs of acrocentric chromosomes. The third pair of chromosomes presents a secondary constriction on the long arms near the centromere. The X chromosome is a metacentric and the Y chromosome a small acrocentric, as shown in Plate 3. As in the other cases mentioned, one of the smallest pairs of chromosomes resembles a pair of metacentric chromosomes found in many of the preparations. The fundamental number is 70.

DISCUSSION

The karyotype of the different subspecies of *Propithecus verreauxi* are similar, and our results confirm the findings of Chu and Bender (1961) for *P. v. verreauxi* and *P. v. coquereli*.

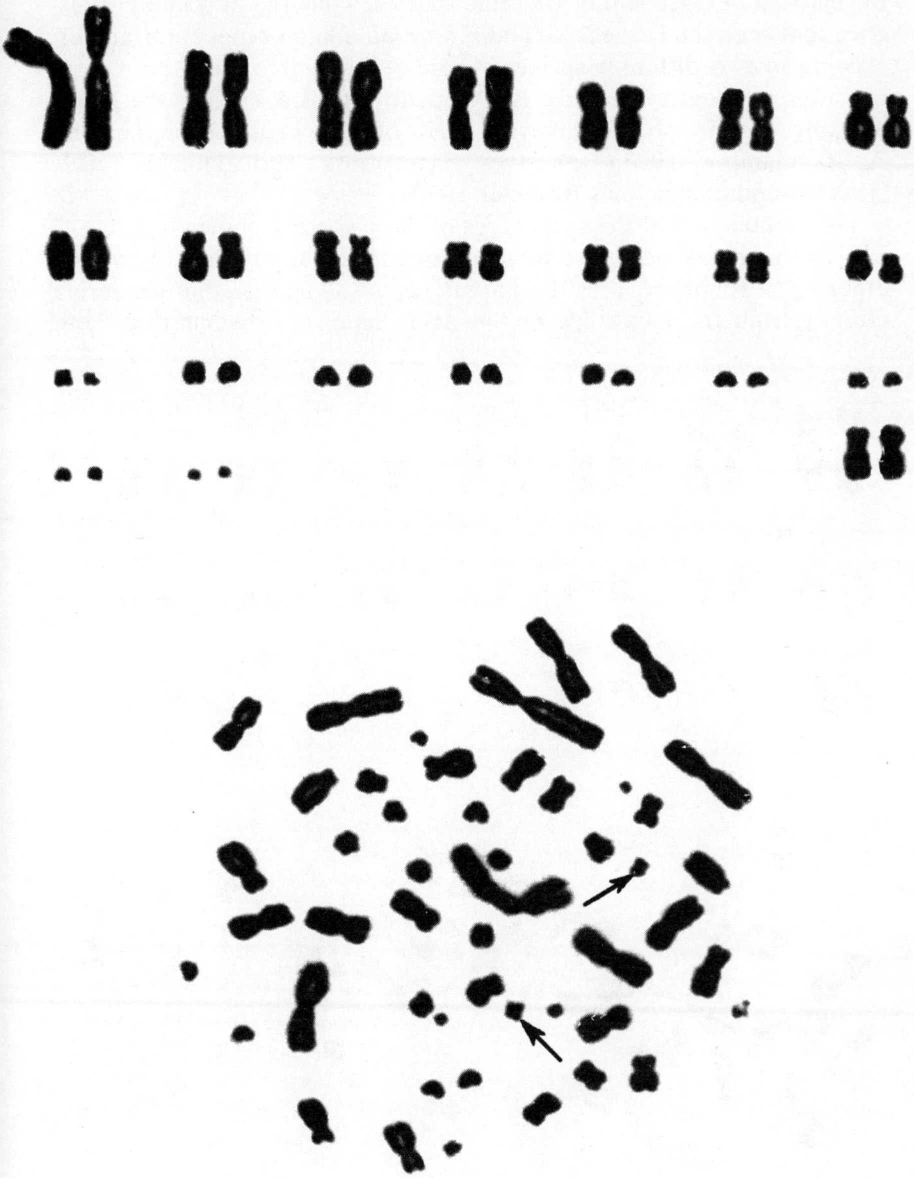

Plate 1. Karyotype of a female *Propithecus verreauxi* and the metaphase that produces it. Note the presence of the small chromosomes (arrows) which are difficult to interpret

The karyotype of *P. d. diadema* and *P. d. perieri* are similar, confirming the classing of these two in the same species, while the important differences between the *P. diadema* and *P. verreauxi* karyotypes confirm their classing in two different species. These results agree with those from other morphological characteristics recently studied, such as the scent-marking glands of the anterior wall of the neck (Rumpler and Andriamiandra 1971) or the fingerprints and knuckle pads (Rakotosamimanana and Rumpler 1970).

The comparison of the karyotypes of these species, shown in Table 1, reveals that many chromosome pairs correspond approximately to each other in morphology; thus the karyotypes of all species did not derive directly from the karyotype of the *Avahi laniger*. It seems that these

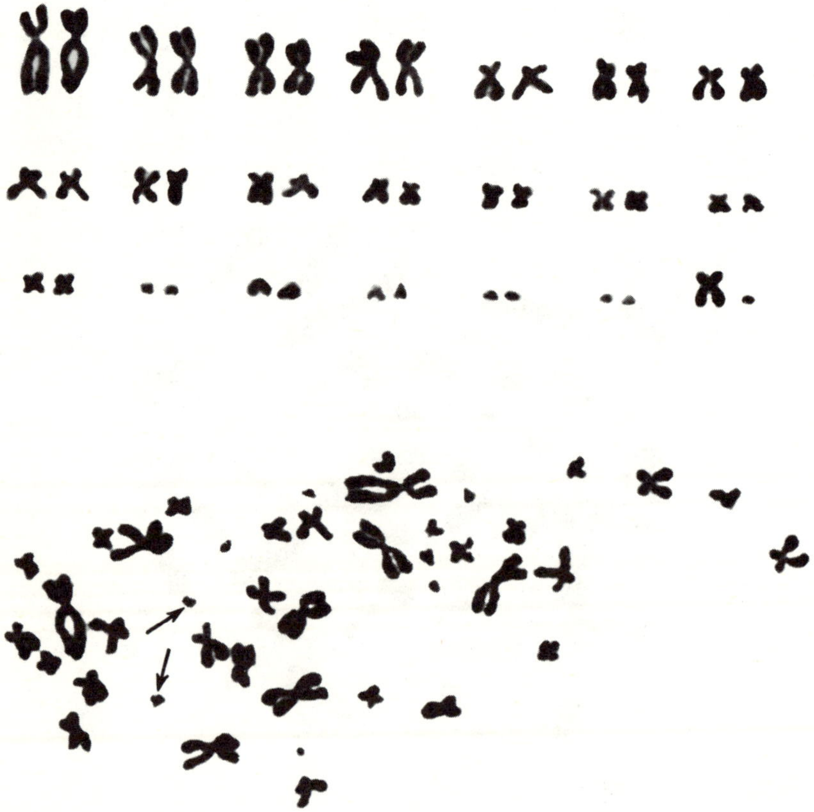

Plate 2. Karyotype of a male *Propithecus diadema* and the metaphase that produces it. Note again the presence of the small chromosomes (arrows) which are difficult to interpret

different karyotypes have developed from common intermediate forms.

Further studies after the capture of new *A. laniger* will permit us to confirm its karyotype, and the use of banding techniques would certainly demonstrate the manner in which the different karyotypes have branched off from the common stem.

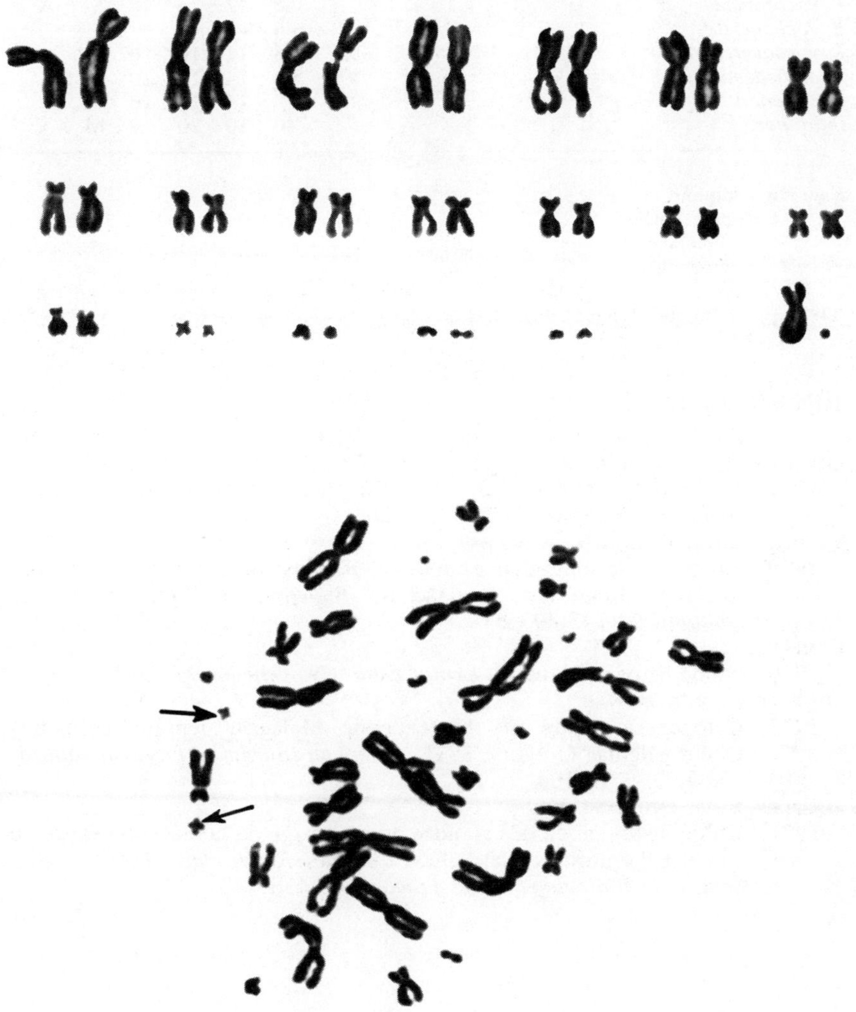

Plate 3. Karyotype of a male *Indri indri* and the metaphase that produces it. Note once again the presence of the pair of chromosomes (arrows) which are difficult to interpret

Table 1. Chromosome number and karyotype of the Indridae

Species	Number and specimen		Chromosomes					
	male	female	2n	M	S	A	X	Y
Avahi laniger	1	–	64	–	4	60	M	A
P. v. verreauxi[a]	1	–	48	–	–	–	–	–
P. v. verreauxi	2	2	48	12	16	18	S	A
P. v. deckeni	1	1	48	12	16	18	S	A
P. v. coronatus	1	1	48	12	16	18	S	A
P. v. coquereli[a]	1	2	48	–	–	–	–	–
P. v. coquereli	1	–	48	12	16	18	S	A
P. d. diadema	3	–	42	14	16	10	M	A
P. d. perieri	1	–	42	14	16	10	M	A
Indri indri	1	–	40	10	20	8	M	A

Key: 2n = diploid.
M = metacentric.
S = submetacentric.
A = acrocentric.

[a]Data from Chu and Bender 1961.

REFERENCES

CHU, E. H. Y., M. A. BENDER

1961 Cytogenetics and evolution of primates. *Annals of the New York Academy of Sciences* 92(1):253–266.

RAKOTOSAMIMANANA, B., Y. RUMPLER

1970 Etude des dermatoglyphes et des coussinets palmaires et plantaires de quelques Lémuriens malgaches. *Bulletin de l'Association des Anatomistes* 148:493–510.

RUMPLER, Y.

1970 Etude cytogénétique du *Lemur catta*. *Cytogenetics* 9:239–244.

RUMPLER, Y., R. ALBIGNAC

1973 Cytogenetic study of the endemic Malagasy lemurs, subfamily Cheirogaleinae Gregory, 1915. *American Journal of Physical Anthropology* 38:261–264.

RUMPLER, Y., A. ANDRIAMIANDRA

1971 Etude histologique des glandes de marquage de la face antérieure du cou des Lémuriens malgaches. *Comptes Rendus des Séances de la Société de Biologie et de ses Filiales* 165:436–442.

ABSTRACT:

C-Bands in Chromosomes of Presbytis entellus entellus *(Dufresne)*

T. SHARMA and G. S. GARG

Study of the C-banding pattern using a cytological DNA denaturation–reassociation technique has revealed an interesting result in the somatic metaphase chromosomes of the Indian common langur, *Presbytis e. entellus*. As seen in Plate 1, C-bands are present in the kinetochore regions of only three pairs of autosomes and in the distal portion of the small acrocentric Y in the total complement of 44 chromosomes of the

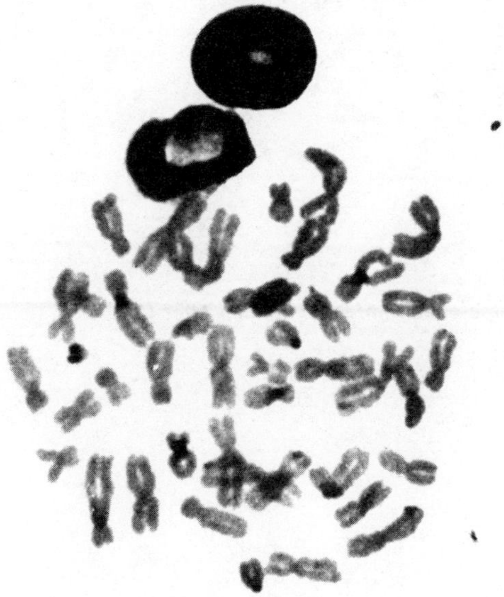

Plate 1. Metaphase plate of *Presbytis e. entellus* after C-banding treatment

male studied. The treatment of slides for denaturation in NaOH for different durations has shown C-bands either in the three pairs of autosomes and the Y only, or not at all. This peculiarity is observed to be consistent in all the 149 metaphases exhibiting the C-band reaction. Moreover, the slides of a female rhesus monkey, *Macaca mulatta* (Zimmerman), treated simultaneously with that of *Presbytis* and shown in Plate 2, show C-bands exclusively in the kinetochore regions of all 42 chromosomes including the X chromosomes. The distinctive feature observed in *Presbytis*, illustrated in Plate 3, will be discussed in the light of presumptive roles of centromeric heterochromatin as well as the heterogeneity of satellite DNA.

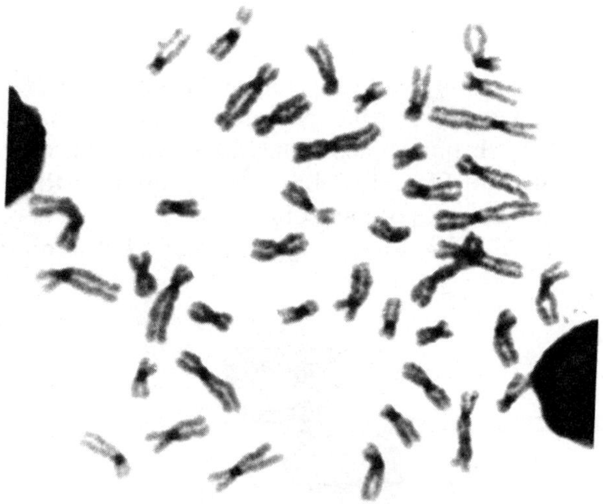

Plate 2. Metaphase plate of *Macaca mulatta* after C-banding treatment

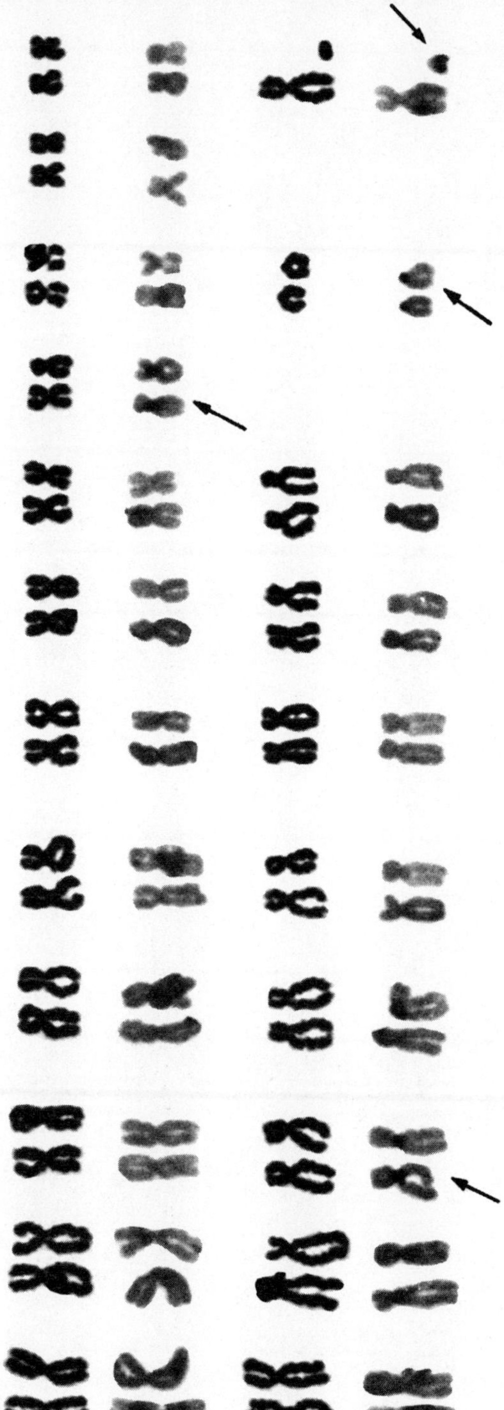

Plate 3. The chromosomes of *Presbytis entellus* after normal staining (upper row) and after C-banding treatment (lower row). The arrows indicate the pairs with C-bands

Chromosome Evolution in the Genus Cercopithecus

ROBERT B. ECKHARDT

Of all the fields contributing new data of potential use in the reconstruction of primate evolution, karyology is among the most rapidly developing at this time. The existence at the IXth Congress of Anthropological and Ethnological Sciences of a special conference, of which this volume is a record, supports this statement, as does the explosive expansion in literature on primate chromosomes over the past decade. It is probably fair to say that most of the contributors to this volume are not interested in the evolution of chromosomes *per se,* or even in the evolution of primate chromosomes in particular, but rather in what the evidence provided by this system can tell us about the biological relationships of the present populations of living organisms that belong to our order.

Reasoning from a part of the organism — even such a basic part as the spatially organized genome represented by the karyotype — to the whole is often not easy. There are several reasons for this. Two are of particular importance:

1. Organisms are mosaics of a number of different systems, and these systems can evolve at different rates.

2. Even the same system (whether it be the dentition, the structure of some protein such as albumin, or the karyotype) can evolve at widely varying rates (a) during different periods in the same phyletic line, and (b) at the same time in separate but closely related lines. Therefore phylogenies based on the assumption of constant rates of evolution may not represent the course of evolutionary events that actually occurred.

For these reasons it is always desirable to compare reconstructions based on one system with those based on others. Traditionally the starting point or standard of reference has been the fossil record, but the evidence that this record provides is as open to differences in interpretation as that afforded by any other approach. Furthermore, for some important subdivisions of our order the fossil record is tantalizingly sparse. This is especially true in the case of guenons, African monkeys of the genus *Cercopithecus*, which makes the increasingly abundant knowledge about their chromosomes of particularly great value. Before turning to the karyotypic data, however, it is desirable to review briefly what is known about the antecedents of this genus from the evidence contained in geological strata.

There is no appreciable fossil record of unequivocal members of the genus *Cercopithecus*. Pilbeam and Walker (1968) discuss one tooth that they consider to be an upper molar (M^1 or M^2) of a cercopithecine. This was recovered from deposits approximately nineteen to twenty-two million years old at Napak, Uganda. Neither, however, does this genus completely lack fossil antecedents, unless it be insisted that all members of the line leading to it appear indistinguishable from modern forms. Such insistence would, of course, be an implicit denial of evolution, in which the possibility for change is understood.

Fossil remains are known of some other African Miocene catarrhines such as *Prohylobates* and *Victoriapithecus* (Simons 1969; 1972:192–195). Specimens attributed to both these genera lack the bilophodont molars typical of living cercopithecoid monkeys but share other features with them, namely, (1) molars which are (a) hypsodont (i.e. have basal crowns high relative to the cusps which rise from them), and (b) waisted (i.e. have marked vertical grooves which separate the anterior trigonid from the posterior talonid); (2) premolars (P_3 and P_4) in which the long axis of the tooth is set at an angle oblique to the line formed by the tooth row; and (3) posterior premolars (P_4) which have an anterolateral projection. As pointed out by Simons (1969), these features are also shared with a still older primate, *Parapithecus*, whose remains come from Oligocene deposits in the Fayum region of Egypt. It is therefore possible to trace the ancestry of cercopithecoid monkeys back that far with some degree of assurance. This introductory review of the fossil record makes evident the value of the data bearing on phylogenetic interpretations provided by the karyotypes of living species. These may be of help in differentiating between several alternative hypotheses as to the course of chromosome evolution in the genus *Cercopithecus*.

Fifteen years ago Chu and Bender (1962) and Bender and Chu (1963) suggested that chromosome numbers in cercopithecines had evolved from higher to lower through centric fusion of the Robertsonian type. This proposal seemed quite plausible at the time since diploid chromosome numbers then known fell into several groups: *C. l'Hoesti* and *C. mitis* with 72 chromosomes; *C. mona* and *C. nictitans buttikoferi* with 66 chromosomes; *C. aethiops, C. diana,* and *C. neglectus* with 60 chromosomes; and (as tentatively suggested then but apparently not yet confirmed) *C. cephus* with 54 chromosomes. The better-known groups were separated by regular differences, the 72-chromsomes species having 18 acrocentrics, the 66-chromosome species having 12 acrocentrics, and the 60-chromosome species possessing 6 acrocentric chromosomes.

In 1968 Chiarelli reviewed the now-more-abundant evidence. There are examples of species with 2n chromosome numbers of 58, 62, 68, and 70. These fall between the groups discussed above and suggest that mechanisms of chromosome evolution other than centric fusion may have been operative here. Because of this, and citing as further support Matthey's conclusion (1964) that there is a very low probability that diploid chromosome numbers higher than 58 represent the primitive condition in eutherian mammals, Chiarelli (1968) suggested that chromosome numbers in the genus *Cercopithecus* have instead evolved from lower to higher numbers.

Since Chiarelli's review one of the major new additions to our knowledge of chromosome numbers in the genus *Cercopithecus* was made by Wurster and Benirschke (1969). These investigators reported a 2n chromosome number of 48 for *Cercopithecus (Allenopithecus) nigroviridis* (Allen's swamp guenon), thus extending downward by a significant amount the lower range of diploid numbers in the genus. The findings of Wurster and Benirschke have been confirmed more recently by Ardito and Mortelmans (1975), who note further that their more detailed studies of chromosome-banding patterns confirm a certain affinity between the chromosomes of *Cercopithecus nigroviridis* and *Macaca mulatta*, the rhesus monkey or rhesus macaque. This Asian taxon belongs to a group of large semiterrestrial and terrestrial Old World monkeys (encompassing those commonly referred to as macaques, baboons, and mangabeys) that manifests a diametrically opposite pattern from that of the genus *Cercopithecus*. The *Cercopithecus* monkeys are karyotypically diverse but relatively uniform morphologically as well as predominantly arboreal and restricted in geographic range to sub-Saharan Africa. The more terrestrial Old World monkeys represented more characteristically by macaques and baboons are, in contradistinction, highly uniform karyotypically but far more varied in visible anatomical features, in addition to being — not coincidentally — more widely distributed (with populations assigned to different taxonomic levels found from North Africa through far-eastern Asia).

The recent karyotypic evidence on the affinities of *C. nigroviridis* is an important step toward confirmation of suggestions made long before on the basis of external anatomical features. Lang (1923) considered this taxon (which he referred to as *Allenopithecus nigroviridis*) to be widely different from other African groups such as *Cercopithecus (Erythrocebus)* and *C. (Miopithecus)* as well as from *Macaca sylvana* of North Africa, though he felt that certain dental features pointed in the direction of the macaques. Hill (1964, 1966) noted that some characteristics link *C.*

nigroviridis with other members of the genus *Cercopithecus*, while other traits suggest alignment with *Macaca* and *Cercocebus*. Included among the latter features of the swamp monkey are its face, which is guenonlike in immature specimens but more macaquelike in its possession of a more prominent muzzle in mature animals; the overall habitus and robust, stocky limbs, which are baboonlike, as is the short, broad manus (though the pes is less different from the condition seen in other guenons); and even the external ear, which has a sharp morphological apex like that of macaques. Among the macaque–baboon–mangabey subgroup of Old World monkeys, *Cercocebus* shows the greatest ecological overlap with *Cercopithecus nigroviridis*. Members of the *Cercocebus albigena* group are wholly arboreal in swampy areas of low-lying forest, while members of the *torquatus* group are more terrestrial, ascending into trees chiefly to feed and sleep.

On balance it appears that the swamp monkey may be a form linking the other guenons characterized by higher chromosome numbers with the 42-chromosome group of semiterrestrial and terrestrial Old World monkeys. Put in other terms, *C. nigroviridis* seems to be — karyotypically, morphologically, and ecologically — a living embodiment of many characteristics close to those of the hypothetical stem cercopithecine. According to Murray (1975) the stem cercopithecines utilized arboreal and some terrestrial resources within forest habitats, with adaptive radiation in cercopithecines proceeding from an intermediate, generalized forest adaptation to extreme arboreality on the one hand and extreme terrestriality on the other. Other hypothetical characteristics of stem cercopithecines are discussed by Murray and also by Delson (1975).

After *Cercopithecus nigroviridis* the next lowest diploid chromosome number in the genus as it is currently delimited (Kuhn 1967), 54, is known at present for two species: *C.* (formerly *Erythrocebus*) *patas*, known by a variety of common names including patas monkey, red hussar monkey, and military red grass monkey; and *C.* (formerly *Miopithecus*) *talapoin*, usually referred to as the talapoin monkey or dwarf guenon. It is of interest that both were formerly recognized as being closely related to other *Cercopithecus* monkeys but were excluded from membership in that genus on the basis of anatomical characteristics. They were placed in separate genera because each diverged from the main group of *Cercopithecus* monkeys in a distinctive — and almost opposite — direction. Patas monkeys are quite large, reaching a weight of sixty pounds in adult males (Tappen 1960). They have limbs that are proportionately longer and more slender than those of other *Cercopithecus* monkeys, and these limbs terminate in compact hands and feet which bear short, stubby digits. All these features represent adaptations to rapid quadrupedal locomotion on the ground (Hill 1966:654).

Talapoin monkeys differ from other guenons chiefly in their small size. They attain a maximum weight of about four pounds (Tappen 1960). The only other unusual features seem to be the occurrence in the female of a high degree of catamenial swelling and the typical (although not universal) occurrence in some adults of tritubercular (three-cusped) third upper molars (Hill 1966:599 incorrectly states that both upper and lower molars are tritubercular). This last feature of the dentition may be explained sufficiently by the small size of the animal. In small jaws dental reductions and simplifications both occur frequently at the posterior part of the dental arch. However, tritubercular molars are also of possible significance because they are a primitive feature. One well-supported and widely-accepted theory of dental evolution traces the highly diverse molar patterns of later placental mammals back to primitive tritubercular types (Gregory 1922:150). This point takes on added significance when it is noted that the jaw of the talapoin matches very closely in size, general shape, and even a number of finer details of tooth morphology with that of *Parapithecus* (Simons 1972:191) from the Oligocene. Also impressive is the observation that the ecological setting reconstructed for the Egyptian Fayum deposits from which *Parapithecus* was recovered is that of gallery forest bordering rivers — the same type of habitat as that in which the talapoin is often found living today (Rowell 1972). Some of the behavioral observations fit with (though they do not require) the hypothesis that the talapoin is a relatively primitive survivor among African monkeys. For example Rowell (1973) stressed that the social organization of the talapoin is unlike that of any other Old World monkey and is paralleled most closely by that of the ecologically similar squirrel monkey (*Saimiri sciureus*) of the New World — a member of a group that is generally recognized as being more primitive phylogenetically than Old World monkeys. These fascinating features of talapoin biology may be mere coincidences, due to parallel or convergent adaptation and evolution. But if they are not, *Cercopithecus talapoin* may well represent the oldest surviving species of its present genus and a close living relative of the basal populations from which other species of the genus have been derived.

With regard to the course of this derivation there are several possibilities. Among these the one that is karyotypically simplest would derive all other *Cercopithecus* species, including the patas and the talapoin, from *C. nigroviridis* by a series of increments in diploid chromosome number. Another somewhat more complex possibility takes into account what may be the most primitive characteristics of the talapoin. Perhaps evolutionary changes including a reduction in 2n chromosome number from 54 to 48 gave rise to the ancestors of *C. nigroviridis*, from which subsequent developments led to the radiation of mangabeys, macaques, and baboons. An independent series of stepwise increases in diploid chromosome number from 2n = 54 could have given rise to other members of the genus *Cercopithecus*.

It does seem probable, overall, that the general progression within the genus *Cercopithecus* has been from lower diploid numbers to higher ones, although perhaps not necessarily in a strictly linear sequence. In terms of the causes for this chromosomal evolution, there seems merit in the arguments of Hamerton (1958) and Egozcue (1969) that the primary evolutionary significance of chromosomal changes is that they produce genetic and reproductive isolation. Tappen (1960) has emphasized the strong factors at work to produce speciation among African forest monkeys. The rather discontinuous distribution of many species suggests that populations have in the past frequently expanded over wide areas when ecological conditions were favorable, then secondarily become eliminated in many areas as climate changed and forest cover disappeared. The resulting isolation of small groups would have favored the rapid establishment of changes in the chromosome complement. Moreover, these differences would have kept the populations distinct when the next climate fluctuation made possible the spread of forests and contact between previously isolated faunal components. Indeed, in areas of overlap, selection for chromosomal differences between populations, as well as for other isolating mechanisms at all levels, might have been intensified. This would occur because gametes expended in matings between individuals from different populations might have resulted in the production of hybrids which were either less fertile or less well adapted than members of either parent population.

It might be noted in closing here that this general explanation for the diversification of chromosome numbers in the genus *Cercopithecus* in terms of its value in promoting isolation also provides a basis for understanding the situation seen in the specific case of *C. (Erythrocebus) patas.* This is one of the most widely-distributed species of the genus, and it has — so far as is now known — a uniform diploid chromosome number over the whole range. Somewhat surprisingly, it combines a karyotype rather similar to that of *C. talapoin* with a strikingly different morphological appearance. It is tempting to offer as a working hypothesis the suggestion that all these observations can be accounted for by assuming that the patas monkey has diverged from the ancestral stock represented by the talapoin fairly recently in geological time, perhaps during the Pilocene, when savanna regions became rather more widespread than before.

Adaptations by the ancestors of patas monkeys to this environment would have made them more wide-ranging than other species in the genus, and this would in turn have made it less likely that isolation and resulting chromosomal divergences would have taken place. At the same time, the patas monkey would have become ecologically isolated from forest-living populations of *Cercopithecus* monkeys, including the talapoin. Consequently, genetic isolating mechanisms would have been largely superfluous. If this reconstruction is correct, it would provide us with a case of extremely rapid morphological evolution combined with little change at all at the chromosomal level.

The present discussion has touched briefly on just a few of the specific reconstructions of the course of primate evolution that can be made when the data on chromosome differences are integrated into the general framework provided by primate evolutionary biology. Much more synthetic work of this sort remains to be done, and it can be accomplished with the results now at hand.

REFERENCES

ARDITO, G., J. MORTELMANS
1975 Cytology of the Cercopithecidae: the banding pattern of *Cercopithecus nigroviridis* Pocock 1907. *Journal of Human Evolution* 4(5):377–381.

BENDER, M. A., E. H. Y. CHU
1963 "The chromosomes of primates," in *Evolutionary and genetic biology of primates*, volume one. Edited by J. Buettner-Janusch. New York: Academic.

CHIARELLI, B.
1968 Chromosome polymorphism in the species of the genus *Cercopithecus*. *Cytologia* 33:1–16.

CHU, E. H. Y., M. A. BENDER
1962 Cytogenetics and evolution of primates. *Annals of the New York Academy of Sciences* 102:253–266.

DELSON, E.
1975 "Paleoecology and zoogeography of the Old World monkeys," in *Primate functional morphology and evolution*. Edited by Russell H. Tuttle, 37–64. The Hague: Mouton.

EGOZCUE, J.
1969 "Primates," in *Comparative mammalian cytogenetics*. Edited by K. Benirschke, 357–389. New York: Springer.

GREGORY, W. K.
1922 *The origin and evolution of the human dentition*. Baltimore: Williams and Wilkins.

HAMERTON, J. L.
1958 Problems in mammalian cytotaxonomy. *Proceedings of the Linnaean Society of London* 169:112.

HILL, W. C. O.
1964 XXI. The external anatomy of *Allenopithecus*. *Proceedings of the Royal Society of Edinburgh*, Section B: Biology 68(21):302–326.
1966 *Primates, comparative anatomy and taxonomy. VI. Catarrhini (Cercopithecoidea, Cercopithecinae)*. New York: Inter-Science.

KUHN, H.-J.
1967 "Zur systematik der Cercopithecidae," in *Neue Ergebnisse der Primatologie*. Edited by D. Starck, R. Schneider, and H. -J. Kuhn, 25–46. Stuttgart: Gustav Fischer.

LANG, H.
1923 A new genus of African monkey, *Allenopithecus*. *American Museum Novitates* 87:1–5.

MATTHEY, R.
1964 Evolution chromosomique et spéciation chez les *Mus* du sous-genre *Leggada* Gray 1837. *Experientia* 20:657–666.

MURRAY, P.
1975 "The role of cheek pouches in cercopithecine monkey adaptive strategy," in *Primate functional morphology and evolution*. Edited by Russell H. Tuttle, 151–194. The Hague: Mouton.

PILBEAM, D. R., A. C. WALKER
1968 Fossil monkeys from the Miocene of Napak, northeast Uganda. *Nature* 220:657–660.

ROWELL, T. E.
1972 Toward a natural history of the talapoin monkey in Cameroon. *Annales de la Faculté des Sciences du Cameroun* 10:121–134.
1973 Social organization of wild talapoin monkeys. *American Journal of Physical Anthropology* 38(2):593–597.

SIMONS, E. L.
1969 A Miocene monkey (*Prohylobates*) from northern Egypt. *Nature* 223:687–689.
1972 *Primate evolution*. New York: Macmillan.

TAPPEN, N.
1960 Problems of distribution and adaptation of the African monkeys. *Current Anthropology* 10:91–120.

WURSTER, D., K. BENIRSCHKE
1969 Chromosomes of some primates. *Mammalian Chromosome Newsletters* 10:3–5.

Comparison of Banding Patterns in Man and Primates and Their Evolutionary Significance

CATHERINE TURLEAU and J. DE GROUCHY

Conventional techniques have shown that the chromosome number is 46 in man and 48 in the three hominoid primates, and that there is some degree of similarity between the karyotypes of these species.

Precise comparisons were limited, however, due to the absence of any criteria other than size and centromeric index of the chromosomes (Bender and Chu 1963; Chiarelli 1961, editor 1971; De Grouchy et al. 1972; Lejeune et al. 1973; Turleau and De Grouchy 1972). New banding techniques have permitted a more precise study of the karyotypes of these species (Bobrow and Madan 1973; Chiarelli and Lin 1972; Dutrillaux and Lejeune 1971; Hamerton et al. 1963; Pearson 1971; Ruffié et al. 1970; Turleau and De Grouchy 1973; Turleau, Ruffié et al. 1972; Vallois 1955).

MATERIAL AND METHODS

Four chimpanzees (two males and two females), two gorillas, and two orangutans (one male and one female of each species) were studied.

Karyotypic preparations obtained from whole-blood cultures were submitted to controlled heating denaturation according to Dutrillaux and Lejeune (1971) and stained with Giemsa (R-bands). The karyotype of one of the chimpanzees was studied after enzyme digestion with trypsin (G-bands).

RESULTS

The banding techniques used for man are applicable to the study of the karyotype of the hominoid apes. Banding patterns comparable to those observed in man are obtained.

According to the phylogenic tree proposed by Vallois (1955) and shown in Figure 1, the following comparisons were made: man–chimpanzee, in Plate 1; man–orangutan, in Plate 2; and chimpanzee–gorilla, in Plate 3. For each comparison, the karyotype of the first-mentioned species was established. Then the best matching chromosome from the second species was selected on the basis of the banding pattern.

The homologous chromosomes thus recognized were then compared in an attempt to establish the phylogenic evolution of each chromosome in the primate complement and to identify the ancestral chromosome (or the closest chromosome possible) from which the others could be derived.

CHROMOSOME 1. The homologous chromosome of the human 1 is similar in the three apes. It has the same banding pattern as the human 1, but the subcentromeric constriction present in man is not present in the apes. The

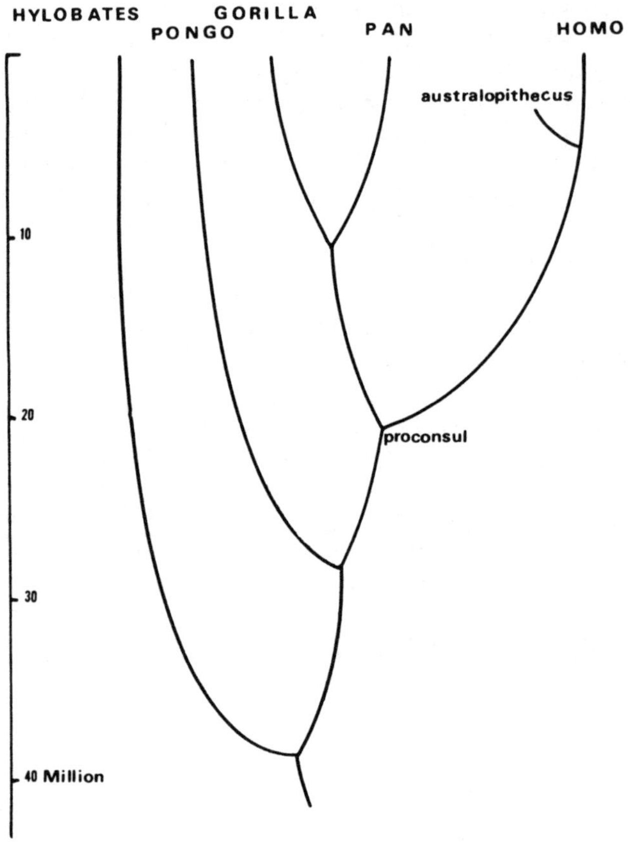

Figure 1. Phylogenic tree of man and the anthropoid apes

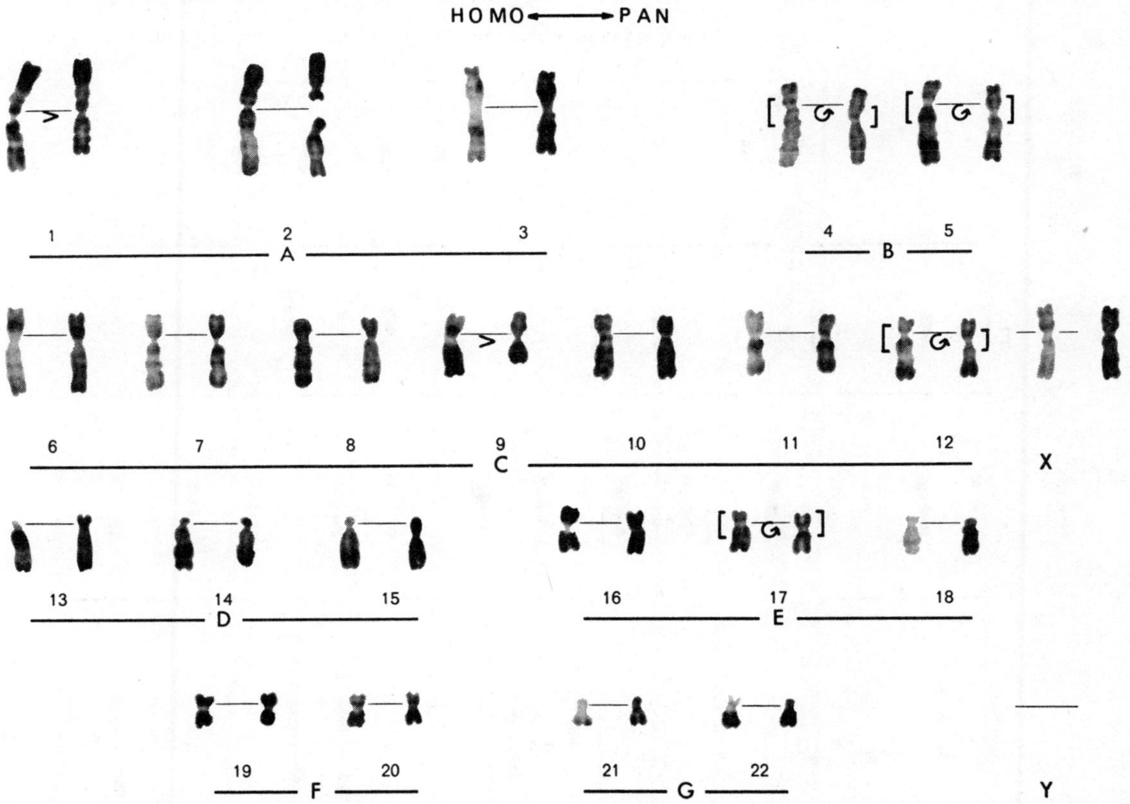

Plate 1. Comparison of human and chimpanzee karyotypes after heat denaturation (R-bands). A curved arrow indicates a pericentric inversion. The signs > and < indicate loss or acquisition of chromosomal material

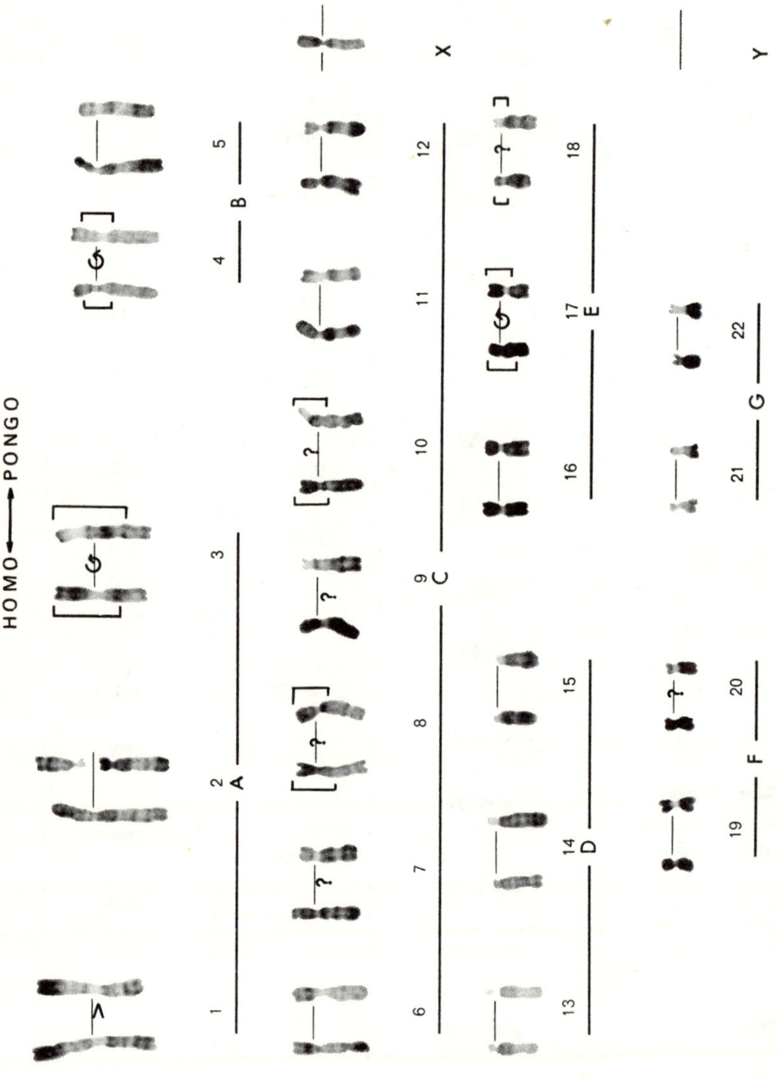

Plate 2. Comparison of human and orangutan karyotypes after heat denaturation (R-bands)

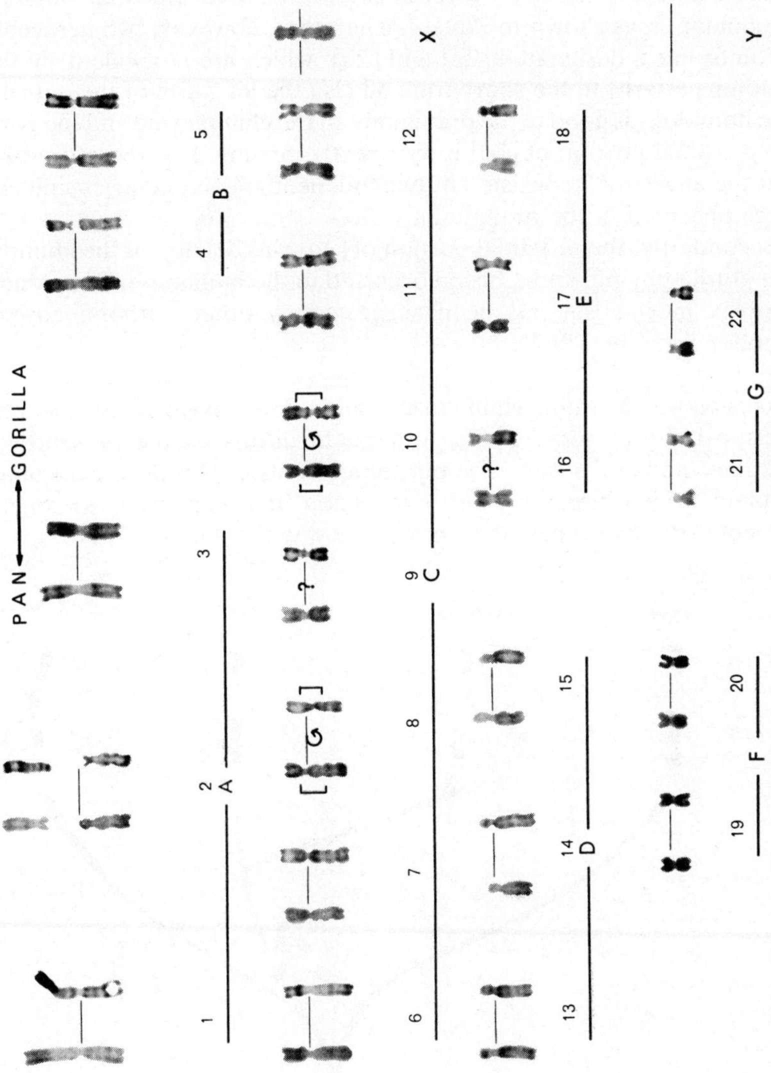

Plate 3. Comparison of chimpanzee and gorilla karyotypes after heat denaturation (R-bands)

ancestral chromosome 1 must have been similar to that observed today in the apes and the secondary constriction acquired in the human phylum.

CHROMOSOME 2. None of the three apes possesses a chromosome similar to the human 2, as shown in Plate 4. There are, however, two acrocentric chromosomes, designated [2p] and [2q], which are equivalent by their banding patterns to the short arm and also the long arm of the human 2. The homology is good in the orangutan. In the chimpanzee and the gorilla the proximal portion of [2q] is apparently missing. It is therefore likely that the ancestral 2 consisted of two independant acrocentrics similar to those observed in the orangutan.

Secondarily, the proximal portion of [2q] was deleted in the chimpanzee–gorilla phylum and a fusion occurred in the human phylum. A more complex mechanism has been suggested by other authors, however (Bender and Chu 1963).

CHROMOSOME 3. Man, chimpanzee, and gorilla possess an apparently identical chromosome 3. The orangutan differs by the presence of a chromosome of the same size but subtelocentric. This difference may be explained by a large pericentric inversion. It is impossible to conclude which of the two types of chromosome was the ancestral one.

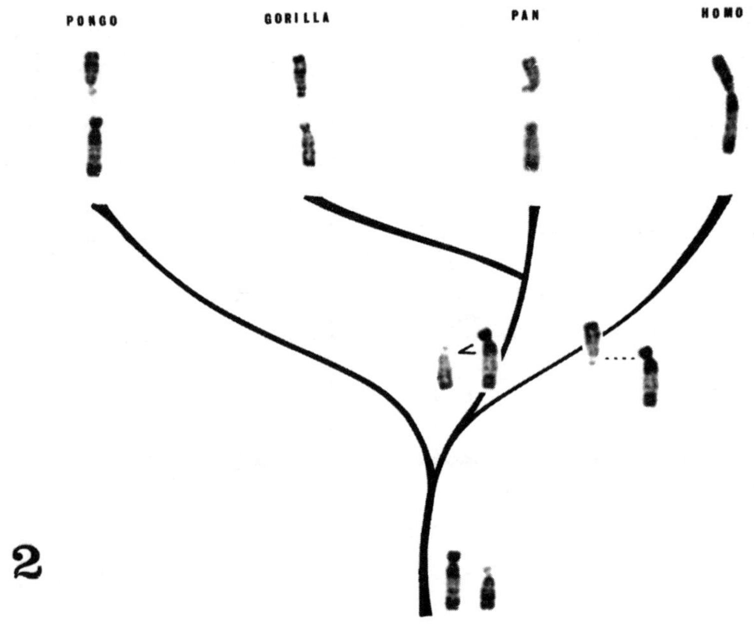

Plate 4. Phylogenic tree of chromosome 2

CHROMOSOME 4. Chimpanzee, gorilla, and orangutan possess a chromosome equivalent to the human 4 but more mediocentric, as shown in Plate 5. This difference may be explained by a pericentric inversion having occurred in the human phylum, the ancestral 4 being similar to that observed in the apes.

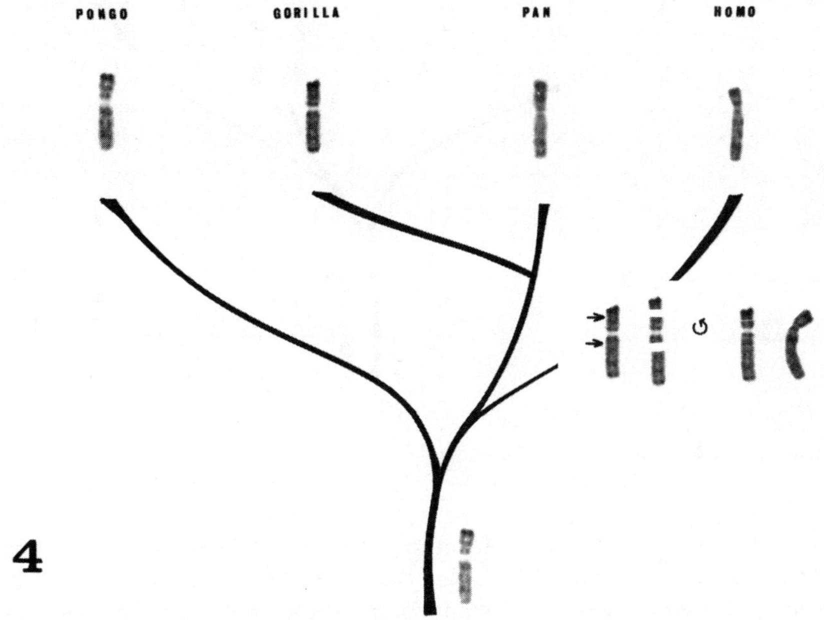

Plate 5. Phylogenic tree of chromosome 4

CHROMOSOME 5. Man and orangutan possess the same 5, while gorilla and chimpanzee possess a more mediocentric chromosome similar to the 4 observed in the apes; see Plate 6. This may be explained by a pericentric inversion having occurred in the chimpanzee–gorilla phylum, the ancestral 5 being of the human–orangutan type.

CHROMOSOME 6. This chromosome appears to be most stable. It is similar in all four species and has probably not varied since the common ancestor.

CHROMOSOME 7. This chromosome is similar in man, chimpanzee, and gorilla. There is no obvious equivalent in the orangutan. The chromosome in Plate 2 was chosen because it is always heavily stained like the human 7. Its banding pattern is similar, mostly for the distal part of the

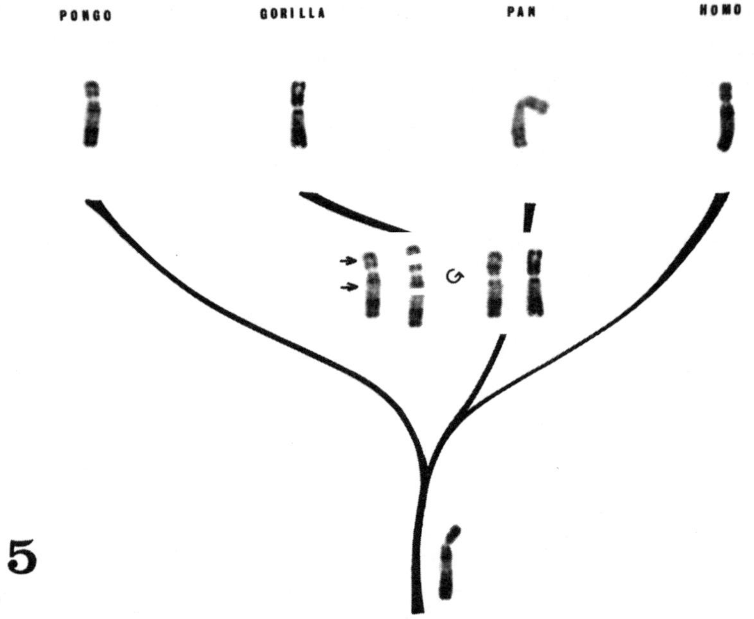

Plate 6. Phylogenic tree of chromosome 5

long arm. It differs markedly from the short arm. The overall size of this chromosome is apparently smaller than the human 7. It would not be meaningful to suggest a rearrangement responsible for the observed differences.

CHROMOSOME 8. This is similar in man and chimpanzee and not very much different in the orangutan. It is mediocentric in the gorilla. It is likely that the ancestral 8 was of the human or orangutan type and that a pericentric inversion occurred in the gorilla phylum.

CHROMOSOME 9. There is no obvious equivalent to the human 9 in the three apes. Chromosome 9 is different in all four species. It is acrocentric in the orangutan; mediocentric and small in the gorilla, with a secondary constriction; also mediocentric in the chimpanzee, but with no constriction and with a different banding pattern. The most economical choice for the ancestral chromosome would be a chromosome similar to that of man or the orangutan.

CHROMOSOME 10. This chromosome is similar in man and chimpanzee. It is mediocentric in the gorilla and acrocentric in the orangutan. The ancestral chromosome may have been of the human or orangutan type. A pericentric inversion occurred in the gorilla.

CHROMOSOME 11. Like chromosome 6, chromosome 11 appears to have been remarkably stable during evolution.

CHROMOSOME 12. This chromosome is similar in man and orangutan. It is identical in the chimpanzee and the gorilla but more mediocentric than that of man or orangutan. Therefore, the ancestral chromosome must have been of the human–orangutan type, and a pericentric inversion (Plate 1) must have occurred in the gorilla–chimpanzee phylum.

D-GROUP CHROMOSOMES. Within the limits of accuracy of the method, the D-group chromosomes appear to have been stable during evolution.

CHROMOSOME 16. This chromosome is more or less mediocentric in all four species. It carries in man a secondary constriction which is not present in the chimpanzee or the orangutan. In the gorilla it is more telocentric and carries a secondary constriction. Like chromosome 1, it probably acquired a secondary constriction in man. The ancestral 16 must have been of the chimpanzee–orangutan type.

CHROMOSOME 17. This is mediocentric in the three species of ape. A pericentric inversion must have occurred in man, the ancestral chromosome being of the simian type.

CHROMOSOME 18. Similar in the three apes, chromosome 18 is slightly more mediocentric in man. A small pericentric inversion may have occurred in the latter. The ancestral chromosome was therefore of the simian type.

CHROMOSOME 19. This is similar in all four species.

CHROMOSOME 20. Similar in man, chimpanzee, and gorilla, chromosome 20 is more telocentric in the orangutan. The ancestral chromosome cannot be determined.

CHROMOSOME 21. This is similar in all four species.

CHROMOSOME 22. This too is similar in all species.

CHROMOSOME X. Once again this chromosome is similar in all species.

CHROMOSOME Y. Apparently smaller in the chimpanzee than in the three other species, the Y chromosome may have lost the distal part of its long arm in the former.

THE CASE OF THE 49, XY, MAR + GORILLA. The karyotype of the male gorilla studied by us showed the presence of a supernumerary marker chromosome. This marker is the size of a G and carries in good metaphases up to three satellites, the medium one being more conspicuous than the two others. The denaturation banding pattern is compatible with trisomy 21.

DISCUSSION

Our results emphasize the major role played in primate evolution by pericentric inversions and by chromosome fusions.

These rearrangements are considered to be harmful in the human population, but only in the heterozygous state; they should not be in the homozygous state.

Considering their frequencies in the human population today, it is probable that they were not rare in ancestral primate populations. The rearrangements established during evolution represent an extremely small fraction of those which have actually occurred. The high degree of incestuous matings in primate populations and a weak advantage, associated by chance with the chromosomal rearrangement, could have established the rearrangement in the homozygous state in few generations. Groups of individuals thus isolated may then have evolved separately, due to newly occurring genic mutations, and become the seeds of new species.

REFERENCES

BENDER, M. A., E. H. Y. CHU

1963 The chromosomes of primates," in *Evolutionary and genetic biology of the primates*, volume one. Edited by J. Buettner-Janusch, 261–310. New York: Academic.

BOBROW, M., K. MADAN

1973 A comparison of chimpanzee and human chromosomes using the Giemsa-II and other chromosome banding techniques. *Cytogenetics and Cell Geneticis* 12:107–116.

CHIARELLI, B.

1961 Chromosomes of the orangutang (*Pongo pygmaeus*). *Nature* 192:285.

1962 Comparative morphometric analysis of primate chromosomes. I. The chromosomes of anthropoid apes and of man. *Caryologia* 15:99–121.

1973 *Evolution of the primates*. New York: Academic.

CHIARELLI, B., *editor*

1971 *Comparative genetics in monkeys, apes and man,* volume 1. New York: Academic.

CHIARELLI, B., C. C. LIN

1972 Comparison of fluorescence patterns in human and chimpanzee chromosomes. *Genen en Phaenen* 15:103–106.

DE GROUCHY, J., C. TURLEAU, M. ROUBIN, F. CHAVIN-COLIN

1973 Chromosomal evolution of man and the primates (*Pan troglodytes, Gorilla gorilla, Pongo pygmaeus*). *Nobel Symposium* 23 (Stockholm, September 25–27, 1972):124–131.

DE GROUCHY, J., C. TURLEAU, M. ROUBIN, M. KLEIN

1972 Evolution caryotypique de l'homme et du chimpanzé. Etude comparative de topographies de bandes, après dénaturation ménagée. *Annales de Génétique* 15:79–84.

DUTRILLAUX, B., J. LEJEUNE

1971 Sur une nouvelle technique d'analyse du caryotype humain. *Comptes Rendus Hebdomadaires des Sciences de l'Académie des Sciences* 272:2,638–2,640.

HAMERTON, J. L., H. P. KLINGER, D. E. MUTTON, E. M. LANG

1963 The somatic chromosomes of the Hominoidea. *Cytogenetics* 2:240–263.

LEJEUNE, J., B. DUTRILLAUX, M. O. RETHORE, M. PRIEUR

1973 Comparaison de la structure fine des chromatides d'*Homo sapiens* et de *Pan troglodytes. Chromosoma* 43:423–444.

PEARSON, P. C.

1971 Evolutionary studies in the hominoid apes by chromosome fluorescence. *Proceedings of the IV International Congress of Human Genetics, Paris.* Amsterdam: Excerpta Medica.

RUFFIÉ, J., P. COLOMBIES, C. GINOUX-MOUNIE, E. CARLES-TROCHAIN

1970 Etude cytogénétique de 4 espèces de primates. Comparaison avec le caryotype humain. *Annales de Génétique* 13:3–10.

TURLEAU, C., J. DE GROUCHY

1972 Caryotypes de l'homme et du chimpanzé. Comparaison de la topographie des bandes. Mécanismes évolutifs possibles. *Comptes Rendus Hebdomadaires des Sciences de l'Académie des Sciences* 274 (series D):2,355–2,357.

1973 New observations on the human and chimpanzee karyotypes. Identification of breakage points of pericentric inversions. *Journal of Human Genetics* 20:151–157.

TURLEAU, C., J. DE GROUCHY, M. KLEIN

1972 Phylogénie chromosomique de l'homme et des primates hominiens (*Pan troglodytes, Gorilla gorilla,* et *Pongo pygmaeus*). Essai de reconstitution du caryotype de l'ancêtre commun. *Annales de Génétique* 15:225–240.

VALLOIS, H.

1955 "Ordre des primates," in *Traité de zoologie*, volume XVII/II. Edited by G. Grasset. Paris: Masson.

ABSTRACT:

A Comparison of the Fluorescent Karyotypes of the Chimpanzee (Pan troglodytes) *and Man*

C. C. LIN and B. CHIARELLI

Individual chimpanzee chromosomes from twenty-two individuals (seven males, fifteen females) have been identified by their characteristic banding revealed by quinacrine fluorescent staining. A fluorescent karyotype of this species was set up to be compared with the standard human fluorescent karyotype. It was found that chromosomes 1, 3, 11, 12, 14, and X of the chimpanzee appear to have banding patterns similar to the equivalent human chromosomes. Chromosomes 6, 7, 8, 10, and 13 also had a fluorescent pattern corresponding to the human chromosomes of the same number, particularly in the long arm. Remarkable variation in intensity and/or size of fluorescent regions was frequently found in the short arm of satellited acrocentric chromosomes 13, 14, 15, 22, and 23. Variations occurred between homologues and between individuals. Such variable fluorescence in a specific chromosomal region of an individual animal is a reproducible characteristic. Unlike that in its human counterpart, the distal segment in the long arm of the chimpanzee's Y chromosome is not brightly fluorescent.

A full report on these results has been published in the *Journal of Human Evolution* (1973) 2:311–321.

Related Satellite DNA in Man and Higher Primates

K. W. JONES

The DNA of constitutive heterochromatin contains highly repetitive nucleotide sequences which may be isolated on isopycnic gradients of caesium salts by virtue of their unique buoyant densities. Such DNA classes are referred to as satellite DNAs because of their separation from the bulk of DNA under these conditions of purification. Satellite DNAs afford readily available and pure fractions of DNA and as such have been intensively studied in recent years. Their function is unknown and they appear to be transcriptionally inert. From *in situ* hybridization experiments (Jones, this volume) it has been shown, first in mouse (Pardue and Gall 1970), and subsequently in man (Jones and Corneo 1971), that satellite DNA is predominantly found in heterochromatin situated close to the centromeres of chromosomes. In man there are five well-authenticated satellites known at present (reviewed in Jones, et al. 1973), two of which have been mapped chromosomally.[1] Satellite II is found predominantly near the centromeres of chromosomes 1 and 16 with lesser amounts on chromosome 9 and minor amounts on the D, F, and G group (Jones and Corneo 1971). Satellite III is prominently localized on chromosome 9 with minor locations on the D-, F-, and G-group chromosomes (Jones, et al. 1973). Human satellite III hybridizes prominently with the chromosomes of the chimpanzee, the gorilla, and the orangutan, indicating the presence of related heterochromatic nucleotide sequences

This work was supported by grants from the Medical Research Council, the Cancer Campaign, and the Muscular Dystrophy Group of Great Britain. Plate 1 is reproduced with the permission of the editors of *Chromosoma*; Plates 4, 5, and 7 are reproduced with the permission of F. K. Schattauer Verlag, Stuttgart-New York. The provision of chromosome preparations of gorilla by Dr. Mary Lucas is greatfully acknowledged.

[1] Since this manuscript was compiled we have found clear evidence of the location of human satellite I on a restricted region of the Y chromosome.

and a common ancestry in this group as briefly demonstrated and discussed here. A previous report has appeared describing some of these studies (Jones et al. 1972).

CHROMOSOMAL LOCATION OF HUMAN SATELLITE DNA III

Radioactive RNA synthesized on a template of pure satellite DNA III *in vitro* and hybridized to human chromosomes according to the methods detailed in a previous paper (Jones, this volume) becomes concentrated mainly in the centromeric heterochromatin of chromosome 9, as shown in Plate 1. Minor locations are also seen in the centromeric heterochromatin of the chromosomes of the D, F, and G groups. Interphase nuclei show concentrations of grains predominantly in two locations, as well as some minor locations, consistent with the condensed nature of heterochromatin at this stage of the cell cycle in the cells examined. In the karyotype shown, chromosome 9 is identified by means of a pericentric inversion.

CROSS-HYBRIDIZATION OF SATELLITE III WITH PRIMATE CHROMOSOMES

Satellite III cRNA was annealed with chromosome preparations of chimpanzee, orangutan, and gorilla to study the possible existence of a related

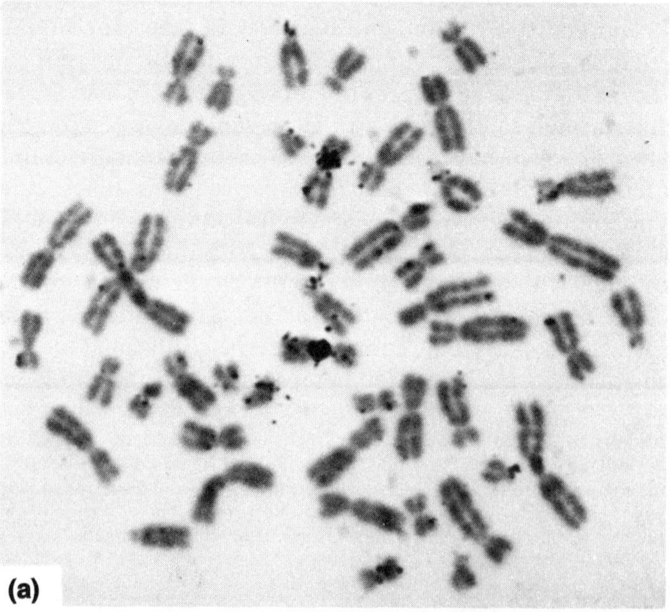

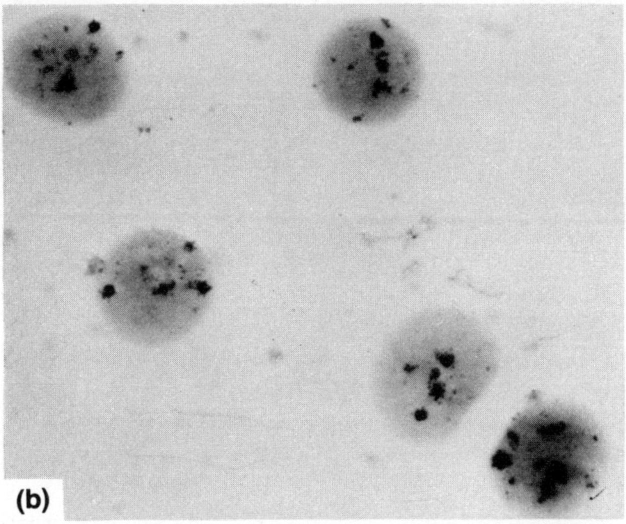

Plate 1. (a) Human chromosomes hybridized *in situ* with human satellite III cRNA. Concentrations of silver grains can be seen over the centromeric heterochromatin of chromosome 9. The homologues are identified by a pericentric inversion of the hybridized region on our member. Minor concentrations of grains are associated with centromeric regions of D, F, and G group chromosomes

(b) Interphase nuclear preparation of the same material showing concentrations of grains over condensed chromatin, indicating the heterochromatic nature of the regions hybridized (adapted from Jones et al. 1973)

fraction. The results obtained are illustrated in Plates 2, 3, and 4. It is immediately apparent that multiple sites exist in all three ape species which cross-hybridize with human satellite cRNA III. The cross-reacting sites in apes in fact exceed the homologous (human) sites by a significant number in that about twenty strongly hybridized locations are seen, compared with only two in the human karyotype. It is clear from this result that a heterochromatic DNA exists in the higher primates with base sequence complementarity to human satellite III DNA and that, probably, related satellite DNAs are involved. It is striking that all the apes studied exhibit rather similar extents of cross-hybridization as judged from these results, and are in this respect uniformly distinct from man.

THE ISOLATION OF RELATED SATELLITE DNAs FROM NONHUMAN PRIMATES

The implication of the foregoing result, namely the existence of related DNA satellites, has been followed up in work aimed at isolating these satellites for further study. This work has been hampered to a great extent by the difficulty of obtaining tissues for DNA extraction, but it was

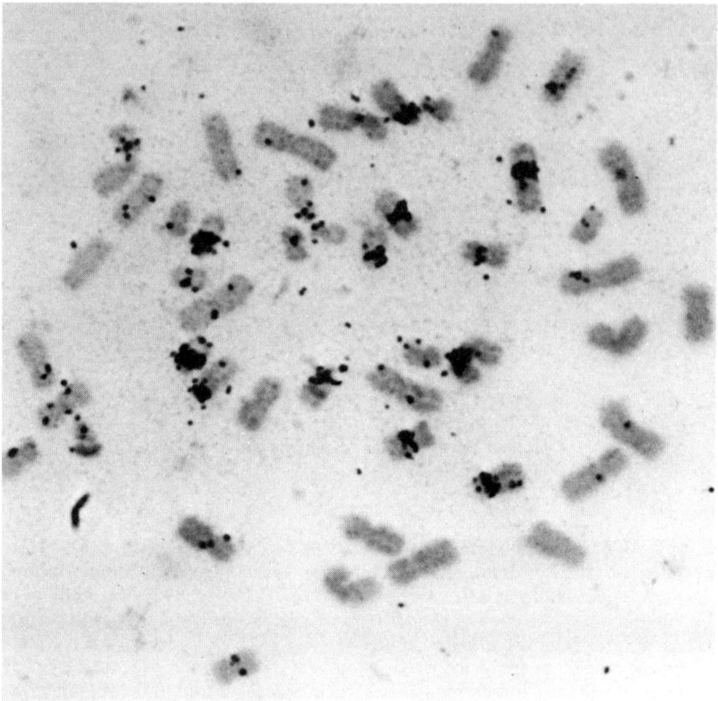

Plate 2. Human satellite III cRNA hybridized *in situ* with chimpanzee chromosomes. Note the numerous pericentric locations of silver grains indicating the formation of heterologous hybrid molecules with DNA sequences in constitutive heterochromatin in these regions, and suggesting the presence of a related satellite DNA in the chimpanzee. This satellite has been isolated and called satellite A (see text)

markedly advanced by the acquisition of a sample of chimpanzee liver from which DNA of reasonable quality was obtained. From this, two satellite DNAs were isolated (Prosser et al. 1973) and named satellites A and B. Satellite A bore a strong likeness to human satellite III in its physical properties (Table 1) and was selected for further study. The cRNA synthesized on a template of this DNA was hybridized to chimpanzee chromosomes, where it localized in centromeric heterochromatin on eight to ten pairs of chromosomes, as shown in Plate 5. These had a size distribution similar to those which reacted with human satellite III cRNA, but they were not classified exactly because of the difficulty of employing chromosome-banding methods using previously hybridized chromosome preparations. Satellite A cRNA, however, localized on exactly the same human chromosomes as did human satellite III cRNA, as shown in Plate 6, thus confirming the molecular homology indicated in the initial work.

When hybridized to orangutan chromosomes, chimpanzee satellite A

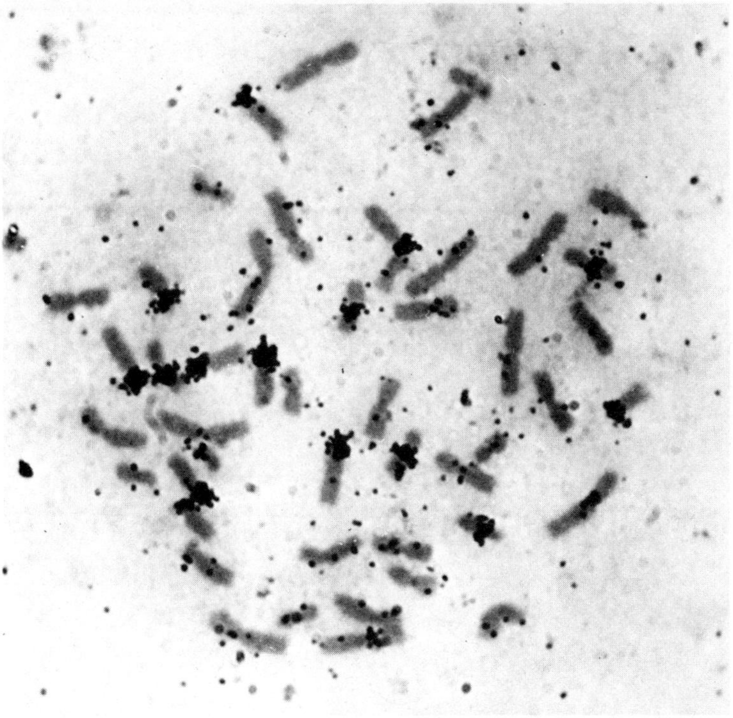

Plate 3. Human satellite III cRNA hybridized with chromosomes of the gorilla. As in the chimpanzee, many strongly hybridizing pericentric regions are seen, implying similar conclusions. The satellite DNA of the gorilla has not been isolated

Table 1. Physical properties of chimpanzee satellite A and human satellite III

	Buoyant density	T_m(°C)	Percent increase in hyperchromicity
Chimpanzee satellite A			
native	1.696	69	32
denatured	1.713		
renatured	1.706	60	18
Human satellite III			
native	1.696[a]	69[b]	30
denatured	1.715[a]		
renatured	1.703[a]	62[b]	–

[a] Taken from Corneo et al. 1972.
[b] Derived from Corneo et al. 1972.

cRNA also localized on the same chromosome group (predominantly the acrocentric group) as did the human satellite III complementary RNA, as shown in Plate 7. This cross-reaction is strongly indicative of the relatedness of the heterochromatic DNA sequences of the three species and

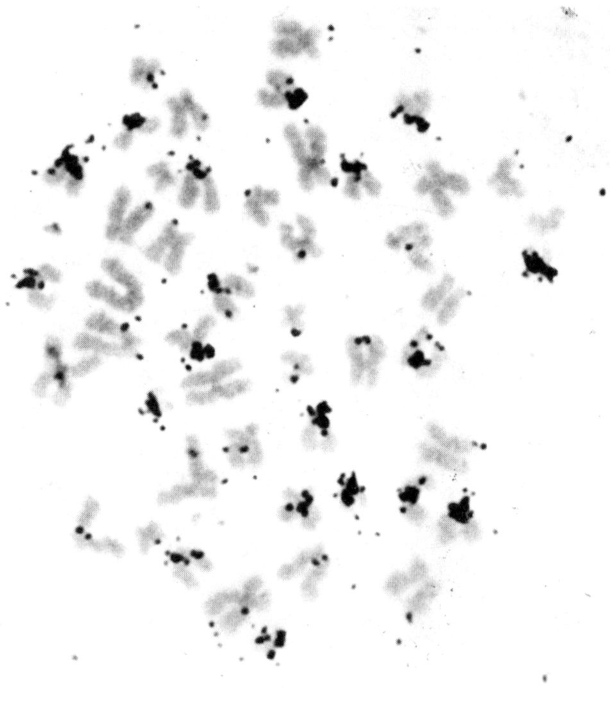

Plate 4. Human satellite III hybridized with chromosomes of the orangutan. Many sites, mainly on the acrocentric chromosomes, have hybridized strongly, indicating related DNA. A satellite fraction has been found in the orangutan (see text) but not characterized as yet

initiated a search for a related satellite DNA in the orangutan. Placenta was obtained from Dr Kurt Benirschke, the DNA of which yielded a rather indistinct satellite fraction. The fractionated orangutan DNA was hybridized with human satellite III cRNA and shown to possess a homologous fraction, but the nature of the fraction hybridized is not yet established. Enough data exists, however, to corroborate the *in situ* result in showing that DNA related to satellite DNA III in humans is present in the orangutan.

Gorilla DNA has not yet become available to us in order to isolate and characterize the related satellite DNA implied by the *in situ* hybridization results.

DO MONKEYS CONTAIN SATELLITE DNA RELATED TO HUMAN SATELLITE DNA III?

The question of whether the related satellite DNAs of the higher primates are exclusive to that group has not been examined in detail. *In situ*

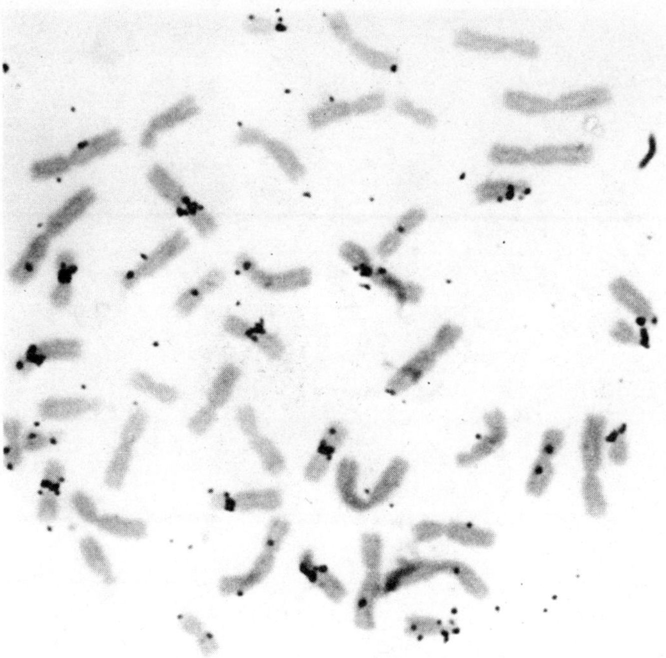

Plate 5. Chimpanzee satellite A cRNA hybridized with chimpanzee chromsomes. A similar number of chromosomes with similar size distribution have hybridized as was found after hybridization with human satellite III cRNA. This is consistent with the existence of related DNA sequences in the satellites of man and chimpanzee suggested by the previous data

hybridization of human satellite III cRNA to chromosomes of the capuchin (*Cebus capuchinus*) and the baboon (*Papio cynocephalus*) has yielded negative results. A satellite fraction has been isolated from the baboon (Prosser et al. 1973). Baboon DNA, however, does not hybridize above background with human satellite III cRNA. Within the compass of these techniques, therefore, it seems likely that the baboon does not possess detectable amounts of DNA homologous to satellite III, and the same is probably true of the capuchin monkey. It therefore appears quite likely that satellite III and its relatives are confined to the great apes and man. The position of the siamang and of the gibbon has not been examined. These two species are of great interest in view of their intermediate position in the suggested evolutionary tree of the most advanced primates (Simons 1964). Chromosomes and tissues of these species are currently sought for investigation.

The related satellite sequences in the higher primates are descended from a DNA sequence of a common ancestral species. It is suggested that this ancestral DNA gave rise to a satellite DNA subsequent to the divergence of the lineages of apes and monkeys and that this original

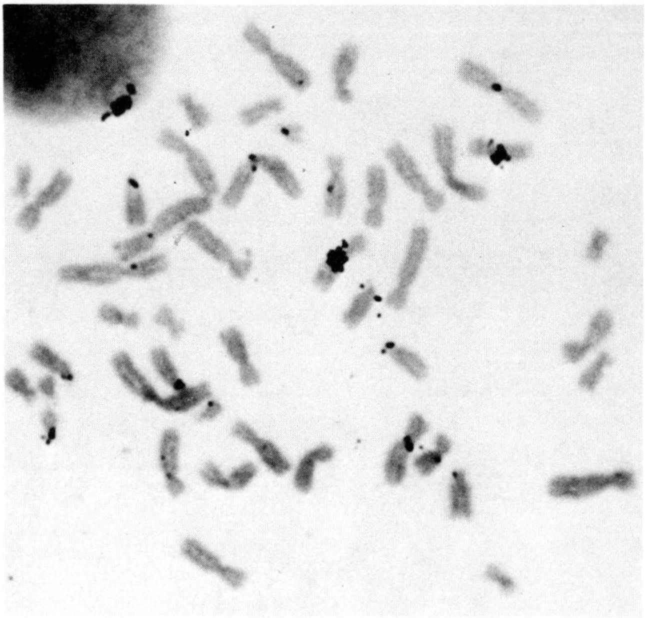

Plate 6. Human chromosomes hybridized with chimpanzee satellite A cRNA. The hybridization pattern is indistinguishable from that obtained with human satellite III cRNA, the number 9 chromosomes being the most strongly involved. This confirms the molecular relatedness of the two satellite fractions of man and chimpanzee

satellite DNA was inherited by all the living branches which subsequently diverged. According to some authorities (see Simons 1964), the gibbon diverged as a separate line close to the time of divergence of apes and monkeys, so that the question of whether or not it possesses the related satellite DNA is of great interest for future studies.

ARE THERE OTHER RELATED SATELLITE DNAs IN THE HIGHER PRIMATES?

Man has at least five authenticated satellite DNAs, one of which (satellite III) is related to satellite DNAs in the higher apes. Are the others similarly related? According to Southern (1970), satellite DNAs may evolve by a process in which part of an original repetitive sequence becomes tandemly repeated. This process is envisaged as being continued at various times in evolution so that, in principle, all satellite DNAs could be related or descended from preexisting satellite DNAs. Whether the "progeny" satellites will cross-hybridize depends upon whether or not a representative portion of the "parental" sequence became amplified.

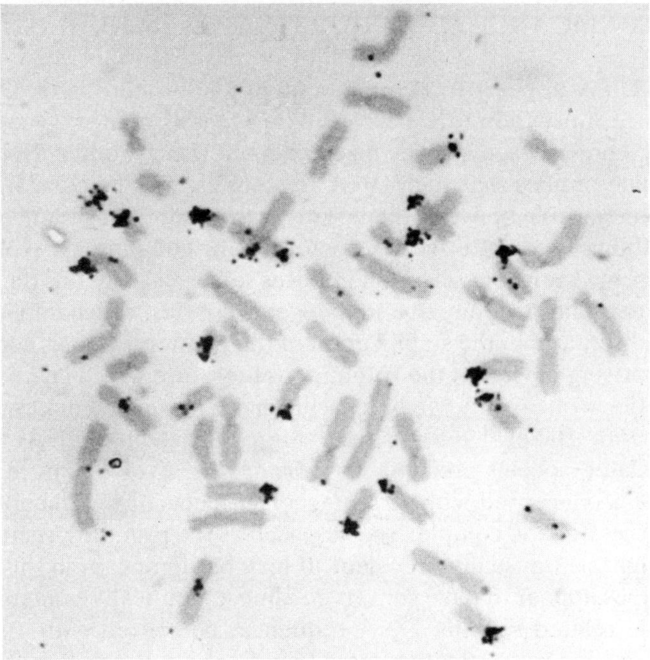

Plate 7. Chimpanzee satellite A cRNA hybridized to orangutan chromosomes. The same group of acrocentrics has hybridized as with human satellite III cRNA, suggesting the existence of satellite DNA sequence homology in all three species

Moreover, the relatedness is further complicated by the accumulation of mutations in the original population. If such is the mechanism of satellite DNA generation in evolution, it is clear that absence of hybridization between satellites may not mean that they are not strictly speaking related. By the hybridization method we are therefore limited to looking at the relation between populations of nucleotide sequences and thus at the time of their amplification in relation to species divergence. Bearing this restriction in mind, the absence of hybridization above background between chimpanzee DNA and human satellites I and II suggests that these two DNA populations arose subsequent to the chimpanzee–human divergence. In the case of satellite II, the centromeric heterochromatin, in which it is concentrated in man, is missing from the otherwise homologous chimpanzee chromosome I (Bobrow and Madan 1973), suggesting that heterochromatin and satellite were contemporaneous in evolution. Consistent with this, the melting behavior of these two satellites (see Jones et al. 1972) suggests that they are relatively younger in evolution than is satellite III.

THE SIGNIFICANCE OF SATELLITE DNA IN EVOLUTION

Satellite DNA appears to be an essentially heterochromatic DNA. Its significance must therefore be considered in this context. As far as is presently known, each species has a characteristic satellite DNA spectrum, which implies that such DNA remains essentially constant within species but variable between them. This in turn implies similar constancy and variability in constitutive heterochromatin. The simplest explanation for such behavior is that selection imposes severe restrictions on changes in heterochromatin within, but not between, species, which implies a role for such chromatin in the stabilization of the phenotype. This conclusion is not surprising in view of the voluminous literature which testifies to the genetic effects of constitutive heterochromatin (reviewed in Lewis 1950; Brown 1966; Baker 1968). Conservation of satellite DNA between closely related species presumably indicates its involvement in genetic mechanisms concerned with advantageous aspects of the phenotype. In highly successful and complex species, such as the primates, such conservation might be unusually prevalent. It may be significant in this respect that the apes appear to have conserved almost identically numerous sites rich in the related satellite DNA sequences compared with man, who exhibits a considerable relative reduction. Such a pattern correlates well with the considerable gap in evolutionary performance which characterizes the two groups. Similar shifts in the amount of satellite DNA correlated with alterations in evolutionary specialization have been documented in the case of the kangaroo rat (Mazrimas and Hatch 1972). The actual role of satellite DNA in the function of heterochromatin is unknown, since this function itself is, largely, only empirically recognized. Since satellite DNA documents the evolutionary cycles of heterochromatin, however, its disappearance and renewal during speciation may testify to the occurrence of important alterations in phenotypic expression, in which case the addition of satellites I and II to the human genome and the relative reduction in satellite III may echo critical past events in human evolution. Such events may have involved the divergence of modern man from those less-successful hominoid forms which are known to have existed from the fossil record.

REFERENCES

BAKER, W. K.

1968 Position effect variegation. *Advances in Genetics* 14:133–169.

BOBROW, M., K. MADAN

1973 A comparison of chimpanzee and human chromosomes using the Giemsa-11 and other chromosome banding techniques. *Cytogenetics and Cell Genetics* 12:107–116.

BROWN, S. W.
1966 Heterochromatin. *Science* 151:417–425.

CORNEO, G., L. ZARDI, E. POLLI
1972 Elution of human satellite DNA on a methylated albumin kieselguhr chromatographic column. *Biochimica et Biophysica Acta* 269:201–204.

JONES, K. W., G. CORNEO
1971 Location of satellite and homogeneous DNA on human chromosomes. *Nature New Biology* 233:268–271.

JONES, K. W., G. CORNEO, E. GINELLI, J. PROSSER
1973 The chromosomal location of human satellite DNA 111. *Chromosoma* 42:445–451.

JONES, K.W., J. PROSSER, G. CORNEO, E. GINELLI, M. BOBROW
1972 Constitutive heterochromatin in man. *Modern Aspects of Cytogenetics:* 45–61.

LEWIS, E. B.
1950 The phenomenon of position effect. *Advances in Genetics* 3:73–115.

MAZRIMAS, J. A., F. T. HATCH
1972 A possible relationship between satellite DNA and the evolution of the kangaroo rat (genus *Dipodomys*). *Nature New Biology* 240:102–105.

PARDUE, M. L., J. G. GALL
1970 Chromosomal localisation of mouse satellite DNA. *Science* 168:1,356–1,358.

PROSSER, J., M. MOAR, M. BOBROW, K. W. JONES
1973 Satellite sequences in chimpanzee (*Pan troglodytes*). *Biochimica et Biophysica Acta* 319:122–134.

SIMONS, E. L.
1964 The early relatives of man. *Scientific American* 211:50–62.

SOUTHERN, E. M.
1970 Base sequences and evolution of guinea-pig α satellite DNA. *Nature* 227:794–798.

Evolutionary Significance of the Chromosome Numbers in the Mammals: A Point of View

MADHURI DATTA

Although the evolution of mammals has long been a subject of interest to many biologists, it is only in recent years that attempts have been made to determine their relationships through the study of chromosomes. Since the chromosome number and morphology are characteristic of a species, analysis of karyotypes of different species has an obvious importance in taxonomic and evolutionary studies. In fact, the information regarding relationships within natural groups of species can be considered more complete in the evolutionary sense if the cytotaxonomic data reinforce the conclusions based on morphological criteria.

In the last decade or so, mainly due to refinement of techniques for handling mammalian chromosomes, a large number of mammalian species has been studied. There are still, however, more unsolved problems than solved ones. The present work deals with the systematic analysis of the mitotic chromosome number and morphology of twenty-five species of mammals, including several primates. The results have been generalized, however, to include different groups of mammals, and an attempt has been made to establish a correlation showing trends of chromosomal evolution in the eutherian mammals with special reference to primates.

Cytological studies on a wide range of species by different authors throughout the world have shown some trends of karyotype evolution in mammals. Painter (1925), from his study of ten species of mammals of seven orders, suggested 48 as the basic number of chromosomes in the primitive eutherian mammals. Makino (1949, 1952), however, on the basis of his study of forty-two mammals, proved that such a conclusion is not substantive.

Matthey (1949, 1959) attempted to arrange the basic chromosome numbers of mammals by equating the diverse numerical types through a simple Robertsonian relation (the evolutionary equivalence of one atelomitic to two telomitic chromosomes). Applying statistical technique, he argued from the reports of 240 eutherian mammals that autherians have developed from a pool of species with 40 to 56 chromosomes; the modal number 48 should be considered as primitive. This view was further confirmed by Matthey (1973) with the data of chromosome complements of 1,165 eutherian species; starting with the primitive number 48, or near it, chromosomal evolution has occurred through a variety of mutations, the most frequent of which are fusions; less frequent are pericentric inversions and fissions.

The diversified numbers in different groups of mammals have been correlated by Sachs (1954) as due to polyploidy. That polyploidy may have taken place in the course of mammalian evolution is evidenced by multiples of some basic chromosome number, e. g. in ungulates with 60 (cow), 48 (buffalo), 54 (sheep) and 66 (horse). Muller (1925) and White (1954), however, raised serious objections to the theory of wide occurrence of polyploidy in the animal kingdom, since this would constitute a serious handicap to proper meiotic division.

Chu and Bender (1961, 1963), in their studies of karyotype evolution in primates, demonstrated a marked correlation between low chromosome number and low number of acrocentrics and between high chromosome number and high number of acrocentrics. These correlations suggest that a Robertsonian type of chromosome evolution by centric fusion may have played an important role in primate evolution.

Todd (1967) has suggested a theory of karyotypic fissioning and genetic potentiation for eutherian chromosome evolution. Following Romer's theory (1966) of classical mammalian phylogeny, that some member of the Pantotheria may have been the ancestor of marsupials and eutherians, Todd postulated that this stock had a diploid number of 14 chromosomes. This number has been accepted principally because of the data for marsupials. During the course of evolution, this "pre-eutherian" karyotype was potentiated by fissioning to yield a protoeutherian karyotype having a diploid number of 24 (14 = 10 ST + 2 AC + XX/XY = 22 AC + XX/XY = 24)[1]. From this potentiated karyotype, different adaptive forms — Insectivora, Chiroptera, Carnivora, etc. — were established as a result of primary genetic potentiation. Accepting this view, an attempt has been made to explain the origin of the chromosome complements of twenty-five species, including primates (*Homo sapiens* among them), and to show phylogenetic relationships on the basis of chromosomes.

[1] ST = submetacentrics + metacentrics; AC = acrocentrics.

MATERIALS AND METHODS

The twenty-five species, belonging to eight orders and eighteen families, which I have studied for their number and morphology of chromosomes are listed in Table 1. In most cases, karyological studies were done on bone marrow and corneal cells following the methods of Ford and Hamerton (1956) and Beatty (1957), with colchicine administered *in vivo*. *In vitro* colchicine treatment was used in the cases of the nine-banded armadillo, the chimpanzee, and man.

Table 1. Chromosome numbers and morphology

		Number studied			Autosomes		Sex chromosomes	
Genus	Species	2n	♂	♀	ST[a]	AC[b]	X	Y
Suncus	*murinus*	40	1		18	30	ST	AC
Pteropus	*giganteus*	38	1	1	30	6	ST	ST
Pipistrellus	*mimus*	38	1		26	10	ST	AC
Dasypus	*novemcinetus*	64	1		18	44	ST	AC
Oryctologus	*cuniculus*	44	1	1	34	8	ST	AC
Rattus	*norvegicus*	42	1	1	22	18	AC	AC
Rattus	*rattus*	42		1	22	20[c]		
Mus	*musculus*	40	1	1	0	38	AC	AC
Cavia	*porcellus*	64	100	100	36	26	ST	AC
Funambulus	*pennanti*	54	1	1	16	36	AC	AC
Felis	*chaus*	38	1	1	32	4	ST	ST
Felis	*domestica*	38	1	1	32	4	ST	ST
Canis	sp. (interbreed)	78	1	1	0	76	ST	AC
Herpestes	*auropunctatus*	35/36	1	1	30	4	ST	0
Lutra	*perspicillata*	38		1	28	10[c]		
Canis	*aureus*	78	1	1	0	76	ST	AC
Vulpes	*bengalensis*	60	1		10	48	ST	AC
Ovis	*orientalis*	4	1		6	46	AC	ST
Bos	*indicus*	60	1		0	58	ST	ST
Capra	*hircus*	60	1	1	0	58	AC	ST
Sus	*scrofa*	38	1	1	24	12	ST	ST
Loris	*tardigradus*	62	1	1	38	22	ST	AC
Macaca	*mulatta*	42	1	1	40	0	ST	ST
Pan	*troglodytes*	48	1		34	12	ST	AC
Homo	*sapiens*	46	1	1	34	10	ST	AC

[a] ST = metacentrics + submetacentrics.
[b] AC = acrocentrics.
[c] Sex chromosomes not separately identified.

RESULTS

Results of the karyotype analysis are given in the table. As far as could be determined, all species possessed XX/XY sex chromosomes except *Herpestes auropunctatus*, which has X0/XX sex chromosomes.

DISCUSSION

The results on twenty-five species will be analyzed according to Todd's theory (1967) of karyotypic fissioning and genetic potentiation, in order to test its validity for primate evolution. Among the twenty-five species in this study, a few Carnivora and Artiodactyla species already examined by Todd (1967) were reinvestigated for confirmation. These have been suitably indicated in Figure 1.

As regards the order Carnivora, Todd proposed that it originated from insectivore stock by Miacidae, which culminated in the separation of most of the recent families into two classes; Feloidea and Canoidea. Since it was postulated that his Insectivora was potentiated by fissioning of the pantotherian karyotype from 14 to 24, it would be supposed that miacid radiation followed a secondary fissioning of the primitive Insectivora karyotype. This fissioning was thought to be from 24 to 38 chromosomes, as 14 ST + 8 AC + XY = 36 AC + XY, since the most frequently occurring diploid number among carnivores is 38. Among seven species studied by me in this order, members of Feloidea retained the same 38 chromosomes, as exemplified here by *Felis chaus* and *Felis domestica* of the family Felidae, with 2n = 32 ST + 4 AC + XY = 38 chromosomes; and by *Lutra perspicilliata* of the family Mustelidae, with 2n = 28 ST (including XX) + 10 AC = 38 chromosomes. Sometimes closely related numbers are most prevalent, as is found (2n = 30 ST + 4 AC + X0/XX = 35/36 chromosomes) in *Herpestes auropunctatus* of the family Viverridae. But among Canoidea, there is a conspicuous example of tertiary karyotypic fissioning, as 38 AC + XY = 38 ST (by inversion) + XY = 76 AC (by fission) + XY = 78. This chromosome complement is found in many species of the family Canidae as well as in the *Canis* sp. and *Canis aureus* of the present study. But in *Vulpes bengalensis* little further karyotypic modification is observed, since the diploid number is 2n = 10 ST + 48 AC + XY = 60 chromosomes. Secondary fusion of 20 acrocentrics to form 10 meta- and submetacentrics from 76 acrocentric chromosomes can explain it to some extent, but not quite correctly.

In the order Artiodactyla, four species have been studied. From the protoeutherian karyotype it was expected to originate by secondary fission: 22 AC + XY = 14 ST (by inversion) + 8 AC + XY = 36 AC + XY. Artiodactyla is also divided into two groups, Suina and Ruminantia. *Sus scrofa* of the Suina (family Suidae) group, with 2n = 24 ST + 12 AC + XY = 38 chromosomes, proves that no further fission has occurred. But among the Ruminantia, tertiary fissioning is assumed, as 36 AC + XY = 22 AC + 14 AC + XY = 22 ST (by inversion) + 14 AC + XY = 44 AC (by fission) + 14 AC + XY = 58 AC + XY = 60. This is exemplified here by the karyotypes of *Bos indicus* and *Capra hircus*, with 2n = 58 AC + XY = 60 chromosomes. In *Ovis orientalis*, 2n = 6 ST + 46 AC + XY =

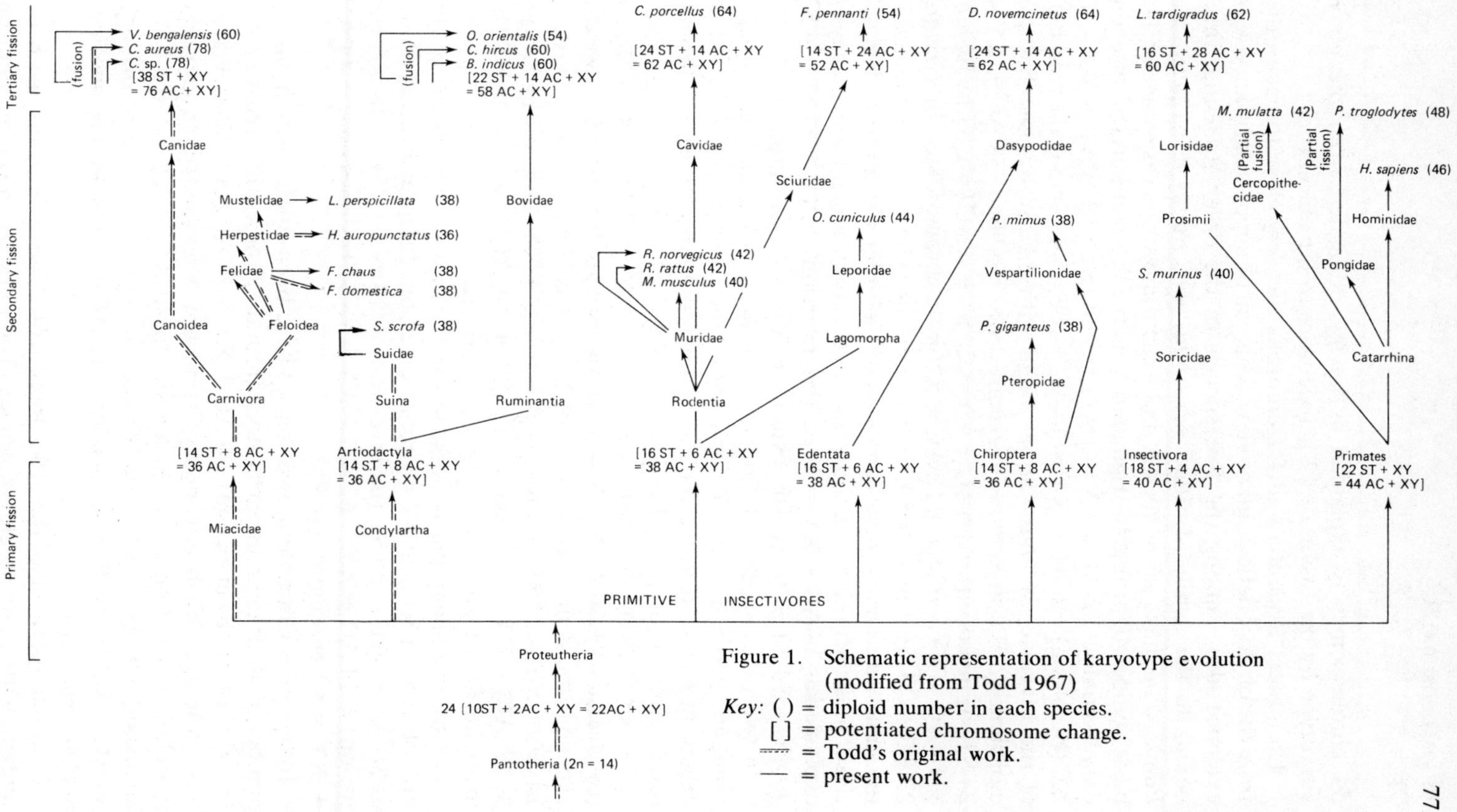

Figure 1. Schematic representation of karyotype evolution (modified from Todd 1967)

Key: ( ) = diploid number in each species.
[ ] = potentiated chromosome change.
(double dashed line) = Todd's original work.
— = present work.

54 chromosomes is explained as having originated from the above karyotype by the fusion of 12 acrocentrics to 6 metacentrics.

The orders Rodentia and Lagomorpha are very close to each other from the phylogenetic point of view. This can be supported by karyotype evidence also. Among the five species of the order Rodentia, three belong to the family Muridae: *Mus musculus, Rattus norvegicus*, and *Rattus rattus* have 2n = 40, 42, and 42 chromosomes respectively. The basic number 40 probably originated from the Protoeutheria after secondary fissioning: 22 AC +XY = 16 ST (by inversion) + 6 AC + XY = 32 AC (by fission) + 6 AC + XY = 38 AC + XY = 40. Among the members of the Muridae, *M. musculus* is found with exactly this number and morphology. But in *Rattus norvegicus*, it was modified: 38 AC + XY = 24 ST (by inversion) + 14 AC + XY = 22 ST + 4 AC (= 2 ST by fission) + 14 AC + XY = 22 ST + 18 AC + XY = 42 chromosomes. In *R. rattus*, 22 ST + 20 AC (including XX) = 42 chromosomes can be explained in the same way. *Oryctologus cuniculus* of the order Lagomorpha is much closer to the Muridae, and the origin of its chromosomal formula may be explained as 38 AC + XY = 38 ST (by inversion) + XY = 34 ST + 4 ST + XY = 34 ST + 8 AC (by fission) + XY = 44 chromosomes.

In two other species of Rodentia, *Cavia porcellus* and *Funambulus pennanti*, partial tertiary fissioning must be assumed to explain their origin. In *F. pennanti*, the diploid number and morphology of chromosomes probably originated as 38 AC + XY = 14 ST (by inversion) + 24 AC + XY = 28 AC (by fission) + 24 AC + XY = 52 AC + XY = 16 ST (by inversion) + 36 AC + XY = 54 chromosomes. In *C. porcellus*, we can explain in the same way the origin of 2n = 64 as 38 AC + XY = 24 AC + 14 AC + XY = 24 ST (by inversion) + 14 AC + XY = 48 AC (by fission) + 14 AC + XY = 36 AC + (12 AC + 14 AC) +XY = 36 ST (by inversion) + 26 AC + XY = 64 chromosomes.

A similar explanation is possible for *Dasypus novemcinectus* of the order Edentata, with 2n = 64 chromosomes, but different in morphology, as 38 AC + XY = 24 ST (by inversion) + 14 AC + XY = 48 AC (by fission) + 14 AC + XY = 62 AC + XY = 18 ST (by inversion) + 44 AC + XY = 64 chromosomes.

In the order Chiroptera, two species have been studied with a diploid number of 38. If their ancestral stock from the Protoeutheria had 22 AC + XY = 14 ST (by inversion) +8 AC + XY = 28 AC (by fission) + 8AC +XY=36 AC+XY by secondary fission, only alteration of chromosome morphology need be assumed, leading to 36 AC + XY = 30 ST (by inversion) + 6 AC + XY = 38 chromosomes in *Pteropus giganteus*, and 36 AC + XY = 26 ST (by inversion) + 10 AC + XY = 38 chromosomes in *Pipistrellus mimus*.

Only one specimen of the order Insectivora, *Suncus murinus*, has been studied, with 2n = 40 chromosomes. Todd (1967), collecting data on large numbers of insectivores, remarked that the ancestral stock had 2n = 42 chromosomes, which probably originated by fission secondarily: 22 AC + XY = 18 ST (by inversion) + 4 AC + XY = 36 AC (by fission) + 4 AC + XY = 42. The origin of 2n = 40 = 8 ST + 30 AC + XY of *Suncus murinus* can be explained from this stock if both fusion and inversion are considered: 40 AC + XY = 6 ST (by inversion) + 2 ST (= 4 AC by fusion) + 30 AC + XY = 8 ST + 30 AC + XY = 40 chromosomes.

Lastly, the karyotypes of the four species of the order of primates is explained by Todd's fission mechanism (1967). Staton (1967a, 1967b) studied mainly the primate karyotypes following Todd's model and suggested two types of secondary fissioning from 22 AC + XY: 22 AC + XY = 22 ST (by inversion) + XY = 44 AC (by fission) + XY in Platyrrhina; and 22 AC + XY = 20 ST (by inversion) + 2 AC + XY = 42 AC + XY in Catarrhina. Among the four primate species studied by me, three species (*Macaca mulatta* [2n = 42], *Pan troglodytes* [2n = 48], and *Homo sapiens* [2n = 46]) belong to the catarrhine group, and *Loris tardigradus* (2n = 62) is a prosimian primate. Staton (1967b) explained that from 42 AC + XY, *M. mulatta* attained 40 ST + XY = 42 chromosomes by inversion and fusion. But in the other two species, partial tertiary fission occurred: 42 AC + XY = 4 ST (by inversion) + 38 AC + XY = 46 AC + XY; in *P. troglodytes* 46 AC + XY was transformed into 34 ST + 12 AC + XY = 48 chromosomes by inversion; and fusion led to 34 ST + 10 AC + XY = 46 chromosomes in *H. sapiens*. For the prosimian primate *L. tardigradus* there was no explanation.

But this can be otherwise explained. If secondary fission can be assumed (22 ST + XY = 44 AC + XY instead of 42 AC + XY), the karyotype of *L. tardigradus* originated as 44 AC + XY = 16 AC + 28 AC + XY = 16 ST (by inversion) + 28 AC + XY = 32 AC (by fission) + 28 AC + XY = 60 AC + XY = 38 ST (by inversion) + 22 AC + XY = 62 chromosomes. The karyotypes of three catarrhine species can also be explained without the assumption of tertiary fission from 44 AC + XY, as in *M. mulatta:* 44 AC + XY = 36 ST (by inversion) + 4 ST (= 8 AC by fusion) + XY = 40 ST + XY = 42 chromosomes; in *P. troglodytes*: 44 AC + XY = 36 ST (by inversion) + 8 AC + XY = 34 ST + 4 AC (= 2 ST by fission) + 8 AC + XY = 34 ST + 12 AC + XY = 48 chromosomes; and in *H. sapiens*: 44 AC + XY = 34 ST (by inversion) + 10 AC + XY = 46 chromosomes.

Thus the application of Todd's formula in the present study has successfully demonstrated the origin of the diploid chromosome number and complement of twenty-five eutherian mammals from primitive pantotherian stock with 14 chromsomes by fission mechanism, followed by pericentric inversion and sometimes fusion; and this correlates the seven eutherian orders with the primates on the basis of chromosomes. But recently Matthey (1973) has criticized Todd (1967, 1970) and Staton

(1967a, 1967b), commenting that these theoretical studies are without cytological basis. Present study supports the karyotypic fissioning theory, contrary to Matthey's remark, so far as the origin of variations in different types of autosomes in the chromosome complement are concerned. But this theory does not consider the origin of variation in the structure of sex chromosomes, like that of autosomes, as shown in Table 1.

The exclusion of sex chromosomes from the fission/fusion mechanism was later explained by Todd (1969); the fissioning of sex chromosomes may result in (1) deleterious sex-chromosome–autosome imbalance; (2) difficulty in synapsis during meiotic pairing; (3) possibility of partial monosomy and trisomy; and (4) certain percentages of lethality of zygotes facing selective disadvantage in the production of progeny. Modification of the karyotypic fissioning theory eliminates the necessity of invoking a special behavior for the sex-chromosome centromere by postulating a simple mechanism for selective retention within the population of ancestral sex chromosomes. That the ultimate establishment of ancestral sex chromosomes and fissioned autosomes within a population is entirely dependent on meiotic and natural selection may be logically correct, but it cannot explain the evolution of the structure of sex chromosomes in twenty-five mammalian species as it can that of autosomes.

SUMMARY

An attempt to explain the origin of the karyotypes of twenty-five mammalian species from the basic pantotherian stock with 14 chromosomes is more or less successful following Todd's suggestion (1967) of fission mechanism with subsequent changes by pericentric inversion and translocation. It is possible to demonstrate the phylogenetic relationship of the primates with other orders of eutherian mammals on the basis of chromosomes. The main drawback of this hypothesis, however, is that it ignores the variation in the structure of sex chromosomes, which is observed in these twenty-five species to resemble the variation in autosomal chromosomes.

REFERENCES

BEATTY, R. A.

1957 Chromosome constancy in the corneal epithelium of the mouse. *Chromosoma* 8:585–596.

CHU, E. H. Y., M. A. BENDER

1961 Chromosome cytology and evolution in primates. *Science* 133:1,399–1,405.

1963 "The chromosomes of primates," in *Evolutionary and genetic biology of primates*. Edited by J. Buettner Janusch. New York and London: Academic.

FORD, C. E., J. L. HAMERTON

1956 A colchicine hypotonic citrate, squash sequence for mammalian chromosomes. *Stain Technology* 31:247–251.

MAKINO, S.

1949 A review of the chromosomes of domestic mammals. *Japanese Journal of Zootechnological Science* 19:5–15.

1952 A contribution to the study of the chromosomes in some Asiatic mammals. *Cytologia* 16:288–301.

MATTHEY, R.

1949 *Les chromosomes des vertebris*. Lausanne; Rouge.

1959 Chromosomal evolution in eutherian mammals. *Nature* 184:1,329–1,330.

1973 "The chromosome formulae of eutherian mammals," in *Cytotaxonomy and vertebrate evolution*. Edited by B. Chiarelli and E. Capanna. London and New York: Academic.

MULLER, H. J.

1925 Why polyploidy is rarer in animals than in plants. *American Naturalist* 59:346–353.

PAINTER, T. S.

1925 A comparative study of chromosomes of mammals. *American Naturalist* 59:385–409.

ROMER, A. S.

1966 *Vertebrate palaeontology* (third edition). Chicago: University of Chicago Press.

SACHS, L.

1954 Chromosome evolution in mammals. *Caryologia* (supplementary of IXth Congress of Genetics):668–689.

STATON, D.

1967a Karyotypes and the phytogeny of New World primates. *Mammalian Chromosome Newsletter* 8(3):203–219.

1967b A preliminary analysis of karyotypic evolution in catarrhines. *Mammalian Chromosome Newsletter* 8(3):220–223.

TODD, N. B.

1967 A theory of karyotypic fissioning, genetic potentiation and eutherian evolution. *Mammalian Chromosome Newsletter* 8(4):268–279.

1969 Sex chromosomes and selection in fission–fusion controversy. *Mammalian Chromosome Newsletter* 10(1):15–18.

1970 Chromosomes and equine evolution in the light of recent palaeontological evidence. *Mammalian Chromosome Newsletter* 11(4):136–140.

WHITE, M. J. D.

1954 *Animal cytology and evolution* (second edition). Cambridge: Cambridge University Press.

SECTION THREE

Chromosome Studies in Humans

*Blood-Group Determinations and Chromosome Aberrations Among the Jicaque Indians (Honduras)**

J. CONSTANS, J. C. QUILICI, L. MALASPINA,
and P. COLOMBIES

At the end of the eighteenth century, in the area shown in Map 1, northeast of the Honduran capital, Tegucigalpa, which corresponds to the modern department of Yoro, there lived a group of Indians: the Tolupan–Jicaque. Up to that time, this group had preserved a large part of its pre-Conquest cultural traditions and social structures.

But at the beginning of the nineteenth century, this group came under political and economic pressures probably connected with the problem of labor on the large plantations as well as the struggle to appropriate their lands.

As a result, around the 1870's, seven individuals left the group. After traveling for some time, they decided to settle in a fertile, wooded area of 800 square meters, with abundant game. In this region of Montaña de la Flor, the group multiplied rapidly. Their relationship is shown in Figure 1. The group stayed as isolated as possible, avoiding contact with the Ladinos [mestizos, non-Indians]. It seems that for a certain time the group grew and maintained its original way of life (von Hagen 1943).

A river crossed the site of their settlement, and this very quickly became a barrier between the two halves: a structure which still exists. They hunted a great deal with blowguns and darts. But gradually, with the appearance of firearms among the Ladinos who surrounded them, game became more scarce, forcing the Jicaque to rely increasingly on agriculture for their main source of food and to expand their barter trade

This research project was supported in part by the Centre d'Hémotypologie of the Centre National de la Recherche Scientifique under the direction of Professor J. Ruffié and in part by a grant from the International Biological Program.

We would like to thank A. Jacquard, Director of the Department of Genetics at the Institut National d'Etudes Démographiques, Paris, who entrusted us with this project.

*Please see Addendum for developments since 1973.

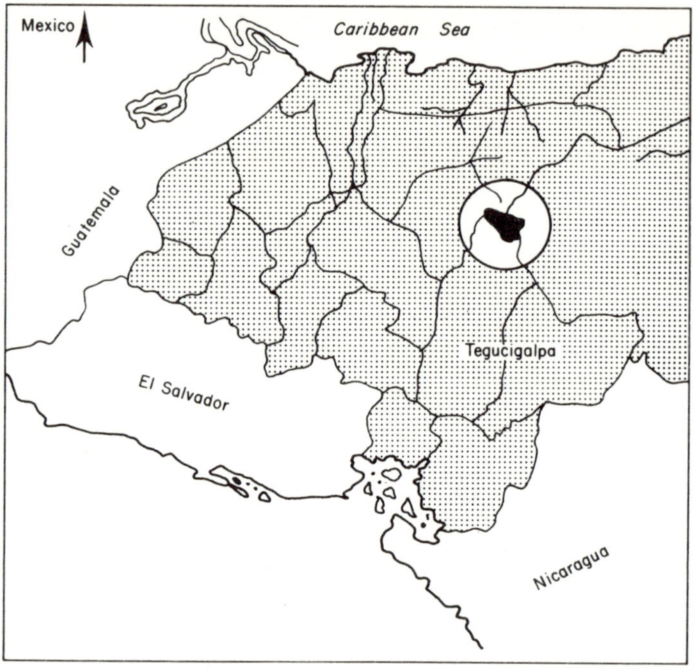

Map 1. Jicaque territory, Montaña de la Flor, Honduras

(Chapman 1962). As a result, there developed an irreversible process of more and more frequent contact with the Ladino, who was a necessary intermediary in their trade (Chapman 1961).

This brief historical and geographical account of the group was provided by the courageous and astute work of the ethnologist, Mrs. A. Chapman (1961, 1962, 1970). For almost fifteen years, she lived with the group during their fascinating final years. She was thus able to trace the complete ancestry of the Jicaque group from the time of the founders. We are grateful to her for sharing her experience with us.

With the genealogical data that had been collected, Chapman and Jacquard (1971) calculated the coefficient of parentage among the members of the fifth generation, i.e. the present generation. The figure they obtained for the mean coefficient of consanguinity had until then never been claimed for any human community.

The small size of the group greatly affected the mathematical valuation of the consequences of their relationships. Furthermore, the data provided by Chapman (1961, 1962, 1970) revealed that from the third and fourth generations mixed marriages took place between Jicaque and Ladino or Jicaque and Indian. In the opinion of Chapman and Jacquard (1971), the effect of these marriages on the fifth generation is that 40 percent of the genes in the population's gene pool are of mestizo origin.

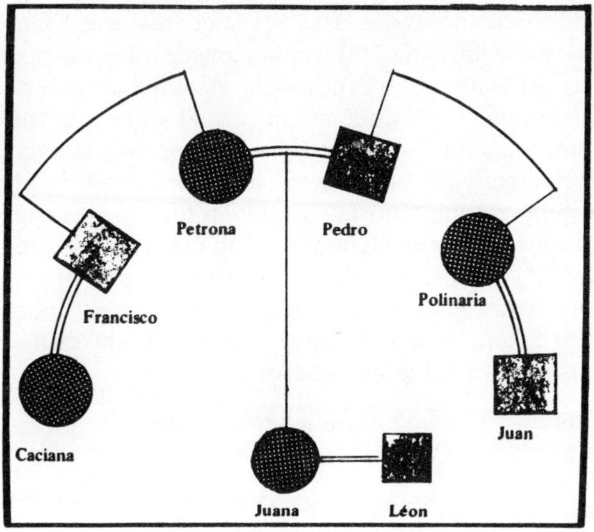

Figure 1. Family relationship between the eight founders at the formation of the Jicaque group

At the time of our investigation, it seems that the group consisted of only 300 individuals, all descended from consanguineal couples. There were several aspects to our research. It was medical in the respect that we carried out a complete physical examination. At the same time, we made an epidemiological survey, collecting a very large number of samples. In addition, blood and saliva samples were taken in order to study the distribution of blood groups in the population.

MATERIAL

Blood samples were taken in sterile tubes, Becton Dickinson vacutainers, and sent for analysis in containers cooled to +4°C. A heparinized sample was taken in order to construct a chromosome map.

RESULTS

In the course of this investigation (Quilici and Constans 1972), we observed neither physical defects nor metabolic anomalies of a congenital type in the population. Following the clinical examination, we identified four individuals who exhibited nervous disorders of an epileptic nature. In addition, several foot deformations were detected. The existence of deaf persons and mutes, however, as noted earlier by Chapman (1970), could not be confirmed. Only one woman and two of her four children were completely mute.

The initial epidemiological data revealed that each individual possessed one or more species of intestinal parasites. There appeared to be a total absence of malaria-type diseases, amoebic dysentery, or bilharziasis, however, which are endemic in coastal areas just 100 kilometers away. Natural selection has played a very important role in this isolate, where all epidemic viral or bacterial infections cause the loss of large numbers of people. Infant mortality remains the most serious scourge for the group. Babies between six months and one year old are the most at risk. See Table 1 for the population distribution in 1965.

During our investigation, we examined eighty-eight individuals, and samples were taken from sixty-three of them. Eight of the sixty-three were the offspring of a Ladino–Indian couple.

Table 1. Distribution of 565 individuals recorded in 1965

Generation	Male	Female	Sex unknown	Total
G 1 (founders)	4	3	–	7
G 2	19	15	–	34
G 3	96	80	–	176
G 4	110	93	44	247
G 5	38	26	37	101
Total	267	217	81	565

Source: Chapman and Jacquard 1971.

Blood Groups

Because the sample is particularly small, our results will be expressed in terms of the phenotypic frequency that we observed. Using this data, we shall be able to make a comparison with the other American Indian populations that we have studied.

1. ERYTHROCYTE SYSTEMS. The data are presented schematically in Table 2. In the ABO system, sixty-one individuals are O, the other two are A_1. A study of the family history shows that one of the two A_1 cases is the progeny of a mestizo (Ladino) couple. In the second case, however, there is a disparity between the subject and the parents, a disparity confirmed in other systems. This clearly poses the problem of the validity of genealogies based on verbal information, particularly as in the present case we are dealing with a perfectly stable couple who have always declared this child to be their own. Other instances of exclusion of paternity have been brought to light by our results, even though the O group is predominant in the isolate.

In the Rhesus system, which is generally the erythrocyte system showing the greatest polymorphism, there are only four phenotypes, distributed as shown in Table 2. The only subject who is R_1R_0 has one Ladino

Table 2. Distribution of phenotype from blood

System	Number of individuals	Total
ABO		
O	61	
A_1	2	
B	–	
AB	–	
Total		63
Rhesus (Rh+ 100 percent)		
R_1R_O	1	
R_1R_1	5	
R_1R_2	20	
R_2R_2	37	
Total		63
Kell (k– 100 percent)		
k+	–	
k–	63	
Total		63
Duffy (Fy [a+] 100 percent)		
Fy (a+)	63	
Fy (a–)	–	
Total		63
P		
P+	27	
P–	36	
Total		63
Kidd		
JK (a+)	35	
JK (a–)	28	
Total		63
MNS		
MMSS	4	
MMss	13	
MMSs	5	
MNS	19	
MNSs	11	
MNss	1	
NNSS	8	
Total		61[a]
MN		
MM	24	
MN	31	
NN	8	
Total		63
Lewis		
Le (a+b–)	–	
Le (a–b+)	30	
Le (a–b–)	24	
Total		54

[a] Two individuals were MMSs negative.

parent, which confirms that the presence of R_O is an indication of white crossbreeding. Usually in American Indian populations, the frequency of the R_1 gene is almost equal to that of the R_2 gene. Here, on the contrary, we observe a predominance of R_2 (over 50 percent). R_z, so frequent in the Andean populations that we studied (Constans and Quilici 1973), is not found here.

In the K, Fy (a), V^w, and I systems, a single gene is present. All the subjects are K-, Fy (a+), V^{w-}, and I+. But a little over 50 percent of them are JK (a+) and P−. These results are frequent among South American Indians.

In the MN system, there is a very high proportion of M genes. Despite the particular constitution of this isolate, polymorphism within the MN system has been maintained together with many heterozygotes.

As was found with the Chipaya group in Bolivia (Constans and Quilici 1973), the individuals examined were 100 percent Le(a−), with approximately 50 percent of them Le(a− b+) and 50 percent Le(a− b−).

2. SERUM SYSTEMS. The data are presented schematically in Table 3.

Immunoglobulins. In the Gm system, the antigen sites detected were 1, 17, and 21 in fifty-four individuals, including all the Jicaque and several Ladinos. The fact that a few subjects have other antigen sites probably signifies crossbreeding with Caucasian characters.

Inv. In the Inv. system, all the Jicaque examined are Inv−, with only a few individuals having an Inv+ antigen site. In the light of our present knowledge, however, they are the only group to manifest such a high percentage of Inv− individuals. We do not know the possible effects of such a phenomenon on the biological defences of the Jicaque in coping with all kinds of aggression.

Haptoglobin. Almost all the subjects are Hp_{1-1}; the Hp_2 gene is only present in the homozygous state in a single individual and in the heterozygous state in four. A study of the genealogies showed that four of these five individuals belong to Ladino–Jicaque families. The only person with Hp_{2-2} also had the Gm 2 site, which, as we have already suggested, may be explained by crossbreeding. Not a single case of HpO was observed. The dosage of haptoglobin is very high: on average 3.70 g/l, signifying a kind of inflammatory reaction, probably in response to the environmental conditions.

Transferrin. All subjects carry the C character, except one who possesses a D-mobility transferrin. Here again, this individual is a mestizo. It should be pointed out that this variant has already been encountered in Central America.

Table 3. Distribution of phenotype from blood

System	Number of individuals	Total
Immunoglobulin Gm		
1, 17, 21	54	
1, 2, 17, 21	6	
1, 2, 3, 5, 10, 11, 17, 21	1	
1, 3, 5, 10, 11, 17, 21	2	
Total		63
Inv		
Inv+	10	
Inv−	53	
Total		63
Haptoglobin		
Hp_{1-1}	58	
Hp_{2-1}	4	
Hp_{2-2}	1	
Total		63
Transferrin		
Tf CC	62	
Tf CD	1	
Total		63
Group component Gc		
Gc_{1-1}	50	
Gc_{2-1}	6	
Gc_{2-2}	–	
Total		56
G6PD B	63	63
6PGD A	63	63
AK 1	63	63
PGM^1_1	63	63
PGM^1_2	63	63
C_5 esterase		
C_5−	25	
C_5+	29	
Total		54
Alkaline phosphatase		
P_O	12	
P+	8	
P++	35	
Total		55
Acid phosphatase		
AA	–	
AB	2	
BB	10	
AC	–	
BC	4	
CC	6	
Total		22

Group Specific Component. The Gc_1 gene is clearly predominant, but six individuals carry the Gc_2 gene in the heterozygous state. This phenomenon is not exceptional, since the Gc_2 gene is usually more frequent than the Gc_1 in American Indian populations.

3. HEMOGLOBINS. No abnormal hemoglobin was encountered, contrary to our findings with the Mayas of Yucatan; to explain this we put forward the hypothesis of black crossbreeding. The test of resistance to alkaline denaturation never revealed F hemoglobin. The dosage of A_2 hemoglobin was always less than 3 percent.

4. ENZYME SYSTEMS. The absence of polymorphism in the systems studied so far holds true in the case of the erythrocyte and serum enzymes, although the number of samples analyzed here is smaller. As in the case of the erythrocyte groups, a certain number of characteristics are present in all the individuals examined (compare Table 2). One subject, it should be pointed out, manifested a PGM_1^{1-2} phenotype, but this is the mestizo already noted in the Gm, Hp, and Gc systems. Other systems, however, show a certain degree of polymorphism. C_5 esterase: twenty-five individuals are $C_5(-)$ and twenty-nine are $C_5(+)$. Alkaline phosphatase: thirty-five of fifty-five subjects are P++, eight are P+, and twelve are P_0. Acid phosphatase: in this system sixteen individuals out of twenty-two carry the B gene, two of them in the heterozygous state AB; four are BC heterozygotes, while ten carry B homozygotes. Of the two AB individuals, one is of mestizo origin. We are therefore led to think that the Jicaque group may be defined in the main as carriers of B and C genes. The peculiarity of this Central American group is the high frequency of the C gene in the form of BC and especially CC.

System HL A. The second investigation permitted the study of HL A polymorphism in twenty-six individuals of this group.

The technique used was the micromethod of deviation of complement on platelets. Fifty-three serum tests were performed. The following antigens were examined in each series:

HL A - A	HL A - B
A_1	B_5
A_2	B_7
A_3	B_{12}
A_9	B_{13}
A_{10}	B_{14}
A_{11}	B_{27}
A_{28}	$B^w{}_{15}$
A_{29}	$B^w{}_{17}$
	$B^w{}_{21}$
	$B^w{}_{35}$
	$B^w{}_{40}$

The following distribution of phenotypes was observed:

HL A phenotype	Number of individuals
A_2, A_9, $B^w{}_{15}$, Y	15
A_2, A_9, $B^w{}_{35}$, Y	1
A_2, X, $B^w{}_{15}$, Y	4
A_2, X, $B^w{}_{35}$, Y	1
A_9, X, $B^w{}_{15}$, Y	2
A_9, X, Y, Y	3

The classical polymorphism of the HL A system is extremely reduced; thus, if one considers the two antigens which have been defined for the products of each locus, one sees that A_2 and A_9 have been found twenty-one times out of twenty-six. Given the high probability of homozygosity indicated by the genealogies, the subjects are probably either A_2/A_2 or A_9/A_9 homozygotes or A_2/A_9 heterozygotes.

In the HL A - B series, the results are not so clear. $B^w{}_{35}$ is very frequent (twenty-one times out of twenty-six), but $B-\pm^1$ is only found five times out of twenty-six. Above all, in three cases, no HL A-B antigen could be found. This indicates that at least one other antigen of this series ought to exist, but it was not detected by our serum tests. In this case, it might be a known gene whose products we were unable to detect because our serum tests did not correspond, or it might be a gene not yet known.

Cytogenetic Study

The initial results of our cytogenetic study concern twenty-four members of the population whom we examined. Using samples of heparinized veinous blood, we cultured leukocytes in accordance with a technique described by Carles-Trochain (1971). This study was undertaken in order to bring to light any chromosomal aberrations which might have appeared in this isolate. In addition, an examination of the chromosome maps helped us establish the nature and frequency of the chromosomal aberrations in this group who live under special ecological conditions. One hundred mitoses were photographed per subject. Chromosome maps were constructed for ten normal cells, each having forty-six chromosomes. Because of transportation difficulties between Montaña de la Flor and Tegucigalpa, there was an interval of seven to eleven days between the taking of samples and the preparation of cultures. We shall return to this point later.

The twenty-four subjects examined are divided equally between the two sexes (twelve men and twelve women) and are between six and forty-eight years of age, twenty-eight being the average age for the men and twenty-six for the women. Table 4 summarizes the distribution of the different characteristics and anomalies.

Table 4. Principal chromosomal aberrations in the Jicaque population

Aberration	Level (percent)
Constrictions	
carried by one no. 9 chromosome	30
carried by two no. 9 chromosomes	61.66
carried by the no. 1 chromosomes	1.66
carried by the no. 2 chromosomes	0.83
carried by the no. 16 chromosome	2.50
Satellites	
large acrocentrics (D)	86.25
small acrocentrics (G)	84.58
Aneuploidy (general average)	19.86
Hyperdiploidy (27 of 2,181 cells studied)	1.24
Structural anomalies	
breakages (126 of 2,181 cells studied)	5.77
various (23 of 2,181 cells studied)	1.17

Constrictions. The very high frequency of a constriction on one or two chromosomes of pair 9 does not strike us as very significant, since it is known to be common in this position. On the other hand, constrictions carried by chromosomes 1, 2, and 16 are much rarer. It should be noted that they exist in normal subjects.

Satellites. These are found on the D and G groups of chromosomes with different distributions depending on the number of chromosomes which carry them, as can be seen in Table 5. The percentages observed coincide with the usual values cited.

Table 5. Distribution of satellites on chromosomes in the Jicaque population

	Number	Percent	Total percent
Satellites on large acrocentrics (D)			86.25
on three pairs	0	0	
on two pairs	53	22.08	
on one pair and one chromosome	45	18.75	
on one pair	78	32.50	
on one chromosome	31	12.91	
Satellites on small acrocentrics (G)			84.58
on two pairs	4	1.66	
on three chromosomes	29	12.08	
on one pair	123	51.25	
one one chromosome	47	19.58	

Translocations and Inversions. None of these chromosomal aberrations was observed in this group.

Aneuploidy. Aneuploidy was discovered in all the subjects under study. The average level of aneuploidy per individual is 19.861 percent, which is one of the highest figures we have encountered. The percentage distribution of aneuploidies by sex is 19.124 percent in males and 20.599 percent in females. Figure 2 shows the histogram of this distribution. Distribution according to age group is as follows:

6 – 15 years: 19.53 percent;
15 – 30 years: 16.77 percent;
30 – 40 years: 18.08 percent;
over 40 years: 22.44 percent.

There is no significant difference between the rates of aneuploidies based on sex or age group.

In the course of our experiments, we examined whether the ageing of

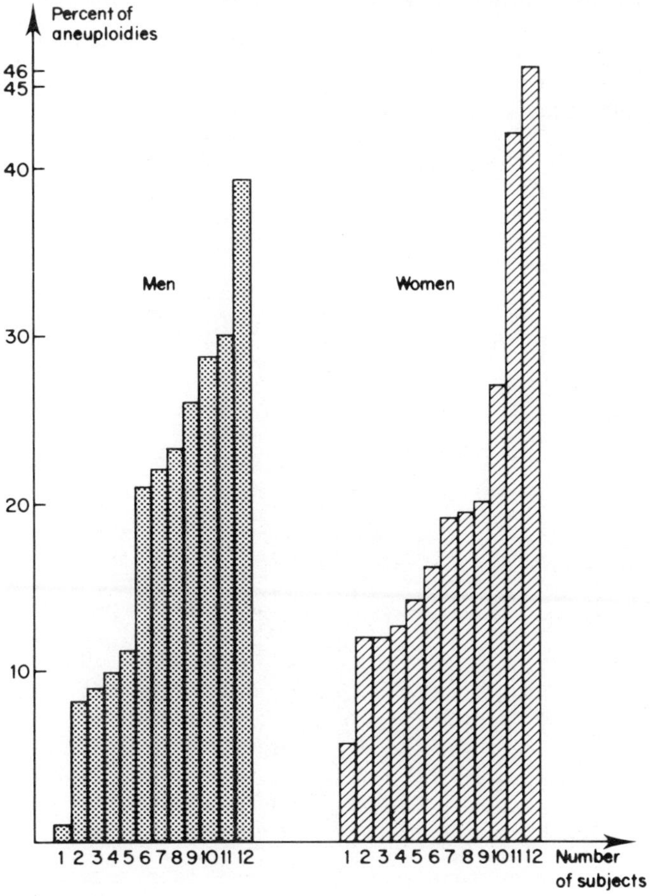

Figure 2. Distribution of aneuploidies among the Jicaque

our blood samples had any effect on the frequency of the chromosomal anomalies that we discovered. We found that the percentage rate of aneuploidies increased very sharply in proportion to the amount of time that had elapsed. The results of our experiments were as follows: after seven days samples from five subjects showed 12.53 percent aneuploidy; after eight days samples from nine subjects showed 23.59 percent aneuploidy; after nine days samples from seven subjects showed 13.63 percent aneuploidy; and after eleven days samples from three subjects showed 35.42 percent aneuploidy. We may thus conclude that the level of aneuploidy among the Jicaque has no particular significance and that the phenomena observed were caused by the ageing of the samples.

We classified by chromosome group the frequency of aneuploidy in the Jicaque population and in a control group of local French people. Results are shown in Figure 3. Table 6 shows that aneuploidy is more frequent in the E, F, and G group of small chromosomes than in the large chromosomes, as was observed by Carles-Trochain in 1971.

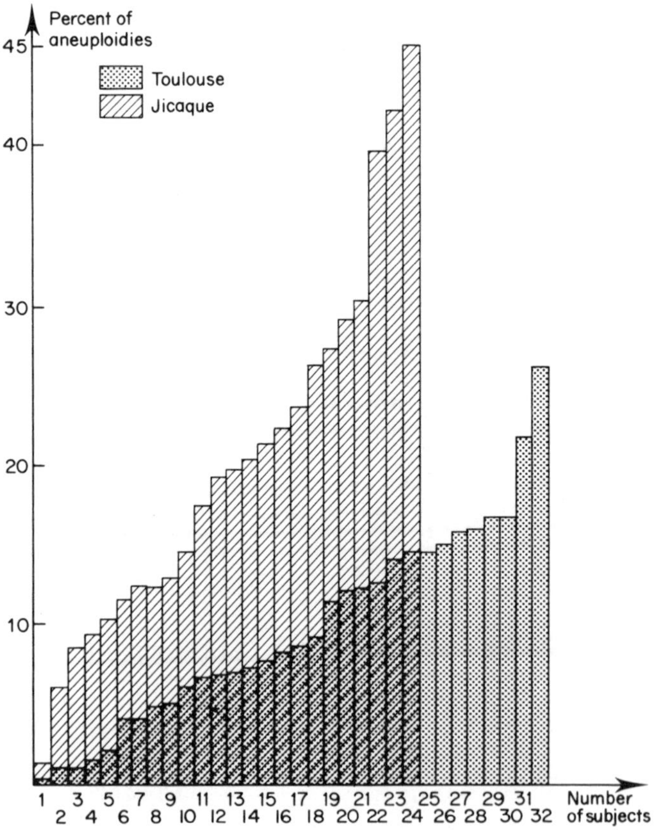

Figure 3. Distribution of aneuploidies in the Jicaque population and a control group

Table 6. Frequency of aneuploidy in the Jicaque population and a control group

Chromosome group	Frequency of aneuploidy (percentages)	
	Jicaque	French
A	0.446	0.145
B	0.672	0.090
C	0.862	0.027
D	0.867	0.237
E	1.016	0.494
F	1.105	0.316
G	1.143	0.555

Hyperdiploidies. Fourteen subjects out of twenty-four studied exhibited one to five cells with supernumerary chromosomes, ranging from 47 to 86 chromosomes per cell. Out of 2,181 cells studied, twenty-seven were hyperdiploids, or 1.24 percent of the cells. The majority (0.81 percent) of these possessed 47 chromosomes, with the following distribution:
0.58 percent C chromosomes;
0.04 percent B chromosomes;
0.04 percent D chromosomes;
0.04 percent E chromosomes;
0.09 percent G chromosomes.
Thus in the majority of cases, C chromosomes were involved.

Structural Anomalies

These are also frequent among the Jicaque Indians.

BREAKS. Of 2,181 cells studied, 126 presented one or more broken chromosomes, a frequency of 5.77 percent of cells exhibiting this alteration. On the other hand, 140 chromosomes (0.013 percent) showed a break, which is considerably higher than the figure obtained by Court Brown in 1967 of 0.003 percent.

Taking into account the number of chromosomes in each of the groups, the following distribution is obtained:

Group	A	B	C	D	E	F	G
Frequencies	56	24	42	7	9	2	0
Percent	0.0092	0.0059	0.0037	0.0006	0.0008	0.0004	

Thus, it is clear that breaks occur most frequently among the large chromosomes, particularly the A-group chromosomes (0.0092 percent). This conclusion confirms other previous findings (Carles-Trochain 1971).

VARIOUS. Twelve subjects — half of those studied – revealed more complex structural anomalies in one or two of their cells. Some of these are shown in Plate 1.

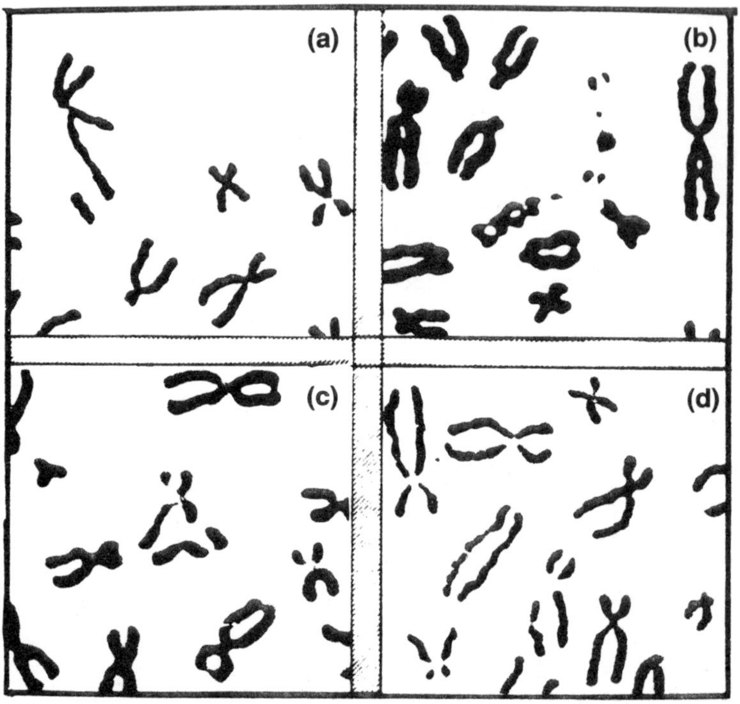

Plate 1. Chromosomic lesions observed among the Jicaques; (a) representative chromosomes with delayed isolocus breaks. Note complete break with total displacement of one chromatid; (b) partial endoreduplication with despiralization chromatids (long arm of a B chromosome); (c) partial endoreduplication (terminal long arm of a C-8 chromosome) with isolocus break; (d) acentric fragment

Deletions. Twelve cells showed a deletion, as follows:
one cell showed a deletion of the lower arm of the A chromosome (Aq−);
one cell showed a deletion of the upper arm of the B chromosome (5p−);
eight cells showed a deletion of the lower arm of the C chromosome (9q−);
one cell showed a deletion of the lower arm of the D chromosome (Dq−);
one cell showed a deletion of the lower arm of the G chromosome (Gq−).

Acentric Fragments. Five cells showed acentric fragments.

Partial Endoreduplication. In the chromosome map of four of our subjects we observed selective endoreduplication of the lower arms of an 8 chromosome in two cases and of one B-group chromosome in the other two cases. In two cases the endoreplicated chromosome was also moniliform in appearance.

COMMENTS

The cytogenetic survey of this population sample did not bring to light any specific systemized anomaly. Two particular points, however, are worthy of note: (1) the very high frequency and marked character of the constrictions on pair 9, verified by the eight cells in which we observed a deletion of the lower arms (9q−); (2) a level of aneuploidy and a breakage frequency higher than those in the populations that we studied previously (the French in Languedoc, Eskimos, Senegalese, and American Indians).

The highly parasitic environment as well as the serious infectious diseases may in part be responsible for the chromosomal aberrations we observed. This hypothesis has already been advanced by Bloom et al. (1970), but in our case we cannot accept it as entirely satisfactory.

CONCLUSION

Our study of the hemotypological characteristics which we have described enables us to draw certain conclusions. First, we are dealing with an unprecedented geographical and cultural isolate. In this group, the marriages have reached a level of consanguinity unknown until now. This phenomenon leads to a uniformity of results which prohibits any calculation of gene frequency.

The results obtained show the characteristic genetic traits of an American (especially Central American) Indian population. The points at which these results tend to coincide with those obtained on a Maya group in Yucatan (Constans and Quilici 1973) confirm this observation.

It seems that the individuals who participated in the formation of the group were not bearers of genetic disorders or unfavorable molecular mutations, since we were unable to detect any of these in any of the subjects. The genetic pool is shown schematically in Figure 4.

We might note that our results compared to those of Matson and Swanson (1963) for a group of Jicaque in the department of Yoro permit us to conclude that the data are very similar. In our sample the gene frequencies are different, which might be explained by the particular makeup of our group.

Among the Jicaque of Yoro, crossbreeding is very obvious. The genetic polymorphism of Amerindian populations is in general rather slight. In the case of the isolate we studied, despite the small number of subjects,

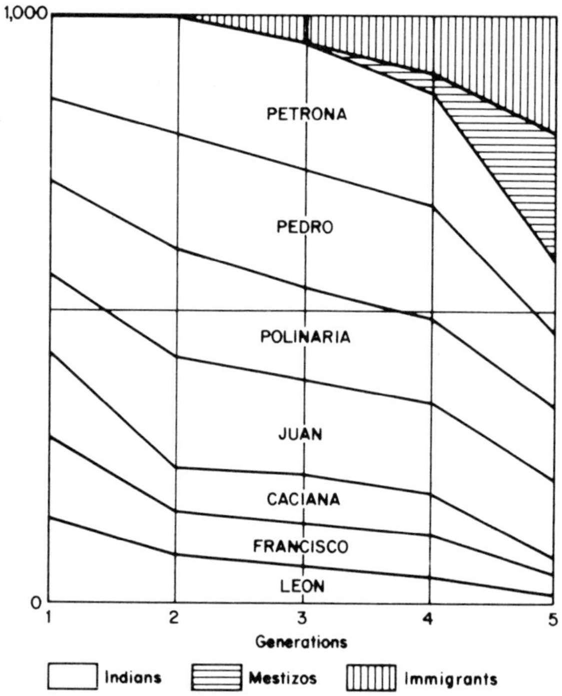

Figure 4. Effect of immigration on the genetic pool of the population (Chapman and Jacquard 1971)

we have found heterozygotes in various systems. The presence of genes such as A (ABO system), $Gm^{3,5}$, R_o and Tf^D allowed us to confirm the exclusions of paternity in the genealogies.

There is a fairly good correspondence between the coefficient of probability of the origin of genes calculated by Pison and Vu Tien (1974) at 20 to 30 percent and the genetic polymorphism observed for the various blood markers.

REFERENCES

BLOOM, A. D., J. V. NEEL, K. W. CHOI, S. IIDA, N. CHAGON

1970 Chromosome aberrations among the Yanomama Indians. *Proceedings of the National Academy of Sciences* 66(3):920–927.

CARLES-TROCHAIN, E.

1971 Etude cytogénétique d'un isolat: le groupe Eskimo d'Angmassalik. *L'Anthropologie* 75(3–4):263–268.

CHAPMAN, A.

1961 Mythologie et éthique chez les Jicaques. *L'Homme* 1(1):95–101.

1962 Survivance de l'organisation dualiste chez les Jicaques. *L'Homme* 2(1):91–101.

1970 Chamanisme et magie des ficelles chez les Tolupan (Jicaques) du Honduras. *Journal de la Société des Américanistes* 59:43–64.

CHAPMAN, A., A. JACQUARD

1971 Un isolat d'Amérique Centrale: les indiens Jicaques du Honduras — génétique et population. *Institut National d'Etudes Demographiques* Booklet 60:163–185. Paris: Presses Universitaires de France.

COLOMBANI, J., J. D'AMARO, B. W. GABB, G. SMITH, A. SVEJGAARD

1974 *Microtechnic of platelet complement fixation. Manual of tissue typing techniques.* National Institutes of Health Manual Number 75/545:86–89.

CONSTANS, J., J. C. QUILICI

1973 "Hémotypologie des indiens Chipaya de Bolivie." Paper presented at the IXth International Congress of Anthropological and Ethnological Sciences, Chicago.

COURT BROWN, W. M.

1967 *Human population cytogenetics.* North Holland Research Monographs: Frontiers of Biology, volume 5. Series edited by A. Newbergers and E. L. Tatum Amsterdam: North Holland.

MATSON, G. A., J. SWANSON

1963 Distribution of hereditary blood antigens among Indians in Middle America IV. In Honduras. *American Journal of Physical Anthropology* 21:545–559.

PISON, G., J. VU TIEN

1974 Les indiens Jicaques du Honduras. Mise à jour des données généalogiques. *Cahiers d'Anthropologie et d' Ecologie Humaine* 2:131–141.

QUILICI, J. C., J. CONSTANS

1972 Informe sobre el estado de salud de la poblacion Jicaque en la Montaña de la Flor, Honduras. *Revista Medica Hondureña* 40:4–6.

VON HAGEN, W. V.

1943 *The Jicaque (Torrupan) Indians of Honduras.* Heye Foundation Indian Notes and Monographs. New York: New York Museum of the American Indian.

ADDENDUM

This text was presented at the Ninth International Congress of Anthropological and Ethnological Sciences held in Chicago in September, 1973. Five years later, it seems necessary to add several comments on the data and analyses used.

An initial genetic study permitted the discovery of a significant number of exclusions of paternity in a sample of sixty-six persons. A second study confirmed this.

After the changes made in the genealogies by Chapman and Jacquard (1971), their calculations were redone by Pison and Vu Tien (1974). The new coefficient obtained shows a diminution of the genetic effects of crossbreeding with the Ladinos.

The second study concerned seventy-six subjects. These results have been published by Quilici (1975) and Vergnes et al. (1976).

We would like to note especially the case of the acid phosphatases, in which we have found a high number of individuals of phenotypes BC and CA. The frequency of the gene P^c reaches a value, never before observed, of 0.392. This gene is practically absent in Amerindian populations.

Spencer et al. (1964) have demonstrated that the activity of the molecule is greater when the gene P^c is present. We do not know what mechanisms may have favored the transmission of this allele in the Jicaque. The result obtained could also be due to a relation between the activity of this phenotype and the environment.

We have also studied the polymorphism of serum albumin in these seventy-six subjects. In one case, we found a slow variant type Mexico in a Ladino subject.

CHAPMAN, A., A. JACQUARD
1971 Un isolat d'Amérique Centrale: les indiens Jicaques du Honduras — génétique et population. *Institut National d'Etudes Demographiques* Booklet 60:163–185. Paris: Presses Universitaires de France.

PISON, G., J. VU TIEN
1974 Les indiens Jicaques du Honduras. Mise à jour des données généalogiques. *Cahiers d'Anthropologie et d'Ecologie Humaine* 2:131–141.

QUILICI, J. C.
1975 "Structure hémotypologique des populations indiennes en Amérique du Sud." Unpublished doctoral dissertation, Department of Human Biology, University of Toulouse.

SPENCER, N., D. A. HOPKINSON, H. HARRIS
1964 Quantitative differences and gene dosage in human red cell acid phosphatase polymorphism. *Nature* 201:299–300.

VERGNES, H., J. C. QUILICI, J. CONSTANS
1976 Serum and red cell enzyme polymorphisms in six Amerindian tribes. *Annals of Human Biology* 3(6):577–585.

*Cytogenetic Study of the Eskimo in Igloolik, N.W.T.**

B. CHIARELLI and RORY DANIEL

The Eskimo population is of particular interest for biological study not only because of its successful adaptation to a stressful climate, but also because of the length of time the group has been separated from the ancestral Asian stock. While the Eskimo can first be identified as a separate group some 5,000 years ago, it is believed that the movement into the Arctic of peoples ancestral to the Eskimo and Aleut began 10,000 to 12,000 years ago (Hughes 1973). A population such as the Eskimo can act as a model in the study of the biological variation occuring in a population under strong selective pressures for a considerable period of time. Morover, if we want to compare the Eskimo population with the European one, we can estimate a biological separation of at least 20,000 years, which means a generation gap of 800 to 1,000 generations, considering generation time as twenty to twenty-five years.

As part of the IBP-HA (International Biological Program-Human Adaptability) research project going on in the Eskimo community of Igloolik, N.W.T. (Canada), we had the opportunity to begin an extensive cytogenetic analysis in this settlement.

The purpose of our work was threefold. First, we wished to detect any gross chromosomal abnormalities which might be present in the study subjects. Second, we wanted to observe whether there were chromosomal polymorphisms present in the population. Third, we intended to look at "break" and "gap" in patterns in the Eskimo chromosomes and compare them with those of a control group.

We were stimulated to include this third aspect in our work by the hypothesis made by some workers that human populations not exposed to the various noxious chemicals and mutagens of the industrial world

*A similar paper has in the meantime been published in *Journal of Human Evolution* (1974) 3:411–416.

would have a lower percentage of chromosomal aberrations than those exposed. Studies of the isolated Yanomama in South America, however, show that these people have an abnormally high rate of somatic chromosome breakage, probably due to environmental factors such as viral disease or to the use of the hallucinogen *ebene* (Bloom and Neel 1970).

To facilitate comparison of the chromosomes of the Eskimo and control (European) populations both directly and with regard to the above hypothesis, over half the cultures were prepared using the blood of an Eskimo and that of an IBP team member of the opposite sex. It was felt that preparation of the cultures in the field would help eliminate artifacts that can appear because of the delay which occurs in sending samples from the field; and that use of mixed cultures would help us identify any artifacts which would appear in our work because of differences in the manipulation of the material during the preparation for the study.

MATERIALS AND METHODS

As mentioned in an earlier article (Daniel and Chiarelli 1971), blood samples from ninety Eskimo individuals were cultured over a two-week period in Igloolik in January, 1971. The collection of our sample was random, though further analysis showed an occasional family grouping. The original sample included forty-six females and forty-four males ranging in age from four to seventy-one years. Two males have since been eliminated because of recording errors.

In total, 104 cultures were prepared, 53 percent of them as mixed cultures. The blood of four individuals of European descent in the IBP team was used for the control cultures. On average, five slides were prepared from each culture. Two complete sets of slides were stained, one with ordinary Giemsa and the second with buffered Giemsa.

Nearly all the cultures harvested contained mitoses; but since we had no distilled water for the hypotonic solution (we had to use melted snow), a number of mitoses proved to be too clumped for adequate investigation. Cultures on two separate days gave such poor results that they had to be eliminated from our data. Thus, the actual sample for which we have material for study constitutes sixty-nine individuals, about 15 percent of the Eskimo population in Igloolik.

In the analysis are included only those individuals for whom we have karyotyped five to twenty well-spread mitoses taken from the two slides stained for each individual. Karyotyping was done according to the Denver Convention. Over 400 Eskimo karyotypes were prepared, another 100 for the controls.

The banding techniques of trypsin and quinacrine fluorescent staining have been applied to a small number of slides in a preliminary analysis.

RESULTS

In none of the sixty-nine individuals for whom karyotypes are available were there found any gross chromosomal abnormalities such as those related to pathological conditions which arise from sex or autosomal aneuploidy or structural aberrations. There were, however, some plates with an aneuploid complement. A brief analysis of these plates showed that in those cases in which the individual could be identified (in mixed cultures) there was no significant difference in the Eskimos and controls in the frequency of aneuploid cells, as shown in Table 1.

Table 1. Aneuploid cells in both Eskimo and control individuals

	Hypomodal	Hypermodal	Total cells	Aneuploid cells (percent)
Eskimo	17	9	417	6.2
Control	4	2	102	5.9
Unidentified plates[a]	8	5	13	

[a] Refers to plates of individuals in mixed cultures which could not be identified with certainty.

The Eskimo karyotype did not differ significantly from the standard of the normal human karyotype nor from that of the cocultured control individuals.

Studies of chromosome banding in this population are still at a preliminary stage, but if small structural changes are present in the Eskimo chromosomes they could be detected by these techniques. The highly fluorescent region of the Y chromosome is detectable in males in this population, as can be seen in Plate 1.

The second aspect of our study involved detection of chromosomal polymorphisms in the population. Using the traditional techniques, we made the following observations:

1. Variation in the size of the Y chromosome is observable, but this chromosome is never as large as those in the F group, as shown in Plate 2. Because of this factor and because of difficulties of obtaining genealogical data among the Eskimo, the study was not carried out.

2. A noticeable variant occurs in the length of the short arm in a G-group chromosome. In a preliminary investigation it was thought that this chromosome was a metacentric Y present in two males. Further analysis showed, however, that this chromosome was present in another male and three females. Family studies showed one of these females to be the mother of the two original males; thus this chromosome belongs to the G group. Family data is not available for the other individuals.

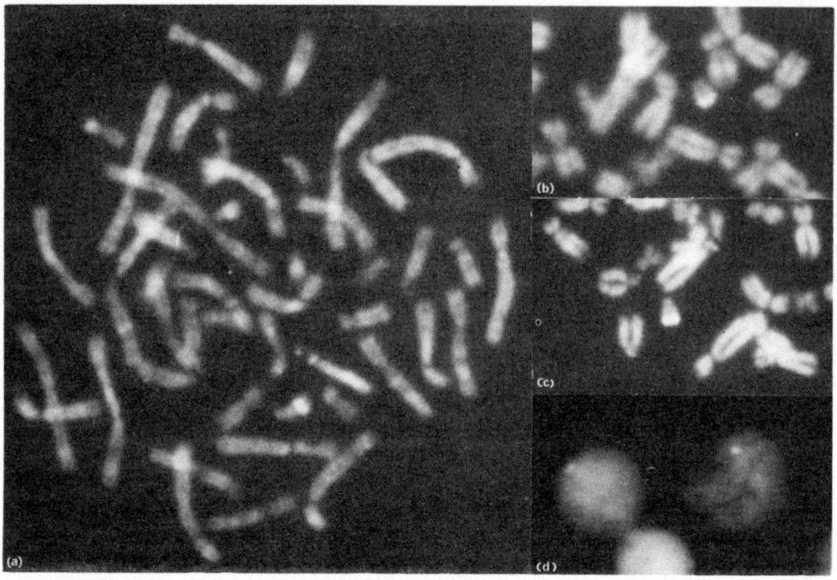

Plate 1. Eskimo chromosomes treated with quinacrine: (a) metaphasic plate of a male; (b) the bright fluorescence of the long arm of the Y chromosome; (d) nuclei of leukocytes after quinacrine treatment; the bright spots indicate the Y chromosome

Plate 2. Variation in the length of the short arm and change in the centromeric index in the 22nd pair

3. Heteromorphism in the size of chromosome 16 occurs unusually consistently in the plates of seventeen Eskimo individuals. The biological significance of this variation is not clear.

4. Variation in the secondary constriction of chromosome 9 is apparent. These constrictions vary their expression between individuals and be-

tween homologues. They were found in both Eskimo and control individuals (see Plate 3).

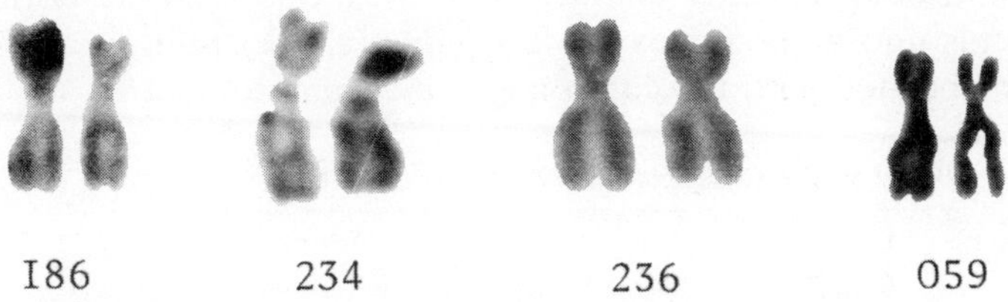

Plate 3. Different types of "gaps" occurring in chromosome 9

Further evidence of chromosomal structural variation in the Eskimo population may appear, as mentioned, as a result of the banding study. The use of mixed cultures will facilitate comparison of the two groups.

The third aspect of our work involved elucidation of frequencies of small morphological differences in the chromosomes of Eskimo and control individuals, differences related to heterogeneity in the staining along the chromosome filament.

This characteristic of chromosomes is traditionally treated under the heading of "chromosome breakage"; we have not regarded it in this traditional way, however. In our analysis we have included all occurring achromatic "gaps" in the stained areas and also very lightly stained areas along the chromosome arms. We have divided these features into two groups: those supposedly randomly distributed, and those which occur in specific locations and on both the arms and/or in homologues of only certain chromosomes.

The data were inadequate to give accurate frequencies of these random "gaps" in the chromosomes, since a minimum of 100 plates per individual would have to be analyzed. In order to obtain a general picture, however, the data were pooled into the Eskimo and control groups and frequencies of random "gaps" in each compared. All chromosomes were considered, and "gaps" occurring on sister chromatids were scored as "2." As Table 2 indicates the frequency of gaps per plate does not differ significantly in the two groups.

Table 2. Random "gaps" observed in Eskimo and control chromosomes

	Total plates studied	Subterminal	Near the centromere	Central area of arm	Total	Frequency of gaps per plate (percent)
Eskimo	417	14	45	32	91	0.22
Control	102	3	7	9	19	0.19

Concerning the second group of nonrandom "gaps," tabulated in Table 3, the following observations were made. In two Eskimo individuals a lightly stained area is found in the subterminal region of the p arm in chromosome 1. This characteristic was observed in both sister chromatids and homologues. A few individuals exhibit an achromatic region associated with the centromere in chromosome 16.

Table 3. Number of plates containing nonrandom "gaps"

Chromosome	Eskimo		Controls	
	homologues	sister chromatids	homologues	sister chromatids
I(A)	1	3	–	1
I(B)	2	10	–	3
9	6	50	1	19
16	5	22	2	6
17	7	7	2	4

An interesting phenomenon occurs in the long arm of chromosome 17 in a number of plates. This phenomenon, an achromatic or lightly stained region in the median area of the long arm, occurs in the plates of five Eskimo and three controls. In mixed cultures in which this characteristic occurs it usually appears in both individuals.

In chromosome 9, "gaps" appear in four different configurations. In some cases the entire secondary constriction is lightly stained or achromatic; in others in which no secondary constriction is apparent, only a small area immediately above the centromere is unstained. In a number of cases the area below the secondary constriction is unstained, while in a few plates the centromere is stained and there is an achromatic region both above and below it. These phenomena appear in both sister chromatids and homologues.

DISCUSSION AND CONCLUSION

A number of cytogenetic analyses have been carried out on semi-isolated and isolated human populations. These include studies of the Greenland Eskimo (Barnicot and Travers 1963; Carles-Trochain 1971); the Ainu, Orok, and Gilyak of Hokkaido and Sakhalin (Hungerford, Makino, et al. 1969); the people of Tristan da Cunha (Hamerton et al. 1965); the Todas of South India (Chandra and Hungerford 1966); New Guinean natives (Hungerford, Giles, et al. 1965); a group of pygmies in the Congo (Cavalli-Sforza et al. 1969); a small number of West Africans (Barnicot and Travers 1963); the Hottentot (Heltne and Singer 1971) and the Yanomama (Bloom and Neel 1970). Chromosome abnormalities found in the Yanomama have already been briefly mentioned. In the Hottentot work it was reported that heteromorphic chromosome pairs and an "extra chromatin fragment" occur consistently in some individuals. It is uncertain whether or not these results are an artifact of the culturing technique.

In all these studies except that of the Hottentot, the culturing was done in a specialized laboratory away from the field after a delay of a few days. The advantage of our work is that it was performed in the field (in the world's most northerly cytogenetic lab!) and involved the use of controlled mixed cultures.

The data we have collected have not supplied us with any very unusual results. That there were no individuals found with gross chromosomal abnormalities is not surprising, mainly because of the limited sample and the fact that all the individuals studied appeared clinically normal. It also seems reasonable that there would be strong selective pressures working against abnormal individuals in this population, though as the Eskimo become more sedentary and acculturated this pressure loses much of its importance.

All the chromosome variations observed in the Eskimo chromosomes are comparable to those found in other human populations. An analysis of the different frequencies of these variations in populations has not been carried out except for the "large" Y chromosome (Soudek 1973).

The problem of chromosome breakage could be of some ecological interest, since chromosome analysis can supply some information on genetic damage that has occurred in a population due to its exposure to deleterious substances in the internal and external environments. It appears that the traditional approach to the problem of breakage may have to be revised, however, on the basis of new concepts arising from increased knowledge of chromosome ultrastructure and relationships involving the chromatid filament and nuclear membrane (Chiarelli: "Bands or chromomeres," this volume). Rather than being actual "breaks" or "gaps" in the chromosome filament, these achromatic and lightly stained areas may be a result of inactivation of protein sythesis in a particular area, or of differences of coiling.

Revision of the breakage concept is also indicated by the work of Caspersson and co-workers (1972), which shows that chromosome "breakage" seems to occur more frequently in lightly stained areas of (banded) chromosomes treated by fluorescent staining procedures.

The most important aspect of our work, we feel, rests on its most obvious finding: that of the homology of the Eskimo karyotype with that of directly comparable European controls. This data supports other data on the seeming stability of the human karyotype, a stability maintained in spite of the differentiation of mankind into various more or less distinct populations.

The more refined techniques of chromosome banding applied to individuals of various populations, especially isolated ones, and in the context of the mixed culture, could add to the reliability of this statement on homogeneity. In any case, results of this kind of study could have major evolutionary and ethnological implications.

REFERENCES

BARNICOT, N. A., P. J. TRAVERS
1963 Comparison of the human karyotype in various populations. *Proceedings of the II Congress of Human Genetics*: 1,164–1,166.

BLOOM, A., J. V. NEEL
1970 Population genetics of the American Indian. *IBP Newsletter* 3(2):2.

CARLES-TROCHAIN, E.
1971 Etude cytogénétique d'un isolat. *L'Anthropologie* 75:263–268.

CASPERSSON, T., ULLA HAGLUND, B. LINDELL, LORE ZECH
1972 Radiation-induced non-random chromosome breakage. *Experimental Cell Research* 75:541–543.

CAVALLI-SFORZA, L. L., L. A. ZONTA, F. NUZZO, L. BERNINE, W. W. DE JONG, *et al.*
1969 Studies on African pygmies: a pilot investigation of Babinga pygmies in the Central African Republic. *American Journal of Human Genetics* 21:252–262.

CHANDRA, H. S., D. A. HUNGERFORD
1966 Chromosome studies of the Todas of South India. *Human Biology* 38:194–198.

DANIEL RORY, B. CHIARELLI
1971 Cytogenetic investigation of chromosomes from Eskimo in Igloolik, N.W.T. *Canadian Review of Sociology and Anthropology* 57:259–262.

HAMERTON, J. L., A. J. TAYLOR, R. ANGELL, V. M. McGUIRE
1965 Chromosome investigations of a small isolated human population: chromosomal abnormalities and distribution of chromosome counts according to age and sex among population of Tristan de Cunha. *Nature* 206:1,232–1,324.

HELTNE, P. G., R. SINGER
1971 Cytogenetic studies in the Hottentot population: count distribution, report of a fragment, and preliminary description of morphology. *American Journal of Physical Anthropology* 34:1–36.

HUGHES, D. R.
1973 Introduction to the "Paper presented to the Fifth Arctic Symposium on the Biology of Circumpolar People" Reykjavik, Iceland June 28, 1973.

HUNGERFORD, D. A., E. GILES, C. G. CREECH
1965 Chromosome studies of East New Guinea natives. *Current Anthropology* 6:107–110.

HUNGERFORD, D. A., S. MAKINO, M. SASAKI, A. ARA, C. B. BALABAN
1969 Chromosome studies of the Ainu population of Hokkaido. *Cytogenetics* 8:74–79.

SOUDEK, D.
1973 Chromosomal variants with normal phenotype in man. *Journal of Human Evolution* 2:341–355.

Observations on Centromeric Heterochromatin

ANN L. KOEN

Although the Giemsa banding techniques are now used almost exclusively for identification of chromosomes or chromosome parts, the C-banding stain for centromeric heterochromatin may also yield useful information about the mechanism of chromosome separation and juxtacentromeric breakage. Comparison of the action of variant chromosomes with their normal homologs in the same environment can be used to determine whether the observed activity is in fact due to the quantitative variation in heterochromatin or to some other undetermined characteristic of individual chromosomes.

The chromosomes of a family exhibiting heterozygosities of centromeric heterochromatin in three chromosome pairs have been studied by means of C-banding techniques. The heterozygous pairs were nos. 1, 9, and 16; each of the variant chromosomes had a reduced amount of centromeric heterochromatin. The family consisted of the father (I-I), mother (I-2), and five siblings (II-I,♂), (II-2,♂), (II-3,♀), (II-4,♂) and (II-5,♂). The siblings II-4 and II-5 were identical male twins. Inheritance of the variant chromosomes by the children from the parents appeared to be consistent with Mendelian laws. All the family members were phenotypically normal.

These chromosomes were used to study the localization of centromeric heterochromatin and the sequence of chromatid separation in late metaphase. In addition, unequal division of the centromere was observed in a few cases.

I thank M. Conley for her competent technical assistance.

METHODS AND MATERIALS

Blood samples were obtained by finger puncture and collected into chromosome medium 1A with phytohemagglutinin (Gibco). Culture time was seventy-two hours and cells were harvested from either hypotonic sodium citrate (0.75 percent) or potassium chloride (0.075 N). Slides were prepared by the conventional air-drying technique. The harvesting procedure was essentially that of Moorhead et al. (1960), except that in some cases 0.1 ml of phytohemagglutinin M (Difco) was added to the culture tubes three hours before harvest and 0.1 ml colcemid (Gibco) was added five minutes before harvest. The addition of the extra phytohemagglutinin M stimulated the lymphocytes to further division; without this extra stimulation very few metaphases were found when the exposure to colcemid was limited to five minutes. This modification made possible the study of chromosomes in late metaphase; the addition of colcemid for only five minutes was sufficient to dissolve the mitotic spindle and thus yield good chromosome spreads with a minimum effect on the division itself. Fifty late metaphase cells with heterozygous pairs nos. 1, 9, and 16, obtained by this method, were examined for sequence of chromatid separation. An additional fifty cells with partially separated chromatids and with the same heterozygosities obtained from cultures exposed to colcemid for three hours were also studied.

Differential staining for centromeric heterochromatin was carried out according to the method of Arrighi and Hsu (1971) and by a modification of this method. The modification consisted of the omission of the RNase treatment and the substitution of saturated $Ba(OH)_2$:0.07 N NaOH (3:1) for 2.5 minutes instead of the 0.07 N NaOH two-minute treatment of the Arrighi and Hsu method. If 0.07 N NaOH is used for two minutes without previous RNase treatment, excessive swelling of the chromosomes occurs. The milder alkali mixture of $Ba(OH)_2$ and NaOH does not cause this swelling. In our hands, however, the original procedure with RNase was superior in delineating the exact placement of centromeric heterochromatin. The substitution of $Ba(OH)_2$ was suggested by Sumner (A. T. Sumner 1973:personal communication).

From twenty to fifty karyotypes were constructed for each subject, using at least four different blood samples from each over a period of two years. Subject I-2 was repeated fifteen times. Patterns of centromeric heterochromatin were entirely repeatable and did not change over the period of the investigation. Consistency of the patterns was also checked by culturing cells from I-1, I-2, and II-2 in the same test tube and identifying the three different chromosome complements of the cells.

RESULTS

Inheritance of the variant chromosomes was consistent with Mendelian laws. As shown in Table 1, the father was heterozygous in chromosomes 1 and 9 and the mother in no. 16, while the children had various combinations of these. Identical twin siblings had the same heterozygosity of nos. 1 and 9. Plate 1 shows a full karyotype of subject II-1; he was heterozygous in all three of the pairs in question, nos. 1, 9, and 16. Plate 2 shows representative pairs of these chromosomes from the rest of the family.

Table 1. Heterozygosities of chromosomes in the family

Subject number	A 1	Zygosity of C 9	E 16	Number of cells karyotyped
I-1	hetero	hetero	homo	28
I-2	homo	homo	hetero	50
II-1	hetero	hetero	hetero	25
II-2	hetero	homo	hetero	20
II-3	hetero	hetero	hetero	35
II-4	hetero	hetero	homo	25
II-5	hetero	hetero	homo	25

When cells from I-1, I-2, and II-2 were cultured in the same test tube, the three different chromosome complements (one heterozygous in nos. 1 and 9, one heterozygous in nos. 1 and 16, and one heterozygous in no. 16 only) could easily be distinguished from each other without reference to the sex chromosomes. Thus it seems certain that the external environment does not play a part in the occurrence of variant forms of centromeric heterochromatin.

Cells in which the chromatids were partially separated from each other were best suited for localization of the centromeric heterochromatin. Plate 3 presents selected chromosomes from a single cell. Although not all the chromosomes can be identified by number, since other banding techniques were not carried out on these slides, nos. 1, 2, 3, 9, 16, and unspecified members of the B, C, D, E, and G groups can be distinguished from each other. Aside from nos. 1, 9, and 16 with their large heterochromatin blocks, it can be seen that the heterochromatin block in the short arm of no. 2 is about three times the size of that in its long arm; that no. 3 has small heterochromatin blocks on each side of the centromere; that one of the B-group pairs has a block of heterochromatin only in the short arms, while the other B-group pair has heterochromatin blocks on either side of the centromere. The placement of the heterochromatin can be distinguished on each of the other representative chromosome pairs. It is interesting that, in those cases in which the areas of heterochromatin can be definitely established, the centromere itself does not appear to be

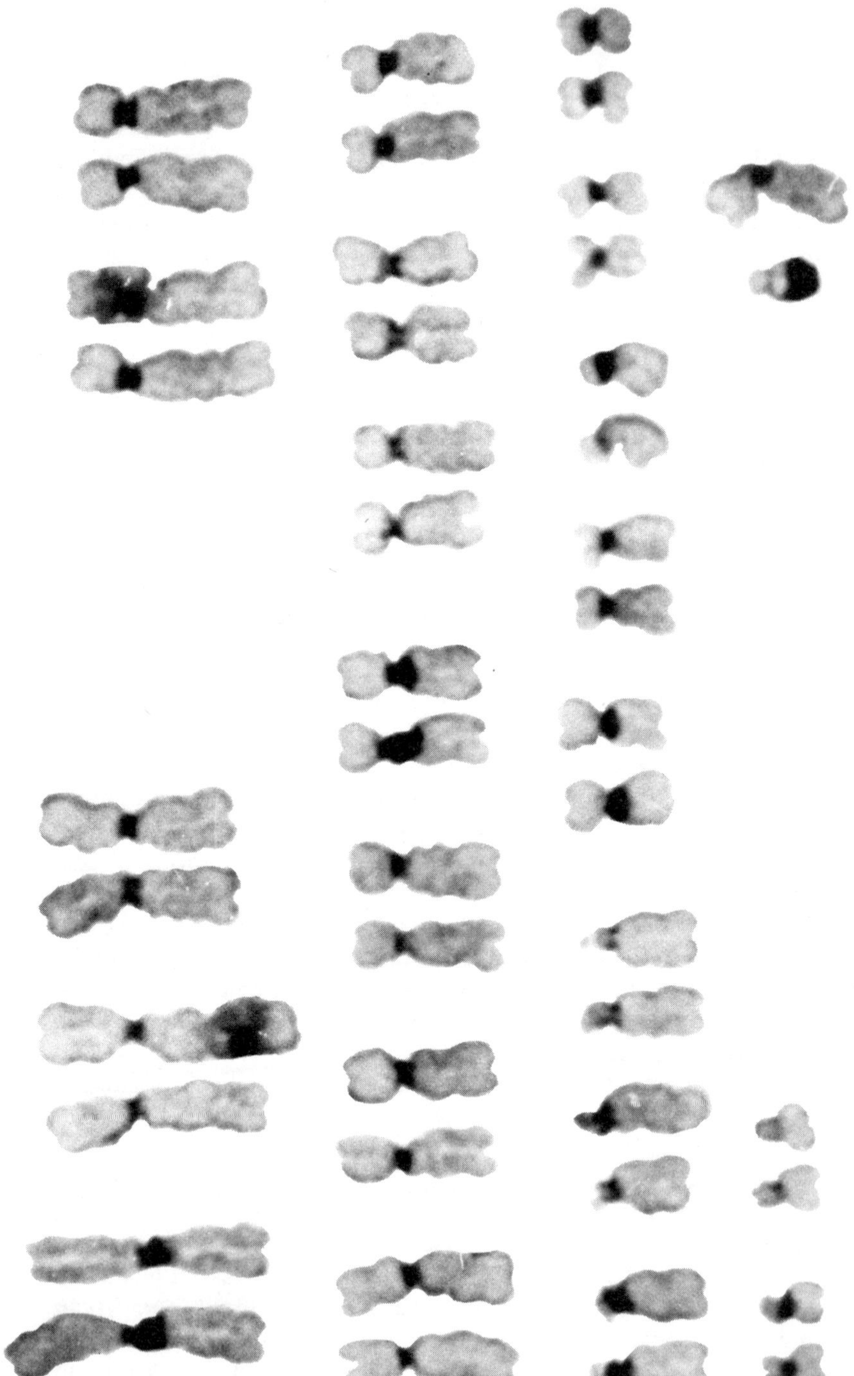

Plate 1. Karyotype of subject II-1. Chromosomes are arranged according to the Denver classification

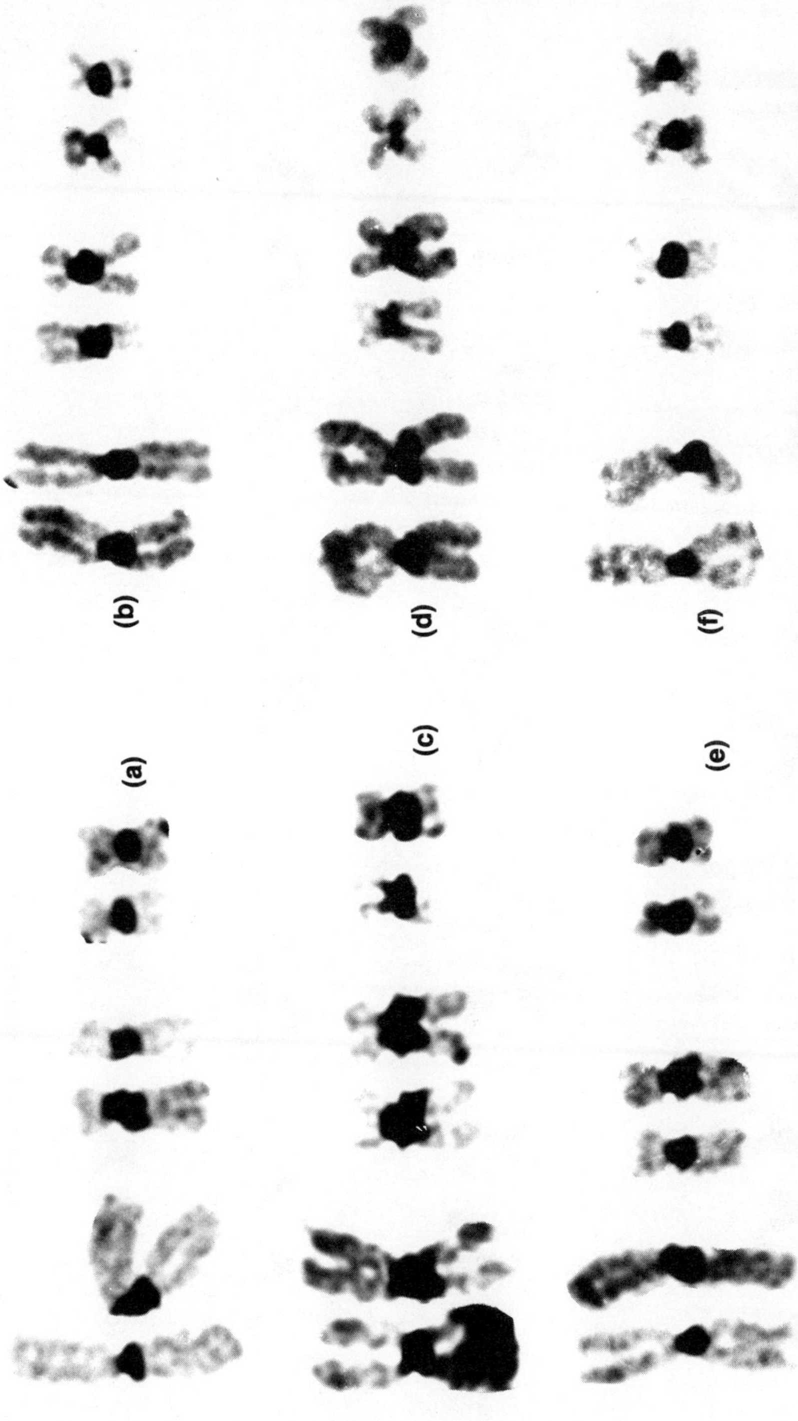

Plate 2. Partial karyotypes (chromosome pairs 1, 9, and 16) of other members of the family: (a) father; (b) mother; (c) brother; (d) sister; (e) and (f) twin brothers

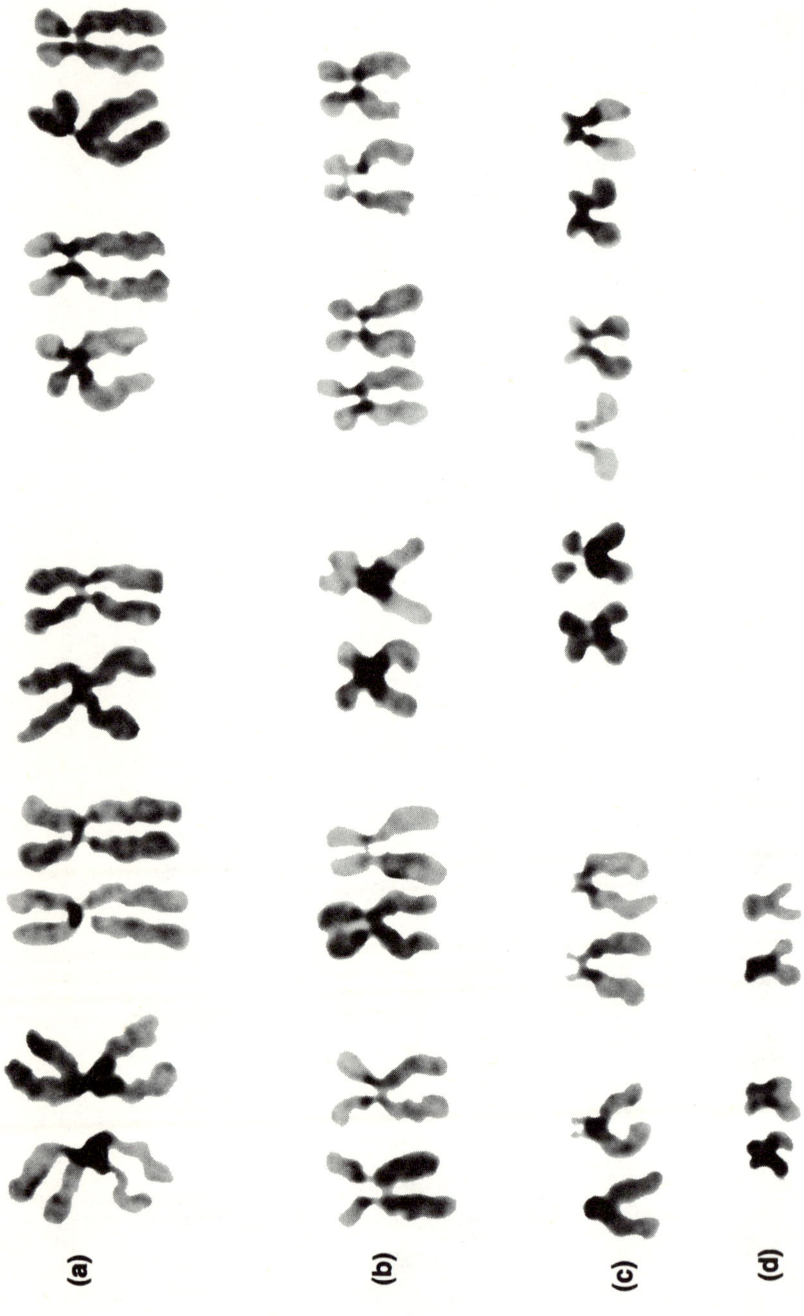

Plate 3. Partial karyotype of late metaphase chromosomes showing placement of heterochromatin. Line a, Groups A and B; Line b, group C; Line c, groups D and E; Line d, group G chromosomes

heterochromatic, but euchromatic. The centromeric heterochromatin is located on one or the other side, or on both sides, of the centromere. In view of the variability found in centromeric heterochromatin, it would not be surprising if members of different families exhibited different placement of centromeric heterochromatin.

The photograph in Plate 3 also indicates that the time of separation of the sister chromatids may be related to the amount of heterochromatin at the centromere; the chromosomes with less heterochromatin are already separated, while those with more are still firmly coupled together. This observation was made in cells which had been exposed to colcemid for three hours as well as in those which had been exposed to it for only five minutes, as shown in Table 2. Therefore, centromeric heterochromatin apparently plays the same role in sequential separation of sister chromatids whether or not the mitotic spindle is present.

Table 2. Relative separation distances of chromatids in heterozygous chromosome pairs

Chromatid separation	Chromosome 1	Chromosome 2	Chromosome 16
Colcemid five minutes			
a[a] > b[b]	43	39	35
a < b	4	4	3
a = b	3	7	12
Total	50	50	50
Colcemid three hours			
a > b	40	38	30
a < b	2	3	4
a = b	8	9	16
Total	50	50	50

[a] Chromosome with reduced centromeric heterochromatin.
[b] Chromosome with normal centromeric heterochromatin.

That this difference in time of chromatid separation is purely a function of the amount of heterochromatin at the centromere rather than of some other individual chromosomal characteristic is indicated by the data in Table 2 and exemplified in Plate 4. In the plate showing heterozygous no. 1 chromosomes (4a), no. 9 chromosomes (4b), and no. 16 chromosomes (4c) from different persons, the left-hand member of each pair, with less heterochromatin, is fully separated, while in the right-hand member, with more heterochromatin, the chromatids are still in contact at the centromere.

Unequal division of the centromeric material in chromosomes containing large blocks of heterochromatin at the centromere was observed in a few cases, as shown in Plate 5. It is not known if the cultured cells containing the unequally divided chromosomes would be viable if the cell had continued through the next division cycle. Since the pattern of

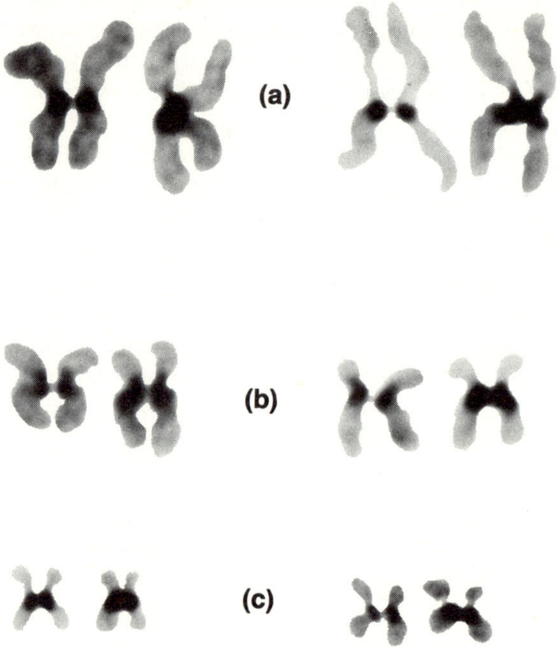

Plate 4. Heterozygous pairs nos. 1, 9, and 16 of subjects I-1 (a), II-2 (b), and II-3 (c), showing difference in separation time of sister chromatids

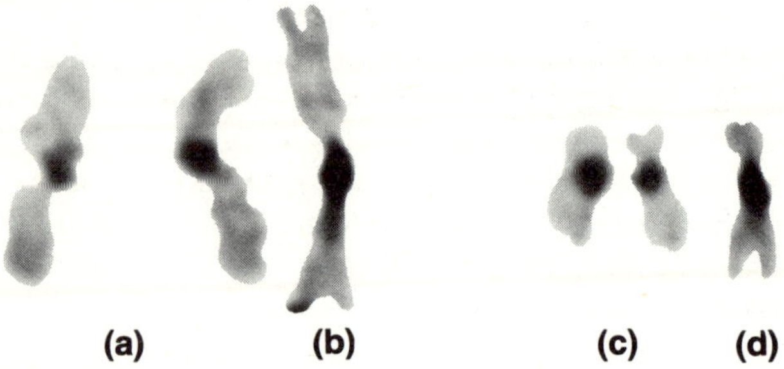

Plate 5. Chromosomes 1 and 9 showing unequal division of centromeres: (a) no. 1 with unequally divided centromere; (b) normal no. 1 from different cell of same person as (a); (c) no. 9 with unequally divided centromere; (d) normal no. 9 from different cell of same person as (c)

centromeric heterochromatin observed in many cultured cells of many persons has been consistent, it is probable that either the anomalous division is very rare, or else the cells in which it occurs do not replicate.

DISCUSSION

Repeatability of the chromosome pattern was very reliable, even in mixed cultures, indicating that the differences in centromeric heterochromatin are intrinsic and not influenced by *in vitro* environmental effects.

The localization of heterochromatin on either side of the centromere, but not at the centromere itself, may perhaps have a relationship to the formation of isochromosomes. It is possible that the chromosome breaks occur in the very small areas of euchromatin between the areas of heterochromatin, rather than in the heterochromatin itself. In clear photographs of good spreads of slightly separated chromatids, fine bridges of chromatin could be discerned linking the two heterochromatin blocks (Plate 3). The small areas of euchromatin between the heterochromatin blocks at the centromere bear a superficial resemblance to the bridges between sister chromatids. The mechanism of the abnormal separation leading to isochromosomes may therefore be similar to that of the normal chromatid separation. This needs further investigation.

The chronological sequence of chromatid separation appears to be related to the amount of heterochromatin at the centromere. That this control of separation time is a function of centromeric heterochromatin only and not of heterochromatin per se is indicated by the difference between the Y chromosome and the autosomes 1, 9, and 16. Although the distal portion of the long arm of the Y contains large amounts of heterochromatin, the separation of Y chromatids does not appear to be influenced by this. The Y chromosome has very little heterochromatin at the centromere (Plate 1), and the Y chromatids separate at approximately the same time as other chromosomes with little centromeric heterochromatin. Many authors have commented on the difference in characteristics of centromeric heterochromatin and Y heterochromatin. Aside from the response to quinacrine stains, the most striking difference seems to be the influence of centromeric heterochromatin on time of chromatid disjunction and the lack of this effect by Y heterochromatin. The large blocks of Y heterochromatin do not seem to influence the distal ends of the Y chromatids to adhere to each other during metaphase.

Although the cases of unequal centromeric division observed occurred in mitotic cells in an artifical environment, it is possible that the same mechanism may have been operative in germ cells which, in the ancestors of this family, gave rise to the observed variant chromosomes. This may represent a case of unequal crossing-over at the centromere.

SUMMARY

Heterozygous chromosme pairs nos. 1, 9, and 16 were studied in members of a human family by means of C-banding techniques. All members of the family were phenotypically normal, and inheritance was apparently in accordance with Mendelian laws. Juxtacentromeric heterochromatin was well defined in good spreads; some chromosomes have single blocks in either the long or the short arm, while others have blocks in both arms. Observation of chromosomes in late metaphase led to the conclusion that time of separation of sister chromatids is correlated with the amount of heterochromatin at the centromere. A few instances of unequal centromere division were observed; this may represent unequal crossing-over, a mechanism by which variant chromosomes with increased or reduced centromeric heterochromatin might arise.

REFERENCES

ARRIGHI, E. E., T. C. HSU
1971 Localization of heterochromatin in human chromosomes. *Cytogenetics* 10:81–86.

MOORHEAD, P. S., P. C. NOWELL, W. J. MELLMAN, D. M. BATTIPS, D. A. HUNGERFORD
1960 Chromosome preparation of leukocytes cultured from human peripheral blood. *Experimental Cell Research* 20:613–616.

The Problems of "Normal" Chromosomal Variants in Man

D. SOUDEK

ABSTRACT

The scientific field of major chromosomal aberrations and their clinical effect seems to be nearly exhausted at present, and much interest is attracted by minor chromosomal variants, which make nearly every human being's chromosome set different from every other. These variants constitute chromosomal polymorphism in man, i.e. the occurrence of chromosomes in two or more alternative structural forms within a population. In man there does not exist a numerical chromosomal polymorphism, that is, several chromosomal numbers, as exists in many animal species. There is a problem linked with human chromosomal polymorphism: are all chromosomal variants entirely without phenotypic effect (as some cytogeneticists suppose), or do some of them possess an adaptive value, i.e. different survival value and reproductive capability for their carriers in comparison with carriers of "standard" chromosomal forms? It has become known that in other human polymorphisms (blood groups, HLA system) some variants are associated with a higher probability of certain diseases.

This review summarizes basic known facts on variants, with some selected references. It is concentrated around the problem of the possible pathogenic influence of some variants.

For nearly half a century, it has been known that the substance forming the main part of the cell nucleus — chromatin — has two components: one (euchromatin) which contains genes, and another which lacks genes. It is estimated at present that only 2 percent of mammalian cell nucleus DNA contains genes (Ohno 1973). The other 98 percent is not transcribed and translated and its functions, if any, are obscure. Heterochromatin belongs in this category of seemingly functionless chromatin. Recently several microscopic methods have been developed which allow that certain fractions of constitutive heterochromatin (i.e. heterochromatin which is permanently in a genetically inactive state) can be observed directly under the microscope *in situ*.

This study was facilitated by the Ontario Mental Health Foundation.

LOCATION OF CONSTITUTIVE HETEROCHROMATIN

Constitutive heterochromatin is located in every human chromosome, in certain chromsomal parts:

1. In metacentric chromosomes, there is a cluster of heterochromatin in the centromeric region. In chromosomes 1, 9, and 16, heterochromatin is located in secondary constrictions.

2. In acrocentric chromosomes, constitutive heterochromatin is located in the centromere (short arms) satellite region. It is important to keep in mind that on secondary constriction (also nucleolar constriction) between short arms and satellites, ribosomal cistrons, containing 18 + 28s rDNA and producing ribosomal RNA (Hsu et al. 1975; Goodpasture et al. 1976), are located. This activity is connected with nucleolus formation; therefore, another term applied to these chromosomal components is "nucleolar organizer" (NOR). These comprise about 400 ribosomal genes which are located only on the ten human acrocentric chromsomes (nos. 13–15, 21, and 22), approximately forty genes to every NOR location. It is to be emphasized that NORs contain special genetically active DNA which is located between segments of heterochromatin but which is not heterochromatic.

3. In the Y chromosome, there is no centromeric block of heterochromatin, but another block forms most of the distal part of the long arm.

4. In rare cases, genetically inactive, small, supernumerary chromosomes (also called accessory or B-chromosomes) are found. Some of these chromosomes contain satellites on one or both arms; they carry in these cases either one or two extra NORs. Supernumerary chromosomes are common in some plant and insect species, but they are very rare in mammals and in man (Soudek 1976).

METHODS FOR STUDY OF VARIANTS IN CONSTITUTIVE HETEROCHROMATIN

At present, there are two basic methods for microscopic study of chromosomal constitutive heterochromatin (Osztovics et al. 1973):

1. USE OF SPECIFIC FLUOROCHROMES: quinacrine (Q-banding), acridinorange, Hoechst 33258. Only Q-fluorescence will be discussed here. The original idea (Caspersson et al. 1968) supposed that more or less intense fluorescence of chromosomal segments reflects different nucleotide base compositions. This hypothesis, however, has not been confirmed, and it seems that the intensity of fluorescence depends not only on DNA bases but also on chromosomal proteins. Anyway, some

heterochromatic chromosomal segments in cells of some individuals show specific but variable fluorescence, e.g. the centromere in nos. 3 and 4, the centromeres, short arms and/or satellites of acrocentrics, the distal part of the long arm of the Y chromosome. It is important to realize that two kinds of polymorphism can be observed after the application of fluorochromes: (a) size of the fluorescent segment; and (b) intensity of the fluorescence, from brilliant fluorescence to no fluorescence at all.

The advantage of the fluorescence method is that the shape of fixed chromosomes is not changed, and other methods can be applied afterwards; there is no destruction of chromosome components involved. The fluorescence phenomenon is only temporary, however, and the slide cannot be stored; a microphotograph is the only permanent proof of the existence of the variant. It should be noted that brilliantly fluorescent heterochromatin is specific for three primate species only: gorilla, chimpanzee, and man.

2. USE OF C-BANDING METHODS. It is interesting to note also that the original theory of DNA renaturation (Arrighi and Hsu 1971), which started C-banding, was later proven not entirely correct. After a period of destructive action by the alkaline solution and saline, or proteolytic enzyme, only constitutive heterochromatin persists and can be stained. The slides can be stored. The method is final, however, and cannot be followed by any other method. Some special modifications have been developed especially for C-heterochromatin of 9qh (Bobrow et al. 1972; Gagne and Laberge 1972). Both fluorescence and C-banding methods demonstrate some identical chromosomal components, e.g. the distal part of the Y chromosome and most cases of polymorphism in acrocentrics. Secondary constrictions in nos. 1, 9, and 16, however, are observable only after C-banding, and after Q-banding only as negatives (as a nonfluorescent gap). Constitutive heterochromatin around the centromere of no. 3 is always present, but it is not fluorescent in all chromosomes (Schnedl 1971). A case has been described in which the long arm of the Y chromosome was nonfluorescent but positive after C-banding (Soudek and Laraya 1976). It should be concluded that these two methods present the same general chromosome components and are somehow complementary, so that both methods should be used for the study of chromosomes of each human subject, because both have certain advantages and disadvantages. The two mentioned methods allow for distinguishing at least four subgroups of constitutive heterochromatin as shown in Table 1.

Jalal et al. (1974) distinguish at least six different C-heterochromatin fractions, according to different staining qualities.

Table 1. Staining methods for constitutive heterochromatin

	Q-bands	C-bands	Special methods[a]
Some 3c, 4c, distal part Yq	+	+	0
Pericentromeric heterochromatin and 1qh, 9qh, 16qh	0	+	0
Intercalary (band) heterochromatin	+	0	0
9qh	0	+	+

[a]Bobrow et al. 1972; Gagne and Laberge 1972.

3. NOR STAINING METHODS. There are special methods for the staining of NORs or N-bands (Funaki et al. 1975; Goodpasture and Bloom 1975).

4. SPECIAL METHODS. More kinds of polymorphism in human chromosomes are on the horizon after the use of special methods (e.g. Angell and Jacobs 1975). The methods are not yet in general use.

CHROMOSOMAL VARIANTS

Even before the use of new cytogenetic methods in the 1970's, it was known that some chromosomal parts are variable (short arms and satellites of acrocentrics, the length of Y). New methods, which permitted the detailed study of constitutive heterochromatin, disclosed high variability of most heterochromatic parts of chromosomes. This is in sharp contrast with euchromatic parts of chromosomes, which contain intercalary heterochromatin and are identifiable by G- or R-banding methods. These bands are preponderantly species-specific, and most of them can be traced also in related species (e.g. those of man can be traced in the chimpanzee, the orangutan, and the gorilla), their analogy being close to 99 percent (Dutrillaux et al. 1975). This corresponds to the fact that, in related species, genes are almost the same: for example, 99 percent of the genes found in man and the chimpanzee are identical (King and Wilson 1975). In contrast, heterochromatin components are so variable that nearly every human being has his own chromosomal characteristics. There are a number of these variables:

1. Q-FLUORESCENCE INTENSITY. This occurs in 3c and 4c and in centromeres, short arms, and satellites of acrocentrics. Yq always has very intense fluorescence, with the rare exception of some "short Ys" which completely lack fluorescence.

2. SIZE OF THE HETEROCHROMATIC SEGMENT. This kind of variability is found especially in 1h, 9h, 16h, and Yq (Craig-Holmes and Shaw 1971; Craig-Holmes et al. 1973; Jalal et al. 1974), and also in some centromeric

heterochromatin: 6, 12 (Sofuni et al. 1974), and acrocentrics. Segments of just a fraction of the average size or of increased volume can frequently be observed. Some authors have suggested that the length of the heterochromatic segment increases only in multiples (e.g. Kim 1973). Variation occurs not only in length, however, but also in the total volume of the heterochromatic segment involved. It seems that the duplication theory is not based on sufficient evidence and that the variation is continuous. In acrocentrics, there are normal, increased, or deleted short arms and increased ("giant"), duplicated, or missing satellites. The length of secondary constrictions ("satellite stalks") is also variable (Orye 1974a, 1974b; Zankl and Zang 1974).

3. POSITION OF CENTROMERIC HETEROCHROMATIN. Secondary-constriction heterochromatin (in chromosomes 1, 9, 16) or centromeric heterochromatin of some metacentrics (i.e. no. 3) is normally located on the long arm of the chromosome involved. In some individuals, however, it is located on the short arms, or just between the arms (intermediary position, or "incomplete" inversion) (de la Chapelle et al. 1974; Mutton and Daker 1973; Soudek, O'Shaughnessy, et al. 1974). This inversion of centromeric heterochromatin is more frequent in some chromosomes than was previously thought and has been referred to as "pericentric" inversion, although this may be an inexact term, which may later be replaced by a more specific expression.

4. VARIABILITY OF NUCLEOLAR ORGANIZERS. As has already been mentioned, there are ten NORs in man. In carriers of Robertsonian translocation, however, there are two NORs missing for every translocation. There are accessory NORs in carriers of trisomies of acrocentrics or in carriers of satellited B-chromosomes. In man there can exist phenotypically normal individuals with any number of NORs between six (Orye and Delire 1967) and sixteen (Sergovich et al. 1967). There was much discussion connected with hypothetical compensation mechanisms to keep the production of nucleolar material constant. It seems at present that the NOR of every individual acrocentric chromosome has a different rDNA level, which is correlated with the morphological variability of the short-arm region. Combinations of special sites of NORs, inherited from parents, determine the individual level of rDNA of every human being (Dittes et al. 1975; Hayata et al. 1977). The question remains, "How broad is the spectrum of normal variability of the human NORs and which types of combinations of variant organizers are compatible with normal development?" (Dittes et al. 1975). The function of NORs is reflected in the fact of frequent associations between short arms of acrocentrics. Extensive but often contradictory literature exists on the subject of associations, which cannot be entered into here. NORs deserve a special review; they can be only briefly mentioned here.

5. RARE SATELLITED VARIANTS OF SOME CHROMOSOMES (e.g. 17 and X, Y). In the morphological variants of this group, chromosomes which do not normally carry satellites are now found carrying them. Satellites on Y are probably satellites originating from an acrocentric-Y translocation. "Satellites" on 17 and X have only a satellitelike shape, not a satellite chemical composition (Lubs 1969; Sandstrom and Jenkins 1973, 1974).

FREQUENCY OF VARIANTS

Chromosomal variants were, for a long time, a very neglected field of the study of human chromosomes, probably because human cytogenetics was influenced by the needs of medicine (chromosomal pathology) and was interested primarily in gross chromosomal errors. Categorization and quantification of variants depends very much on the laboratory involved, because there have been almost no universally accepted norms for the evaluation of variants; it is nearly impossible to compare the results of different laboratories. The experience of the investigator also plays an important role; more experienced persons find more variants within the same sample. Some authors have tried a quantitative classification of chromosomal variants (Patil and Lubs 1977b; Deryaguin et al. 1977). There are still problems, however: for example, it was proven that 1qh does not contract proportionally with the rest of the chromosome during mitosis (Schmiady and Sperling 1976; Baliček et al. 1977), which could cause classification error.

The frequency figures of chromosomal variants are a result of two factors. The first of them is the *subjective* factor, identical with the technical limits of the evaluation process. It was shown how a simple technical factor — the density of the negative film record — can influence the apparent presence or absence of a variant in the resulting photograph (Overton et al. 1976). This factor is dependent on the length of the exposure both in the microphoto camera and in the enlarger; on the length, dilution, and temperature of the developing process; on the kind of developer used; etc. The second factor is an *objective* one, based on real frequency differences of individual studied populations. The published results are a combination of both subjective and objective factors. There will probably be not only racial (Park and Antley 1975, 1976) but also local population differences: for example, the long arm of the Y chromosome is longer in Mediterranean populations than in Scandinavians (Lubs and Patil 1975). There was a difference in the length of the Y chromosome and in the frequency of some variants between two Indian populations studied, Rajputs and Punjabis (Ghosh and Singh 1975). Strongly fluorescent Q-bands and larger C-bands are more common in blacks. Inversion 1h is more common in Caucasians, while inversion 9h is more frequent in blacks (Lubs, Kimberling, et al. 1977). Australian aborigines have a very short fluorescent segment in Y (Angell 1973). Chromosomal variants in normal populations have been summarized and discussed in several papers (Pearson et al. 1973; Soudek 1973; Ferguson-Smith 1974). They were studied usually in samples of newborns (Zankl and Zang 1971; M. L. Lubs and H. A. Lubs 1972; Friedrich and Nielsen 1973; Bochkov et al. 1974; Cohen et al. 1975; Müller and Klinger 1975a, 1975b; McKenzie and Lubs 1975; Lin et al. 1976; Buckton et al. 1976) and of adults (Mikelsaar et al. 1973, 1975; Ghosh and Singh 1975, 1976), or of selected healthy adults (Soudek et al. 1976). Some authors, however, studied chromosomal polymorphism in a mixed sample of normal healthy persons and of patients without chromosomal aberrations, or of their relatives (Geraedts and Pearson 1974; Buckton et al. 1976). Such studies give the frequency of a variant in a part of the general population but do not help to ascertain possible clinical effects of the variant. In a collaborative study (Lubs, Patil, et al. 1977), the health problems of probands and the presence of a variant were recorded and their possible association evaluated. A population cytogenetics conference, held in Albany, New York, in 1975, was to a large extent devoted to the results of the variant study.

As an example of discrepancies in the results of different laboratories, see Table 2, a comparative table of frequency of variants. The results are given in percentages of subjects carrying a variant.

Results given in some more recent papers, using new cytogenetic methods, cannot be compared with results from other laboratories quoted in Table 2, because their evaluation is of a more complex nature. The study of most chromosomal variants is encumbered, as was mentioned, by several methodological sources of error (e.g. lack of natural categories, problems of scoring of variants, experience of the researcher, etc.). "Inversions" of pericentromeric constitutive heterochromatin, however, are a class of variants which is much less accessible to subjective errors. "Complete inversions," especially, can hardly be interpreted erroneously, and one is inclined to ascribe the observed frequencies mostly to the differences in populations. "Inversions" of pericentromeric heterochromatic regions have the frequencies shown in Table 3.

It is necessary to point out that the term "chromosomal variant" has been used rather vaguely. Therefore, let us try to present a tentative definition: *chromosomal variants are atypical, heterochromatic structures in chromosomes, are found in phenotypically normal persons, and do not decrease significantly the reproductive fitness of carriers.* Of course, there is no natural boundary between variants and aberrations. The limits of both categories are arbitrary; an extra Y chromosome in an XYY male, for example, is nearly totally heterochromatic, and the effect of this chromosomal aberration often corresponds to the above-mentioned

Table 2. Comparison of frequencies of chromosomal variants in populations studied by different authors (in percentages of subjects)

Location	Population studied	1qh+ and 1qh−	9qh+ and 9qh−	Dp+ and Ds+	Dp−	16qh− and 16qh−	Gp+ and Gs−	Gp−	Yq+ ≥F19	≥E18	Yq−
Munich, West Germany[a]	280 newborns			6.80		1.50	1.50		2.10		
Denver, U.S.A.[b]	3,476 newborns	0.23		16.41	0.06	2.59	5.16	0	14.87	0.34	0.40
Arhus, Denmark[c]	5,049 newborns	0.06	0.02	0.34	0.04	0.08	0.40	0.02	1.03		0.57
Tartu, Estonia, U.S.S.R.[d]	103 adults	0.90	5.70	11.20	0.90	1.90	8.50	0.90	3.80		5.70
Maryland, U.S.A.[e]	4,635 normal and institutionalized boys	0.24	1.71	5.32	0.45	0.13	4.58	0.04	4.04		0.71
Moscow, U.S.S.R.[f]	2,500 newborns	0.04		0.44	0.45	0.12	0.32			0.36	0.08
Jerusalem, Israel[g]	500 newborns	0.40		3.20		0.80	1.40			5.80	0.40
Winnipeg, Canada[h]	14,069 newborns	0.04		0.26	0.06	0.09	0.54	0.03		0.89	0.11

[a] Zankl and Zang 1971.
[b] M. L. Lubs and H. A. Lubs 1972.
[c] Friedrich and Nielsen 1973.
[d] Mikelsaar, Tüür, et al. 1973.
[e] Rary and Borgaonkar 1973.
[f] Bochkov et al. 1974.
[g] Cohen et al. 1975.
[h] Hamerton et al. 1975.

Table 3. Frequency figures for "inversions" of pericentromeric heterochromatin (in percentages of subjects)

Population	Sample size	Inversion frequency (percentage) 1qh	3c	9qh	10c	16qh	Complete (c) or partial (p) inversion
Orientals[a,b]	150	3.3 single cases					?
Retarded Caucasians[c]	270		3.00				c
Newborns, Bronx[d]	316		0.30				c
Healthy Caucasian males[e]	100		1.50				c
Newborns, Hamilton[f]	930		0.75				c
Newborns, Edinburgh[g]	482		1.70				c
Fourteen-year-olds, Edinburgh[g]	109		0				c
Patients, London, Ontario[h]	180		0.50				?
Retarded, London, Ontario[h]	100		1.00				?
Normal population, Vienna[i]	1,000		0.50				?
Newborns, Tartu, U.S.S.R.[j]	102		5.00 to 6.00				?
Retarded, Tartu, U.S.S.R.[j]	45		11.00				?
Patients[k]	200			1.50			?
Newborns, Grand Junction[l]	600			0.80			c + p
Retarded[l]	68			4.40			c + p
Patients[m]	282			2.80			?
Unspecified[m]	650			0.70			?
Patients[n]	631			0.80			?
Newborns, Grand Junction[o]	77			4.00			c + p
Newborns, Jerusalem[p]	500			0.20			c
Orientals[b]	150			11.30			?
Newborns, Edinburgh[g]	482			1.40			c
Fourteen-year-olds Edinburgh[g]	109			2.00			c
Whites, U.S.A.[q]	3,084			0.16			c
Blacks, U.S.A.[q]	1,780			1.30			c
Healthy Caucasian males[e]	100			4.00			c
Greeks[r]	400			3.50			?
Patients[n]	631				1.3		?
Unspecified[s,t]						single cases	?

[a] Lee et al. 1974; Tüür et al. 1974.
[b] Park and Antley 1975.
[c] Soudek 1975.
[d] Müller and Klinger 1975a, 1975b.
[e] Soudek, Sroka, et al. 1976.
[f] Lin et al. 1976.
[g] Buckton et al. 1976.
[h] Sergovich, personal communication 1975.
[i] Schnedl, personal communication 1977.
[j] Mikelsaar, personal communication 1977.
[k] Mutton and Daker 1973.
[l] H. A. Lubs and M. L. Lubs 1972.
[m] Madan and Bobrow 1974.
[n] de la Chapelle et al. 1974.
[o] McKenzie and Lubs 1975.
[p] Cohen et al. 1975.
[q] Lubs, Kimberling, et al. 1977.
[r] Tsenghi 1976.
[s] del Solar 1974.
[t] Fonatsch 1977.

definition. All supernumerary sex chromosomes are listed as aberrations, however, and consequently, for this reason, all XYY cases will remain in the group of aberrations. "Chromosomal variant" is often equivalent to the term "marker chromsome."

FEATURES OF CHROMOSOMAL VARIANTS

1. PHENOTYPIC EFFECT. In contrast to chromosomal aberrations, which are pathogenic and which invariably have an unfavorable effect upon the carrier or upon some of its progeny (in most cases of unbalanced translocation or inversions), chromosomal variants are found in clinically normal persons. Chromosomal variants cannot be linked directly to any clinical syndrome (for a possible exception, see Gardner et al. 1974). This leads to the concept of "normal" chromosomal variants, which constitute chromosomal polymorphism. Possible indirect phenotypical effects will be discussed later.

2. INHERITANCE OF VARIANTS. At least 90 percent of variants, and possibly many more, are inherited and can be found in one of the parents. This fact has a practical use in many cases of prenatal chromosomal diagnosis because it makes it possible to distinguish between female fetal cells and maternal cell contamination. The fetal chromosomal variants are a combination of maternal and paternal variants. This fact can also be used for paternity exclusion studies. New mutations are extremely rare; inv 3c, for example, was studied in seven families, and in all the cases the inversion was found in one of the parents (Soudek 1976). Craig-Holmes et al. (1975) found ninety-nine C-band variants inherited, while seven variants were not present in the parents at all and an additional seven variants were different from parental variants. The authors propose that the C-band polymorphism arises from mismatching of the repetitive DNA sequences with subsequent unequal crossing over. This can occur during meiosis as well as in somatic cells. Similar results were reported by Craig-Holmes (1977) and Sekhon and Sly (1975). In contrast, Magenis et al. (1977) and Hoehn et al. (1977) found the size and position of C-bands extremely stable, without new mutations. Translocations of fluorescent segments from Yq to the short arms of acrocentrics have been described or, in some cases, presumed (Noël et al. 1971; Friedrich and Nielsen 1972; Fründ et al. 1972; Wahlström 1972; Develing et al. 1973; Lundsteen and Philip 1973; Reitalu 1973). Girnelli et al. (1977) observed translocation of prominent fluorescent satellites from chromosome 22 to a translocation chromosome (10; 22) in the first generation and again to chromosome 22 in the next generation. Possibly such an exchange can occur at meiosis. In one patient 22s+ was noticed; in neither of the parents was this variant observed. The paternity of the father (99.8 percent probability) was proven by genetic markers (Nagakome et al. 1977). Variants are inherited mostly in simple Mendelian ratios; some authors, however, have presented results which do not fit Mendelian proportions. C-band inversions and C-band size polymorphism appear to segregate in a Mendelian manner (with the possible exception of very

large 9qh+, which may be selectively inherited). There is some evidence to suggest a preferential segregation of brilliant Q-band polymorphism (Robinson et al. 1976; Carnevale et al. 1976). In the case of random distribution of variant chromosomes, the homozygote–heterozygote ratio should follow the Hardy–Weinberg law. Deviation from this law means that some selective factor is involved.

CHEMICAL COMPONENT OF CHROMOSOMAL VARIANTS

DNA, which is, from a genetic point of view, the most important component of the cell nucleus, can be divided into several fractions. One of these fractions is characterized by unique sequences of bases, by which genetic information is coded. Other fractions contain repetitive sequences, some of them with very short repeats, which cannot act as a code for anything. In this fraction belongs the so-called satellite DNA (Yunis and Yasmineh 1971; Jones 1973), which forms about 35 percent of human DNA. It is possible to divide human satellite DNA into four subfractions: satellite I is especially concentrated in Yq; satellite II is localized in C-bands in nos. 1 and 16, with less concentration in no. 9; satellite III is located in constitutive heterochromatin in no. 9, D, and G; satellite IV is also found in no. 9. There is no exact agreement on detailed distribution of individual subfractions of human satellite DNA (Jones et al. 1974; Evans et al. 1974). In Q- and G-bands, another subfraction of repetetitve DNA was identified (Sanchez and Yunis 1974). There is an important evolutionary feature in satellite DNA. While DNA-forming genes did not undergo much transformation other than rearrangement for many millions of years (for example, X-chromosome genes did not change for approximately 100 million years), satellite DNA is species-specific or is similar only in related species. Human satellite DNA III can be hybridized with corresponding satellite DNA of the chimpanzee, which is not the case with satellites I and II. A chimpanzee's Y chromosome completely lacks a fluorescent segment; in contrast, short arms of acrocentrics in the chimpanzee have more fluorescent material than those of human acrocentrics. In summary, there is a small difference (interspecies difference) in euchromatin, but very extensive differences in constitutive heterochromatin.

POSSIBLE FUNCTION OF CONSTITUTIVE HETEROCHROMATIN

There is a variety of hypotheses on the function of the repetitive component of DNA. Ohno (1973) holds an extreme opinion that it is only "junk

DNA." He thinks that this DNA lost its function entirely and "the only contribution which these base sequences have, may be stated as the importance of doing nothing." His argument is that the base sequence of "junk DNA" is functionless as a code carrier. The hypothetical function of constitutive heterochromatin, however, need not be based on base sequence code at all. Other roles attributed to the constitutive heterochromatin (Yunis and Yasmineh 1971; Hsu 1975) are organizing chromosome systems, attracting homologues at meiosis, facilitating evolutionary changes, providing raw material for new genes, acting as gene spacers and loci for recombination, protecting euchromatin, etc. The satellite DNA was recently described as an "enigmatic species of DNA." Another theory, diametrically different from that of Ohno, was first presented by Commoner (1964), who suggested that heterochromatin plays a regulatory role in cell metabolism, especially in determining cell size, growth rate, and total rate of the oxidative mechanism. It was recognized long ago that accesssory sex chromosomes, which are heterochromatic, influence some body functions in a negative way. The intensity of the influence is proportional to the number of extra sex chromosomes. Heterochromatin forming genetically inactive X chromosomes is facultative heterochromatin, which can have different qualities when compared with C-heterochromatin. Extra Ys or supernumerary chromosomes, however, are composed mainly of C-heterochromatin. Barlow (1973) summarized the results of the influence of extra chromosomes. He found that accessory heterochromatin affects birth weight, body height, immunoglobulin level, and cell growth rate. This last fact is reflected in affected intelligence: "even small changes in division rate could lead to large changes in the relative proportion of differentiated cells and so result in abnormal brain structure" (Barlow 1973). Forssman and Hambert (1967) and Forssman (1970) see minimal brain damage as the product of the presence of supernumerary sex chromosomes. Vogel (1973) put forward a hypothesis that the main clinical symptoms caused by chromosome aberrations in man are not caused primarily by aberrant numbers of classical structural genes, but by disturbances in the function of DNA with control function. Davidson and Britten proposed an extensive theory on gene regulation, the last modification of which was published by Davidson et al. in 1975. They contribute the regulatory function to interspersed repetitive DNA sequences located between the structural genes.

Pathak et al. (1973) and Yosida (1975) found independently that some rodent species have identical karyotypes, but their constitutive heterochromatin is different. It is presumed that interspecies differences are based mainly on the differences in the size of the organs. If the classical structural genes in related species (e. g. man and the chimpanzee) are identical, one is tempted to attribute species differences to some other factors: either to gene rearrangement (King and Wilson 1975) or to some other chromosomal component without gene activity. Constitutive heterochromatic parts of chromosomes would be an excellent candidate, if not for the regulatory function, at least for the inhibitory action. Their effect would be much smaller than that of entire extra sex chromosomes. Let us imagine that increased C-band would diminish the number of cells in some organ(s) but without pathogenic effect. The safety limit, however, would be smaller, and the organ might be less resistant to some adverse influences. It is possible that some function of heterochromatin does not regard cellular level but organism level, for example the proportion of cells forming different organs.

From this point of view of the selective value of different chromosomal variants, possibly the entire spectrum exists. On the one side are variants which are selectively neutral or even advantageous; on the opposite side variants can be found which carry increased risk for the carriers and their progeny. Some variants can be very rare because they have a potentially damaging effect, but there can exist, also, a relatively new, selectively neutral, variant mutation in a local population. There can be highly subjective factors involved in the apparent incidence of a variant. Is the high incidence of inv 3 in eastern Ontario a result of a founder effect, or is it merely a reflection of the fact that this laboratory paid special attention to this inversion?

Apart from the hypothetical influence of centromeric heterochromatin on some characteristics of the organism, there is another effect regarding the mechanism of heterochromatin attraction. Attraction of heterochromatin segments is obvious not only between acrocentrics (already mentioned association) but also between (peri)centromeric heterochromatin segments and acrocentrics (Schmid et al. 1975; Maguire 1972). Quantitative differences in the "normal" amount of heterochromatin can influence meiosis and mitosis and increase the probability of chromosomal errors, especially nondisjunction of no. 21 (Stalder et al. 1965; Stonova 1968). It is not an easy task to study the possible influence of constitutive heterochromatin imbalance. Many published cases of abnormal variants in a defective person are based on biased selection, because mostly abnormal persons undergo chromosomal examination. The variant is also regularly found in healthy relatives of the proband. Comparison of the incidence of every category of variant in an unselected normal population (e. g. newborns, or, better, health-checked normal subjects) and a selected abnormal population (and their relatives) is probably one of the methods which can be used for tracking risky chromosomal variants.

REPORTED CLINICAL EFFECTS OF CHROMOSOMAL VARIANTS

Probably the first paper focusing attention on the clinical significance of what are today called "chromosomal variants" was that of German (1966). The numerous cases describing an association between mental and/or physical defects and a chromosomal variant (usually familial) in a patient will not be entered and quoted here, in spite of the fact that they are often impressive, especially cases of two different variants in one patient. All these cases, it can be argued, may be chance coincidences based only on biased selection. Therefore only the few papers based on comparative studies or on observation of more unrelated cases of an identical variant will be quoted here.

YQ+. Higher incidence of a longer Y chromosome has been observed several times in the criminal male population (Nielsen and Friedrich 1972; Kahn et al. 1969; Huebner 1971a; Soudek and Laraya 1974; Nielsen and Norland 1975; Martin-Lucas and Abrisqueta 1976; Tajmirova and Ondrejčak 1976; Funderburk et al. 1976). Other authors did not find this difference or even found slightly shorter Y chromosomes in their sample of prisoners (Benezech, Robert, et al. 1973; Benezech, Noël, Travers, et al. 1976; Benezech, Noël, and Manelphe 1975; Urdal and Brogger 1974; Schwinger and Wild 1974). Genest and Dumas (1976) did not find increased criminality in families with a long Y chromosome. Christensen and Nielsen (1974) found increased incidence of Yq+ in patients of a children's psychiatric hospital. Nielsen and Norland (1975) observed increased activity in boys with a longer Y. This observation was not confirmed by Brogger et al. (1977). This putative effect should somehow conform to the principle of racial and local differences in the size of Y. It is supposed that a long Y chromosome in Caucasians has a Mediterranean origin (Lubs and Patil 1975); a north–south length gradient for Y exists. Patil and Lubs (1974, 1977a) found more abortions in families of carriers of long Y chromosomes. Also, some familial cases of mental retardation in males with a very long Y chromosome would indicate a possible adverse effect of extra Y heterochromatin.

YQ−. Incidence of this variant is 0.08 percent (Rary and Borgaonkar 1973). Žižka (1972) and Giraud et al. (1975) repeatedly found Yq− variants in fathers of Down's-syndrome children. Koulischer (1976) concluded that Yq− is directly related to severe oligospermia or azoospermia.

INV Y. Frequency of this chromosomal anomaly is 0.1 percent. There was no increased risk of abnormal phenotype observed (Friedrich and Nielsen 1974).

INCREASED SHORT ARMS AND/OR SATELLITES OF ACROCENTRICS (usually also with intense Q-fluorescence). Increased short arms are more common in cases of Down's syndrome (Dekaban et al. 1963; Šubrt 1970; Starkman and Shaw 1967; Hamerton 1970; and other authors cited in Giraud et al. 1975 and Robinson and Newton 1977). In contrast, Giraud et al. (1975) did not register increased incidence of aberrant short arms in parents of Down's-syndrome patients in his sample. Christensen and Nielsen (1974) found enlarged satellites to be a factor "at risk." Nielsen, Friedrich, Hreidarsson, Noël, et al. (1974), however, found no increased risk in intensely fluorescent enlarged short arms of acrocentrics. On the other hand, Mikelsaar, Kaosaar, et al. (1975) found more intensely fluorescent satellites in oligophrenics than in newborns, and Funderburk et al. (1976) reached the same conclusion.

DELETED SHORT ARMS OF ACROCENTRICS. Their incidence in the Danish population is 0.16 percent (Nielsen, Friedrich, and Hreidarsson 1974a). These authors did not find any deleterious effects from this variant. Ballantyne et al. (1977) came to the conclusion, after reviewing the literature, that the variant may predispose to developmental anomalies or to nondisjunction, or both.

1QH+. Incidence is 0.06 percent according to Nielsen, Friedrich, and Hreidarsson (1974b), 1.9 percent according to Tüür et al. (1974). Nielsen did not find any deleterious effect. Contrary to this opinion, Gardner et al. (1974) summarized cases in which 1qh+ seemed to be pathogenic, causing disease similar to Patau's and Meckel's syndromes. Halbrecht and Shabtay (1976) described seven carrier families with nonspecific malformations. Atkins (1977; Atkin and Pickthall 1977) found chromosome 1 heteromorphosis (i.e. different size of both homologues) in patients with ovarian cancer.

INTENSELY FLUORESCENT 3 CEN. Mikelsaar, Kaosaar, et al. (1975) found increased frequency of this variant in female oligophrenics.

9QH+. Nielsen, Friedrich, Hreidarsson, and Zeuthen (1974b) found the frequency of this variant to be 0.1 percent in newborns, but 2.8 percent among the relatives of probands with chromosome abnormalities. It constitutes, probably, an increased risk of chromosome abnormality. Fitzgerald (1974) noted that there were twice as many carriers as noncarriers of 9qh+ in families he studied, which he explains by zygotic selection of 9qh+ against 9qh. Therefore 9qh+ should be very common, but it is rare. This would suggest a contra-acting disadvantage, but "we have no indication what this might be." For Wahlström (1975), the presence of 9qh+ in one parent constitutes an indication for prenatal chromosome study. Kunze and Man (1975) emphasize the potentially pathogenic effect associated with 1qh+ and 9qh+. Frequency of 9qh+ is increased in parents of chromosomally abnormal abortions (Holbek et al. 1974). Madan and Bobrow (1974) found Mendelian segregation and no adverse effect. In an extensive collaborative study (Lubs, Patil, et al. 1977), 9qh+ was present in about 20 percent of black children with low I.Q. It was concluded that additional studies of this variant are warranted. Funderburk et al. (1976) found an association between the size of 9h, cranial circumference, and mental retardation. Soudek (1977) found 9qh+ more frequent in the mentally retarded than in a healthy control population.

16QH+. Nielsen, Friedrich, Hreidarsson, and Zeuthen (1974a) found no indication for pathogenic effect of 16qh+ (0.10 percent in newborns, 0.12 percent in psychiatric institutions — the increase is not significant). Kelly and Almy (1971), however, stressed the increased heart defects in many cases of this variant.

17P MAR (OR 17PS). The frequency is 0.88 percent, with no proven phenotypic effect (Sandstrom and Jenkins 1973, 1974).

XQS. Harvey et al. (1977) claimed that X-linked mental retardation is associated with a satellite-like marker on the long arm of the X chromosome. These results were confirmed by Sutherland (1977a, 1977b); the phenomenon is dependent on the use of certain cell cultivation media, however, and is not seen in all cases of X-linked retardation.

ACCESSORY GENETICALLY INACTIVE CHROMOSOMES. These are composed of a substance which only reacts to the C-band method. They do not contain euchromatin at all and represent extra satellite DNA. (More information can be found in Soudek and Sroka 1977 and in Soudek 1976.)

INVERSIONS OF C-BAND HETEROCHROMATIN. These inversions are probably very frequent in the general population (0.5 to 5 percent), as shown in Table 3. H. A. Lubs and M. L. Lubs (1972) found the frequency of inversion in no. 9 to be 5.9 percent in retarded or autistic children, but only 0.84 percent in consecutive newborns. Boue et al. (1975) found inv 9qh in ten unrelated cases of reproductive failure. De la Chapelle (1976) studied pericentric inversions in C-group chromosomes. He could identify seven different types in chromosomes 8, 9, 10, and 11. He believes that many of them will fit the definition of inherited polymorphism traits without clinical consequences.

Ballesta and Serra (1976) and Jenkins et al. (1976) found increased frequency of polymorphism in mentally retarded populations and Say et al. (1977) in children with psychiatric disorders.

CONCLUSION

Many scientists are aware that chromosomal variants have an evolutionary importance and can have some influence on the phenotype. Following are quotations of opinions of some of them:

. . . these small structural variants must be regarded as an evolutionary factor even though their significance cannot be established (Eberle 1975).

The phenotypic changes which commonly accompany speciation might be expected to favour changes in constitutive heterochromatin as mechanism of achieving new homeostatic balance (Jones 1976).

Significance of specialized regions, such as 1qh, 9qh, 16qh, would have an important implication for clinical cytogenetics. One is somewhat uncomfortable in assigning no significance for fetal development to the presence of "variant chromosome" (Moorhead 1976).

This newly found and preliminarily described genetic variability, its source, its composition, its racial distribution, its contribution to fitness and its potential utility as a cytogenetic marker provide mammalian geneticists with a wealth of interesting questions (McKenzie and Lubs 1975).

It is difficult to speculate as to the tolerance of the genome for changes involving heterochromatin. The result of this study demonstrates that a considerable amount of variation and change in the C-band heterochromatin can be tolerated with no apparent deleterious effect on the phenotype (Craig-Holmes and Shaw 1971).

As seen in the discussed data, the problem of possible phenotypic effects of chromosomal variants is an unsolved question which could possibly remain unsolved for a considerable time. There was no direct correlation found between specific variant and specific clinical defect, and it is doubtful if such correlation will ever be found. The possibility of an indirect effect, however, resulting in increased vulnerability of some organ(s) of the variant carrier, cannot be excluded according to the present results. Some variants (e. g. some cases of very long Y chromosome, 9hp+) are suspected of producing an unfavorable effect on some proportion of carriers. Detailed anatomical and biochemical studies of normal carriers of these variants are nonexistent but would be very useful. There were not enough comparative studies of the prevalence of variants in normal and abnormal populations. Until results of such studies are available, the problem of the possible phenotypic effect of chromosomal variants will remain an open question.

REFERENCES

ANGELL, R. R.

1973 The chromosomes of Australian aborigines. *Australian Aboriginal Studies No. 44*, Australian Institute of Aboriginal Studies, Canberra.

ANGELL, R. R., P. A. JACOBS

1975 Lateral asymmetry in human constitutive heterochromatin. *Chromosoma* 51:301–310.

ARRIGHI, F. E., T. C. HSU

1971 Localisation of heterochromatin in human chromosomes. *Cytogenetics* 10:81–86.

ATKIN, N. B.

1977 Chromosome 1 heteromorphism in patients with malignant disease; a constitutional marker for a high risk group? *British Medical Journal* 1:358.

ATKIN, N. B., V. J. PICKTHALL

1977 Chromosomes 1 in 14 ovarian cancers; heterochromatin variants and structural change. *Human Genetics* 38:25–33.

BALIČEK, P., J. ŽIŽKA, H. SKALSKA

1977 Length of human constitutive heterochromatin in relation to chromosome contraction. *Human Genetics* 38:189–194.

BALLANTYNE, G. H., M. I. PARSLOW, A. M. O. VEALE, D. H. H. PULLON

1977 Down's syndrome and deletion of short arms of a G-chromosome. *Journal of Medical Genetics* 14:147–150.

BALLESTA, F. M., M. SERRA

1976 Chromosomal polymorphism; frequency and clinical manifestations. *Clinical Genetics* 10:349.

BARLOW, P.

1973 The influence of inactive chromosomes on human development. Anomalous sex chromosomes complement and the phenotype. *Humangenetik* 17:105–136.

BENEZECH, M., B. NOËL, C. MANELPHE

1975 Chromosome Y long (Yq+) et comportement delictueux. *Annales Medico-Psychologiques* (Paris) 2(133):313–321.

BENEZECH, M., B. NOËL, E. TRAVERS, J. MOTTET

1976 Conduite antisociale et longueur du chromosome Y. *Human Genetics* 32:77–80.

BENEZECH, M., G. ROBERT, B. NOËL, R. ZINTILINI

1973 Agressivité, delinquence psychopathique et longueur du chromosome Y. *Nouvelle Presse Medicale* 2:583.

BOBROW, M., K. MADAN, P. L. PEARSON

1972 Differential staining of specific regions in human chromosomes; particularly the secondary constriction of no. 9. *Nature New Biology* 238:122–124.

BOCHKOV, N. P., N. P. KULESHOV, A. N. CHEBOTAREV, V. I. ALEKHIN, S. A. MIDIAN

1974 Population cytogenetic investigation of newborns in Moscow. *Humangenetik* 22:139–152.

BOUE, J., J. L. TAILLEMITE, P. HAZEL-MASSIEUX, C. LEONARD, A. BOUE

1975 Association of pericentric inversion of chromosome no. 9 and reproductive failure in ten unrelated families. *Humangenetik* 30:217–224.

BROGGER, A., T. URDAL, F. B. LARSEN, N. J. LAVIK

1977 No evidence for a correlation between behaviour and the size of the Y chromosome. *Clinical Genetics* 11:349–358.

BUCKTON, K. E., M. L. O'RIORDAN, P. A. JACOBS, J. A. ROBINSON, R. HILL, H. J. EVANS
1976 C- and Q-band polymorphisms in the chromosomes of three human populations. *Annals of Human Genetics* 40:99–112.
CARNEVALE, A., B. B. IBANEZ, V. DEL CASTILLO
1976 The segregation of C-band polymorphism in chromosomes 1, 9, and 16. *American Journal of Human Genetics* 28:412–416.
CASPERSSON, T., S. FARBER, G. E. FOLEY, J. KUDYNOWSKI, *et al.*
1968 Chemical differentiation along metaphase chromosomes. *Experimental Cell Research* 49:219–222.
CHRISTENSEN, K. R., J. NIELSEN.
1974 Incidence of chromosome aberrations in a children's psychiatric hospital. *Clinical Genetics* 5:205–210.
COHEN, M. M., S. DAHAN, M. SHAHAM
1975 Cytogenetic evaluation of 500 Jerusalem newborn infants. *Israel Journal of Medical Sciences* 11:969–977.
COMMONER, B.
1964 Roles of desoxyribonucleic acid in inheritance. *Nature* 202:960–968.
CRAIG-HOLMES, A. P.
1977 "C-band polymorphism in human populations," in *Population cytogenetics.* Edited by E. B. Hook and I. H. Porter, 161–178. New York: Academic.
CRAIG-HOLMES, A. P., F. B. MOORE, M. W. SHAW
1973 Polymorphism of human C-band heterochromatin. I. Frequency of variants. *American Journal of Human Genetics* 25:181–192.
1975 Polymorphism of human C-band heterochromatin. II. Suggestive evidence for somatic crossing over. *American Journal of Human Genetics* 27:178–189.
CRAIG-HOLMES, A. P., M. W. SHAW
1971 Polymorphism of human constitutive heterochromatin. *Science* 174:702–704.
DAVIDSON, E. H., B. R. HOUGH, W. H. KLEIN, R. J. BRITTEN
1975 Structural genes adjacent to interspersed repetitive DNA sequences. *Cell* 4:217–238.
DEKABAN, A. S., M. A. BENDER, G. E. ECONOMOS
1963 Chromosome studies in mongoloids and their families. *Cytogenetics* 2:61–76.
DE LA CHAPELLE, A.
1976 Pericentric inversions of human C-group chromosomes. *Clinical Genetics* 10:353.
DE LA CHAPELLE, A., J. SCHRÖDER, K. STENSTRAND, J. FELLMAN, *et al.*
1974 Pericentric inversions of human chromosomes 9 and 10. *American Journal of Human Genetics* 26:746–766.
DEL SOLAR, C., I. UCHIDA
1974 Identification of chromosomal abnormalities in quinacrine staining technique in patients with normal karyotypes by conventional analysis. *Journal of Pediatrics* 84:534–538.
DERYAGUIN, G. V., O. A. PODUGOLNIKOVA, A. A. PROKOFIEVA-BELGOVSKAYA
1977 System of registration of chromosome polymorphism in man (in Russian). *Tsitologija* 19:293–297.
DEVELING, A. J., F. A. CONTE, C. J. EPSTEIN
1973 A Y-autosome translocation 6,X t(Yq−;7q+) associated with multiple congenital anomalies. *Journal of Pediatrics* 82:495–498.

DITTES H., W. KRONE, K. BROSS, M. SCHMID, W. VOGEL
1975 Biochemical and cytogenetic studies on the nucleolus organizing regions in man. *Humangenetik* 26:47–59.

DUTRILLAUX, B., M.-O. RETHORE, J. LEJEUNE
1975 Comparison of the karyotype of the orangutan to those of man, chimpanzee and gorilla. *Annales de Génétique* 18:153–161.

EBERLE, P.
1975 Chromosome finds in primates and cytogenetical aspects in the evolution of man. *Journal of Human Evolution* 4:435–439.

EVANS, H. J., J. R. GOSDEN, A. R. MITCHELL, R. A. BUCKLAND
1974 Localisation of human satellite DNAs on the Y chromosomes. *Nature* 251:346–347.

FERGUSON-SMITH, M. A.
1974 Autosomal polymorphism. *Birth Defects*. Original Article Series 10(10):19–28.

FITZGERALD, P. H.
1974 The nature and inheritance of an elongated secondary constriction on chromosome no. 9 of man. *Cytogenetics and Cell Genetics* 12:404–413.

FONATSCH, C.
1977 New chromosome polymorphism inv(16)(p11q12 or 13). *Cytogenetics and Cell Genetics* 18:106–107.

FORSSMAN, H.
1970 The mental implications of sex chromosome aberrations. *British Journal of Psychiatry* 117:353–363.

FORSSMAN, H., G. HAMBERT
1967 Chromosomes and antisocial behaviour. *Excerpta Criminologica* 7:113–117.

FRIEDRICH, U., J. NIELSEN
1972 Presumptive Y-15 and Y-22 translocation in two families. *Hereditas* 71:339–342.
1973 Chromosome studies in 5,049 consecutive newborn children. *Clinical Genetics* 4:333–343.
1974 Pericentric inversion of Y in a population of newborn boys. *Hereditas* 76:147–152.

FRÜND, S., T. KOSKE-WESTPHAL, S. FUCHS-MECKE, E. PASSARGE
1972 Quinacrine mustard fluorescence of a second Y chromosome in a Y-autosomal translocation. *Humangenetik* 14:133–136.

FUNAKI, K., S. MATSUI, M. SASAKI
1975 Location of nucleolar organizers in animal and plant chromosomes by means of an improved N-banding technique. *Chromosoma* 49:357–370.

FUNDERBURK, S. J., R. S. SPARKES, D. GUTHRIE, J. R. WESTLAKE
1976 The significance of minor chromosome variants. *Fifth International Congress of Human Genetics*, Mexico (abstracts):125.

GAGNE, R., C. LABERGE
1972 Specific cytological recognition of the heterochromatic segment of no. 9 chromosome in man. *Experimental Cell Research* 73:239–242.

GAGNE, R., R. TANGUAY, C. LABERGE
1971 Differential staining patterns of heterochromatin in man. *Nature New Biology* 232:29–30.

GARDNER, R. J. M., H. R. McCREANOR, M. I. PARSLOW, A. M. O. VEALE
1974 Are 1q+ chromosomes harmless? *Clinical Genetics* 6:383–393.

GENEST, P., L. DUMAS
1976 The length of the Y chromosome and antisocial behaviour in human beings. *Canadian Journal of Genetics and Cytology* 18:560.

GERAEDTS, J. P. M., P. L. PEARSON
1974 Fluorescent chromosome polymorphisms: frequencies and segregations in a Dutch population. *Clinical Genetics* 6:247–257.

GERMAN, J.
1966 The chromosomal structural load in man. *Texas Report on Biology and Medicine* 24:347–364.

GHOSH, P. K., I. P. SINGH
1975 Morphological variability on the human chromosomes in two Indian populations — Rajputs and Punjabis. *Humangenetik* 29:67–78.
1976 Morphologic variability of human chromosomes; polymorphism of constitutive heterochromatin. *Human Genetics* 32:149–154.

GIRAUD, F., J. F. MATTEI, M. G. MATTEI
1975 Etude chromosomique chez les parents d'enfants trisomiques 21. *Lyon Medical* 233:241–251.

GIRNELLI, G., E. PORRO, F. SANTI, S. SCAPPATICCI, O. ZUFFARDI
1977 "Jumping" satellites in three generations; a warning for paternity tests and prenatal diagnosis. *Human Genetics* 34:315–318.

GOODPASTURE, C., S. E. BLOOM
1975 Visualization of nucleolar organizer regions in mammalian chromosomes using silver staining. *Chromosoma* 53:37–50.

GOODPASTURE, C., S. E. BLOOM, T. C. HSU, F. E. ARRIGHI
1976 Human nucleolus organizers; the satellites or the stalks? *American Journal of Human Genetics* 28:559–566.

HALBRECHT, I., F. SHABTAY
1976 Human chromosome polymorphism and congenital malformations. *Clinical Genetics* 10:113–122.

HAMERTON, J.
1970 *Human cytogenetics*, volume two. New York: Academic.

HAMERTON, J. L., N. CANNING, M. RAY, S. SMITH
1975 A cytogenetic study of 14,069 newborn infants. I. Incidence of chromosome abnormalities. *Clinical Genetics* 8:223–243.

HARVEY, J., C. JUDGE, S. WEINER
1977 Familial X-linked mental retardation with an X chromosome abnormality. *Journal of Medical Genetics* 14:46–50.

HAYATA, I., M. OSHIMURA, A. A. SANDBERG
1977 N-band polymorphism of human acrocentric chromosomes and its relevance to satellite association. *Human Genetics* 36:55–62.

HOEHN, H., K. AU, L. E. KARP, G. M. MARTIN
1977 Somatic stability of variant C-band heterochromatin. *Human Genetics* 35:163–168.

HOLBEK, S., U. FRIEDRICH, J. G. LAUTITSEN, A. J. THERKELSEN
1974 Marker chromosomes in parents of spontaneous abortions. *Human Genetics* 25:61–74.

HOOK, E. B., I. H. PORTER, *editors*
1977 *Population cytogenetics; studies in humans.* New York: Academic.

HSU, T. C.
1975 A possible function of constitutive heterochromatin. *Genetics* 79 (supplement): 137–150.

HSU, T. C., S. E. SPIRITO, M. L. PARDUE
1975 Distribution of 18+ 28S ribosomal genes in mammalian genomes. *Chromosoma* 53:25–36.

HUEBNER, H.
1971a The Y chromosome in selected groups of men in the Polish population. *Bulletin de l'Académie Polonaise des Sciences* 6:(19):467–472.
1971b Karyotype 46,XYq+ and its phenotype. *Bulletin de l'Académie Polonaise des Sciences* 6(19):473–477.

JALAL, S. M., R. W. CLARK, T. C. HSU, S. PATHAK
1974 Cytological differentiation of constitutive heterochromatin. *Chromosoma* 48:391–408.

JENKINS, E. C., L. WILBUR, F. M. GIORDANO, R. G. WEED, H. R. SLAWEN
1976 Chromosomal abnormality and 1,9,16qh variation in mentally retarded population. *Fifth International Congress of Human Genetics,* Mexico (abstracts):131.

JONES, K. W.
1973 Satellite DNA. *Journal of Medical Genetics* 10:273–281.
1976 "Repetitive DNA sequences in animals, particularly primates," in *Chromosomes today*, volume five. Edited by P. L. Pearson and K. R. Lewis, 305. New York: J. Wiley and Sons.

JONES, K. W., I. F. PURDOM, J. PROSSER, G. CORNEO
1974 The chromosomal localisation of human satellite DNA I. *Chromosoma* 49:161–171.

KAHN, J., W. I. CARTER, N. DERNLEY, E. T. O. SLATER
1969 "Chromosome studies in remand home and prison populations," in *Criminological implications of chromosome abnormalities*. Edited by D. J. West, 44–48. Cambridge: Cambridge University Press.

KELLY, S., R. ALMY
1971 Chromosome 16 heterology; non-specific variations in karyotype? *New York State Journal of Medicine* 71:2,297–2,299.

KIM, MY. A.
1973 Polymorphismus des konstitutiven Heterochromatins bei menschlichen A 1 Metaphasechromosomen. *Humangenetik* 18:213–217.

KING, M. C., A. C. WILSON
1975 Evolution at two levels in humans and chimpanzees. *Science* 188:107–116.

KOULISCHER, L.
1976 Polymorphism of the Y chromosome in a population of 1,000 infertile male patients. *Clinical Genetics* 10:363.

KUNZE, J., G. MAN
1975 A1 and C9 marker chromosomes in children with combined minor and major malformations. *Lancet* 1:273.

LEE, C. S. N., K. L. YING, P. BOWEN
1974 Position of the Duffy locus on chromosome 1 in relation to breakpoints for structural rearrangements. *American Journal of Human Genetics* 26:93–102.

LIN, C. C., M. M. GEDEON, P. GRIFFITH, W. K. SMINK, *et al.*
1976 Chromosome analysis on 930 consecutive newborn children using quinacrine fluorescent banding technique. *Human Genetics* 31:315–328.

LUBS, H. A.
1969 A marker X chromosome. *American Journal of Human Genetics* 21:231–244.

LUBS, H. A., W. J. KIMBERLING, F. HECHT, S. R. PATIL, *et al.*
1977 Racial differences in the frequency of Q- and C-chromosome heteromorphism. *Nature* 268:631–633.

LUBS, H. A., M. L. LUBS
1972 "New cytogenetic techniques applied to a series of children with mental retardation," in *Chromosome identification*. Edited by T. Caspersson and L. Zech, 241–250. New York: Academic.

LUBS, H. A., S. PATIL
1975 Mediterranean origin of long Y in Caucasians. *American Journal of Human Genetics* 27:60A.

LUBS, H. A., S. A. PATIL, W. J. KIMBERLING, J. BROWN, *et al.*
1977 "Q- and C-banding polymorphism in 7 and 8 year old children: racial differences and clinical significance," in *Population cytogenetics*. Edited by E. B. Hook and I. H. Porter, 133–159. New York: Academic.

LUBS, M. L., H. A. LUBS
1972 "Compilation and analysis of neonatal cytogenetic survey data," in *Perspectives in cytogenetics*. Edited by S. Wright, B. Crandall and L. Boyer, 297. Springfield, Ill.: C. C. Thomas.

LUNDSTEEN, C., J. PHILIP
1973 Y/22 translocation in YY males. *Cytogenetics* 12:53–59.

MADAN, K., M. BOBROW
1974 Structural variations in chromosome No. 9. *Annales de Génétique* 17:81–86.

MAGENIS, R. E., C. G. PALMER, L. WANG, M. BROWN, *et al.*
1977 "Heritability of chromosome banding variants," in *Population cytogenetics*. Edited by E. B. Hook and I H. Porter, 175–188. New York: Academic.

MAGUIRE, M.
1972 Role of heterochromatin in homologous chromosome pairing. *Science* 176:543–544.

MARTIN-LUCAS, M. A., J. A. ABRISQUETA
1976 Yq+ anomaly in a criminal population. *Fifth International Congress of Human Genetics*, Mexico (abstracts): 140.

McKENZIE, W. H., H. A. LUBS
1975 Human Q and C chromosomal variations; distribution and incidence. *Cytogenetics and Cell Genetics* 14:97–115.

MIKELSAAR, A.-V. N., M. E. KAOSAAR, S. J. TÜÜR, M. H. VIIKMA, *et al.*
1975 Human karyotypes polymorphism III. Routine and fluorescence microscopic investigation of chromosomes in normal adults and mentally retarded children. *Humangenetik* 26: 1–23.

MIKELSAAR, A.-V. N., S. J. TÜÜR, M. E. KAOSAAR
1973 Human karyotype polymorphism I. Routine and fluorescence microscopic investigation of chromosomes in normal adult population. *Humangenetik* 20:89–101.

MOORHEAD, P. S.
1976 A closer look at chromosomal inversions. *American Journal of Human Genetics* 28:294–296.

MÜLLER, H., H. P. KLINGER
1975a Chromosome polymorphism in a human newborn population II. *Cytogenetics and Cell Genetics* 15:239–255.
1975b "Chromosome polymorphism in a human newborn population," in *Chromosomes today*, volume five. Edited by P. L. Pearson and K. R. Lewis, 249–260. New York: J. Wiley and Sons.

MUTTON, D. E., M. G. DAKER
1973 Pericentric inversion of chromosome no. 9. *Nature New Biology* 241:80.
NAGAKOME, Y., T. KITAGAWA, K. LINUMA, E. MATSUNAGA
1977 Pitfalls in the use of chromosome variants for paternity dispute cases. *Human Genetics* 37:255–260.
NIELSEN J., U. FRIEDRICH
1972 Length of the Y chromosome in criminal males. *Clinical Genetics* 3:281–285.
NIELSEN, J., U. FRIEDRICH, A. B. HREIDARSSON
1974a Frequency of deletion of short arms satellites in acrocentric chromosomes. *Journal of Medical Genetics* 11:177–180.
1974b Frequency and genetic effect of 1qh+. *Humangenetik* 21:193–196.
NIELSEN J., U. FRIEDRICH, A. B. HREIDARSSON, B. NOËL, *et al.*
1974 Brilliantly fluorescing enlarged short arms D or G. *Lancet* 1:1,049–1,050.
NIELSEN J., U. FRIEDRICH, A. B. HREIDARSSON, E. ZEUTHEN
1974a Frequency and segregation of 16qh+. *Clinical Genetics* 5:316–321.
1974b Frequency of 9qh+ and risk of chromosome aberrations in the progeny of individuals with 9qh+. *Human Genetics* 21:211–216.
NIELSEN, J., E. NORLAND
1975 Length of Y chromosome and activity in boys. *Clinical Genetics* 8:291–296.
NOËL, B., I. EMERIT, J. M. LUCIANI, B. QUACK
1971 A familial Y/autosome in man. *Clinical Genetics* 2:1–6.
OHNO, S.
1973 "Evolutionary reason for having so much junk DNA," in *Modern aspects of cytogenetics in man.* Edited by R. A. Pfeiffer. Stuttgart: F. K. Schattauer.
ORYE, E.
1974a Relative activation and inactivation phenomena between homologous and nonhomologous nucleolus organizers on the normal human G chromosome. *Cytogenetics and Cell Genetics* 23:252–368.
1974b Satellite associations and variations in length of the nucleolar constrictions of normal and variant human G chromosomes. *Humangenetik* 22:299–309.
ORYE, E., C. DELIRE
1967 Familial D/D and D/G translocation. *Helvetica Paediatrica Acta* 22:36–40.
OSZTOVICS, M., E. M. BUHLER, H. MULLER, G. R. STALDER
1973 Banding techniques in the evaluation of human chromosomal variants. *Humangenetik* 18:123–128.
OVERTON, K. M., R. E. MAGENIS, T. BRADY, J. CHAMBERLIN, M. PARKS
1976 Cytogenetic darkroom magic; now you see them, now you don't. *American Journal of Human Genetics* 28:417–419.
PARK, J., R. M. ANTLEY
1975 Racial differences in C-band heterochromatin polymorphism. *American Journal of Human Genetics* 27:70A.
1976 Racial C-band chromosomal polymorphismus. *Fifth International Congress of Human Genetics*, Mexico (abstracts):144.
PATHAK, S., T. C. HSU, F. E. ARRIGHI
1973 Chromosomes of *Peromyscus* IV. The role of heterochromatin in karyotypic evolution. *Cytogenetics and Cell Genetics* 12:315–326.

PATIL, S. R., H. A. LUBS
1974 Spontaneous abortions and long Y chromosomes. *American Journal of Human Genetics* 26:66A.
1977a A possible association of long Y chromosomes and fetal loss. *Human Genetics* 35:233–236.
1977b Classification of qh regions in human chromosomes. *Human Genetics* 38:35–38.

PEARSON, P. L., J. P. M. GERAEDTS, A. G. J. M. VAN DER LINDEN
1973 "Human chromosome polymorphism," in *Modern aspects of cytogenetics in man.* Edited by R. A. Pfeiffer, 201–213. Stuttgart: F. K. Schattauer.

RARY, J. M., D. S. BORGAONKAR
1973 Analysis of minor chromosomal variants in a survey of Maryland boys. *American Journal of Human Genetics* 25:62A.

REITALU, J.
1973 A familial Y/22 translocation in man. *Hereditas* 74:155–160.

ROBINSON, J. A., K. E. BUCKTON, G. SPOWART, M. NEWTON, *et al.*
1976 The segregation of human chromosome polymorphisms. *Annals of Human Genetics* 40:113–121.

ROBINSON, J. A., M. NEWTON
1977 A fluorescence polymorphism associated with Down's syndrome? *Journal of Medical Genetics* 14:40–45.

SANCHEZ, O., J. J. YUNIS
1974 The relationship between repetitive DNA and chromosomal bands in man. *Chromosoma* 48:191–202.

SANDSTROM M., E. C. JENKINS
1973 A 17p marker chromosome familial study. *Annales de Génétique* 16:267–269.
1974 Cytological elucidation of a 17p marker chromosome. *Caryologia* 27:307–313.

SAY, B., N. J. CARPENTER, P. R. LANIER, C. BANEZ, *et al.*
1977 Chromosome variants in children with psychiatric disorders. *American Journal of Psychiatry* 134:424–426.

SCHMIADY, H., K. SPERLING
1976 Length of human C-bands in relation to the degree of chromosome condensation. *Human Genetics* 35:107–111.

SCHMID, M., W. VOGEL, W. KRONE
1975 Attraction between centric heterochromatin of human chromosomes. *Cytogenetics and Cell Genetics* 15:66–80.

SCHNEDL, W.
1971 Unterschiedliche Fluoreszenz der beiden homologen Chromosomen No. 3 beim Menschen. *Humangenetik* 12:59–63.

SCHWINGER, E., H. WEHNER
1976 Frequency of chromosomal fluorescence polymorphism in normal persons and in clinical patients with diagnosed chromosome aberrations. *Human Genetics* 32:115–119.

SCHWINGER, E., P. WILD
1974 Length of the Y chromosome and antisocial behaviour? *Humangenetik* 22:67–70.

SEKHON, G. S., W. S. SLY
1975 Inheritance of Q- and C-polymorphisms. *American Journal of Human Genetics* 27:79A.

SERGOVICH, R. F., A. LUCE, D. CARR
1967 Chromosome aneuploidy in the father of a mongol. *American Journal of Diseases of Children* 114:407–411.

SOFUNI, T., K. TANABE, K. OHTAKI, H. SHIMBA, A. A. AWA
1974 Two new types of C-band variants in human chromosomes (6ph+ and 12ph+). *Japan Journal of Human Genetics* 19:251–256.

SOUDEK, D.
1973 Chromosomal variants with normal phenotype in man. *Journal of Human Evolution* 2:341–355.
1975 Small chromosomal inversions; are they harmless? *Canadian Medical Association Journal* 112:1,289.
1976 "Small accessory inactive chromosomes in man. A review." Unpublished manuscript.
1977 Longer Y chromosome in criminals; a reappraisal. *Clinical Genetics* 11:314–315.

SOUDEK, D., P. LARAYA
1974 Longer Y chromosome in criminals. *Clinical Genetics* 6:225–229.
1976 C- and Q-bands in long arm of Y chromosomes; are they identical? *Humangenetik* 32:339–341.

SOUDEK, D., S. O'SHAUGHNESSY, P. LARAYA, B. McCREARY
1974 Pericentric inversion of "fluorescent" segment in chromosome No. 3. *Humangenetik* 22:343–346.

SOUDEK, D., H. SROKA
1977 C-bands in seven cases of accessory chromosomes. *Clinical Genetics* 12:285–289.

SOUDEK, D., H. SROKA, P. LARAYA
1976 Chromosomal variants in a sample of normal Canadian men. *International Congress of Human Genetics*, Mexico (abstracts):155.

STALDER, G. R., E. M. BÜHLER, U. K. BÜHLER
1965 Possible role of heterochromatin in human aneuploidy. *Humangenetik* 1:307–310.

STARKMAN, M. N., M. W. SHAW
1967 Atypical acrocentric chromosomes in Negro and Caucasian mongols. *American Journal of Human Genetics* 19:162–173.

STONOVA, N. S.
1968 Possible role of heterochromatin in the appearance of human inherent malformation (in Russian). *Genetika* 4:90–99. (Reprinted in 1972, in English: *Soviet Genetics* 4:64–72.)

ŠUBRT, I.
1970 A further example of familial Gp+ associated with trisomy G. *Humangenetik* 9:86–90.

SUTHERLAND, G. R.
1977a Marker X chromosome and mental retardation. *New England Journal of Medicine* 296:1,415.
1977b Fragile sites on human chromosomes; the demonstration of their dependence on the type of tissue culture medium. *Science* 197:265.

TAJMIROVA, O., M. ONDREJČAK
1976 Cytogenetics in forensic genetics. *Fifth International Congress of Human Genetics*, Mexico (abstracts):158.

TSENGHI, C.
1976 Polymorphism of chromosome no. 9 in 400 Greek subjects. *Clinical Genetics* 10:373.

TÜÜR, S., M. E. KAOSAAR, A.-V. N. MIKELSAAR
1974 1q+ variant in a normal adult population. *Humangenetik* 24:217–220.

URDAL, T., A. BROGGER
1974 Criminality and the length of the Y chromosomes. *Lancet* 1:626.

VERMA, R. S., H. A. LUBS
1975 Variation in human acrocentric chromosomes with acridine orange reverse banding. *Humangenetik* 30:225–235.

VOGEL, F.
1973 Genotype and phenotype in human chromosome aberrations and in the minute mutant of *Drosphila melanogaster. Humangenetik* 19:41–56.

WAHLSTRÖM, J.
1972 Identification by fluorescence of apparently extra human F chromosomes as G chromosomes with giant satellites. *Hereditas* 71:154–157.
1975 Three cases of minor chromosomal aberrations discovered by prenatal chromosome determination. *Humangenetik* 27:223–225.

YOSIDA, T. H.
1975 Diminution of heterochromatic C-bands in relation to the differentiation of *Rattus* species. *Proceedings of Japan Academy of Sciences* 51:659–663.

YUNIS, J. J., W. G. YASMINEH
1971 Heterochromatin, satellite DNA and cell function. *Science* 174:1,200–1,209.

ZANKL, H., K. D. ZANG
1971 Structural variability of the normal human karyotype. *Humangenetik* 13:160–162.
1974 Quantitative studies on the arrangement of human metaphase chromosomes. IV. The association frequency of human acrocentric marker chromosomes. *Humangenetik* 23:259–265.

ŽIŽKA, J.
1972 Down's syndrome: a cytogenetic and clinical study of 134 patients with a special reference to translocation cases. *Collection of Scientific Works*, Medical Faculty, Hradec Kralove 15(1).

ADDENDUM

While this review was being prepared for print, several important papers on chromosomal variants were published. Archidiacono et al. (1977) studied the relation of N-bands to chromosomal variants. Kubien and Kleczkowska (1977) described 17ps in a high-risk family. The abstract of Tsenghi (1976) was published as a full-size paper (Metaxotou et al. 1977). Phillips (1977) confirmed a Mendelian segregation and a rareness of mutations of variants. Akesson and Wahlström (1977) did not find longer Y chromosome in patients examined by forensic psychiatrists. Jacobs (1977) in her review on chromosomal variants concludes that there is no convincing evidence of a phenotypical effect of chromosomal variants; however, she is aware of diminished reproductive fitness in the carriers of extreme variants (Jacobs et al. 1975).

AKESSON, H. O., J. WAHLSTRÖM
1977 The length of the Y-chromosomes in men examined by forensic psychiatrists. *Human Genetics* 39:1–5.

ARCHIDIACONO, N., A. DE CAPOA, M. FERRARO, F. PELLICIA, *et al.*
1977 Nucleolus organizer and N-band distribution in morphologic and fluorescence variants of human chromosomes. *Human Genetics* 37:285–289.

JACOBS, P. A.
1977 Human chromosome polymorphism. *Progress in Medical Genetics* 2:251–274.

JACOBS, P. A., A. FRACKIEWICZ, P. LAW, C. J. HILDITCH, N. E. MORTON
1975 The effect of structural aberration of the chromosomes on the reproductive fitness in men II. *Clinical Genetics* 8:169–178.

KUBIEN, E., A. KLECZKOWSKA
1977 Familial occurrence of chromosome variant 17ph+. *Clinical Genetics* 12:39–42.

METAXOTOU, C., A. KALPINI-MAVROU, M. PANAGOU, C. TSENGHI
1977 Polymorphism of chromosome 9 in 600 Greek subjects. *American Journal of Human Genetics* 30:85–99.

PHILLIPS, R. B.
1977 Inheritance of Q- and C-polymorphisms. *Canadian Journal of Genetics and Cytology* 19:406–413.

TSENGHI, C.
1976 Polymorphism of chromosome no. 9 in 400 Greek subjects. *Clinical Genetics* 10:373.

SECTION FOUR

Perspectives on Supra- and Subchromosomal Organization

Position of Male-Determining Gene Loci on the Y Chromosome and Possible Explanation of XX-Male and XY-Female Phenomena in Man

FATHI ABDEL-HAMEED and LINDA M. GLUESING

In humans, the normal male autosomal and sex-chromosome complement is 46,XY, while that of the normal female is 46,XX. The Y chromosome, being a strong male-determining element in man, is necessary to initiate differentiation of the embryonic gonads into testes during early fetal development (Money and Ehrhardt 1973; Laurian et al. 1971). The embryonic testis directs male differentiation through characteristic hormonal secretions, in the absence of which a female develops (Jost 1947). The XX-male and XY-female phenomena, therefore, appear to represent cases in which the normal sex determination mechanism in man somehow fails.

The XX male is normal in appearance, but pubic, axillary, and facial hair may be decreased and of a female distribution (de la Chapelle 1972; Boczkowski et al. 1968; de la Chapelle et al. 1965; Therkelsen 1964). The external genitalia appear normal, although the testes are usually undersized. The XX males are sterile phenotypic males with a male psychosexual orientation (de la Chapelle 1972).

The XY female is grouped under two distinct clinical syndromes: testicular feminization and "pure" gonadal dysgenesis. XY gonadal dysgenetic females possess uterus, tubes, and female external genitalia, but show "streak" gonads that lack germ cells and hormonal function (Dewhurst 1971; Espiner et al. 1970; Hoffenberg and Jackson 1957; Smith and Pepler 1969; Tetsuji et al. 1971; Sternberg and Barclay 1967). The female with testicular feminization syndrome is typically feminine in appearance and, unlike the XY female with gonadal dysgenesis, not characterized by sexual infantilism. These XY females, however, lack uterus, tubes, and ovaries, but possess inguinally or abdominally located testes (Dewhurst 1971; Boczkowski and Teter 1969; Sternberg and Barclay 1967; Morris 1953; Laurian et al. 1971). The testes, in these XY females, produce estrogens and androgens in a pattern characteristic of the testes of normal males (Ismail et al. 1972; Rivarola et al. 1967; Saez et al. 1971; Rosenfield et al. 1971; Southern et al. 1965; French et al. 1965; Wade et al. 1968), and these hormones are biologically active (Morris and Mahesh 1963; Cardiff et al. 1972). The female with testicular feminization fails to respond to these or administered androgens (Rosenfield et al. 1971; Laurian et al. 1971; Rivarola et al. 1967). As a result, the syndrome is often referred to as a case of androgen insensitivity. The exact defect causing this lack of response to androgens is unknown, although it appears to involve a defect in end-organ response rather than production of biologically inactive androgens.

THE X AND Y CHROMOSOMES IN MAN

The X and Y chromosomes in man are morphologically distinguishable, the Y chromosome being distinctly smaller (*Original Article Series* 1972). Disparity in size exists between the X and Y chromosomes, but association must take place during meiosis as it is necessary to prevent random assortment and the subsequent formation of XO and XXY zygotes following fertilization. Association of the X and Y chromosomes occur during prophase of the first meiotic division of the primary spermatocyte. The short arm of the X chromosome and the short arm of the Y chromosome associate terminally via homologous pairing segments (Pearson and Bobrow 1970), and they adjoin without visual evidence of chiasma. A synaptinemal complex has been observed near or at the end-to-end association (Solari and Tres 1970; Pearson and Bobrow 1970). Homologous pairing segments or regions on the X and Y chromosomes indicate that partial linkage should exist between them. No available evidence suggests that these homologous loci are genetically active, as phenotypic effects from crossing-over in this region to establish partial linkage have not been observed (Morton 1956). X and Y chromosomes in man seem to be quite similar to those of the plant species *Melandrium*, which have been studied in great detail and are shown in Figure 1.

The X Chromosome

Approximately sixty X-linked genes are known, and the relative positions of a few have been determined (German 1970), as diagrammed in Figure 2. In the cells of a normal human female, one X chromosome is shorter, darker, and more tightly contracted in interphase nuclei (Eggen 1969). This heteropycnotic X chromosome has been designated the sex chromatin mass or Barr body. In the human male, the single X chromosome is not

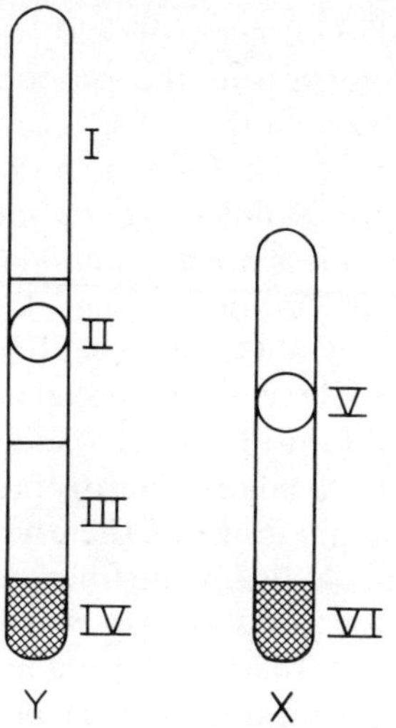

Figure 1. Sex chromosomes of *Melandrium*: I, suppressor of female development; II, male promoting region; III, male fertility; IV pairing segment; V, female development; VI, pairing segment

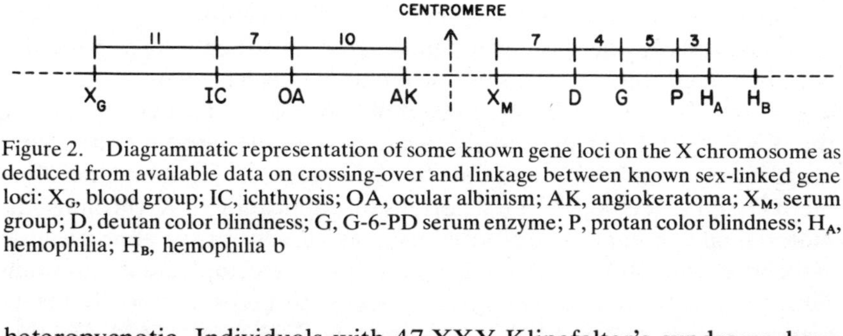

Figure 2. Diagrammatic representation of some known gene loci on the X chromosome as deduced from available data on crossing-over and linkage between known sex-linked gene loci: X_G, blood group; IC, ichthyosis; OA, ocular albinism; AK, angiokeratoma; X_M, serum group; D, deutan color blindness; G, G-6-PD serum enzyme; P, protan color blindness; H_A, hemophilia; H_B, hemophilia b

heteropycnotic. Individuals with 47,XXY Klinefelter's syndrome have one X chromosome tightly coiled in identical fashion to one of the X chromosomes in normal females. Abnormalities associated with the sex chromosome complement 49,XXXXY, in which the cells contain three heteropycnotic X chromosomes, do not differ greatly from those abnormalities associated with the sex chromosome complement 48, XXXY, in which the cells contain two sex chromatin masses. To explain these and other events, Lyon (1962) proposed that dosage compensation in man and mammals, for the X chromosome, is through inactivation, all X chromosomes in excess of one being inactivated. This inactivation is hypothesized to occur at random in early zygotic division stages. On the average, the human female would have the paternal X chromosome inactive in 50 percent of the body cells, while the other 50 percent would have the maternal X chromosome inactive. In such cases, females heterozygous for sex-linked genes would show incomplete penetrance and variable gene expression as the proportion of active normal and mutant chromosomes vary in the cells.

Evidence in support of the Lyon hypothesis comes from study of women heterozygous for X-linked genes such as glucose-6-phosphate dehydrogenase and sideroachrestic anemia. The enzyme glucose-6-phosphate dehydrogenase is an X-linked trait in which the abnormal allele leads to a deficiency of the enzyme. Females heterozygous for the abnormal and normal allele have two populations of red blood cells, one population with normal G-6-PD activity and one with deficient G-6-PD activity (Stern 1973). In X-linked sideroachrestic anemia, the abnormal allele causes severe anemia in which the erythrocytes are hypochromic, microcytic, and morphologically distinct from normal erythrocytes. Women heterozygous for this trait have two distinct erythrocyte populations, those red blood cells that are normal in appearance and those that are microcytic and hypochromic (Lee et al. 1968; MacDiarmid et al. 1967). The presence of two populations of red blood cells is an indication of random inactivation, as a result of which both the normal and the abnormal alleles are being expressed.

Inactivation of the X chromosome takes place at random, but certain loci may become preferentially inactivated. Polani et al. (1964) have found that the X_g locus is always inactivated on structurally abnormal X chromosomes. If the X chromosome is completely inactivated, one would expect the XO Turner female to be a phenotypically normal female and the XXY Klinefelter male to be a phenotypically normal male. Since inactivation takes place in about the third week of embryonic development (Bachman et al. 1971), deletions or duplication of X-chromosome material may be expressed before inactivation, causing the abnormalities associated with XO and XXY sex-chromosome complements. In addition, the inactivated X chromosome seems to be only partially inactivated, showing the so-called "spreading effect" (Russell 1963), and thus explaining XO and XXY abnormalities.

The Y Chromosome

On the basis of several observations, including the fact that individuals with 47,XXY Klinefelter's syndrome are usually phenotypic males, whereas those with 45,XO Turner's syndrome are phenotypic females, it became obvious that the presence of at least one Y chromosome is essential for male differentiation in man. Therefore, it is safe to assume that male-determining gene loci in man are located on the Y chromosome. This view is generally accepted now for other primates and the majority of mammalian species.

The size and shape of the X chromosome are relatively stable, with the exception of the inactivated X chromosome. On the other hand, the Y chromosome is the most variable chromosome in the human karyotype and may vary from the size of the G-group autosomes to that of the D-group autosomes. This size variation is heritable and wholly attributable to the long arm of the Y chromosome, the short arm being relatively stable in size between individuals (Bobrow et al. 1971; Andersson et al. 1972). Extremely long Y chromosomes are, also, not associated with any particular disease state (Sperling and Lackman 1971).

The Y chromosome presumably carries the genes necessary for male development, but apart from these loci, only one trait has been definitely ascribed to this chromosome: that of "hairy ear rim" (Stern 1973). The relative position of these male determining genes has not been determined. Therefore we may assume that they are located on the long or the short arm of the Y chromosome or on both arms. The long arm of the Y chromosome is heterochromatic and unlikely to carry male-determining genes. During early fetal development, however, the long arm of the Y chromosome may be active genetically and become heterochromatic only later in development. The location of these loci on the short or the long arms of the Y chromosome remains a matter of controversy. Structural abnormalities of the Y chromosome produce characteristic phenotypes. Study of these individuals can help to determine the relative position of male-determining loci on the Y chromosome.

THE POSITION OF MALE-DETERMINING LOCI ON THE Y CHROMOSOME IN MAN

Structural abnormalities of the Y chromosome have been reported and can be separated into four general categories: (1) isochromosomes for the long arms of the Y chromosome; (2) long-arm deletions; (3) dicentrics for the short or long arms; and (4) pericentric inversions.

Isochromosomes

Most cases of isochromosomes for the long arm of the Y chromosome are present in association with other cell lines, such as mosaicism (Ferguson-Smith et al. 1969). Mosaicism is a condition in which two genetically different cell lines arise from a single zygote (Hafez 1968). In XO/XY mosaicism, the gonads do not contain testicular tissue unless 46,XY cells are present in the gonads. If the gonad contains approximately one 46,XY cell for each fifty cells, the mosaic patient will develop one histological testis and one gonad that is streak (Starkman and Jaffe 1967).

Ferguson-Smith and co-workers (1969) have reported a mosaic individual with the karyotype 45,X/46,XYqi. One cell line was 45,X while the other was 46,XYqi, the Y chromosome being an isochromosome for the long arms. The individual was phenotypically female and short in stature, lacked breast development, and possessed a shieldlike chest, scant pubic axillary hair, and an infantile uterus. Laparotomy revealed the left gonad as streak and composed only of ovarian stroma; the right gonad was lacking. The clitoris was somewhat enlarged, but the presence of the long arm of the Y chromosome did not produce other masculinization. In cases of mosaicism such as this, it is difficult to determine if the streak gonad resulted from the 45,X cell line. The presence of only ovarian stroma in the streak gonad can indicate either that the gonad was composed only of the 45,X cell line or that the 46,XYqi cell line was incapable of directing testicular differentiation. Gonadal tissue was not karyotyped in this patient. Two female patients have been reported, however, in which mosaicism was not present. Only one cell line was detected with 46,XYqi, the Y chromosome being an isochromosome for the long arms (Jacobs and Ross 1966). The patients were phenotypically similar, exhibiting sexual infantilism, primary amenorrhea, infantile uterus, and streak gonads. In one of these individuals, neck webbing was present. Individuals with Turner's syndrome (45,X) have shieldlike chests, neck webbing, short stature, primary amenorrhea, scant or absent pubic and axillary hair, lack of breast development, and streak gonads. The phenotype of the 45,X Turner female is similar to that associated with the female possessing an isochromosome for the long arms of the Y chromosome. Apparently, long-arm isochromosomes do not contain sufficient genes for male development. The similarities between the phenotypes associated with long-arm isochromosomes and absence of an X chromosome indicate that the long arm of the Y chromosome is largely inert genetically, especially with respect to the sex-determination mechanism.

Long-Arm Deletions

Several individuals with the long arm of the Y chromosome deleted have been reported. Long-arm deletions, at present, cannot be distinguished from isochromosomes for the short arms (Meiser and Inhorn 1972; Telfer et al. 1973). The fluorescent banding pattern of the Y chromosome is characterized by two weakly fluorescing bands, one on each side of the centromere, essentially identical in appearance (Manolov et al. 1971; see Figure 1 in Abdel-Hameed and Gluesing 1979). The Y chromosomes with long-arm deletions are frequently distinguished from isochromosomes for the short arms on the basis of clinical similarities between these patients and XYY males (Meiser and Inhorn 1972; Telfer et al. 1973). As isochromosomes for the long arm appear to contain few male-determining loci, one would expect isochromosomes for the short arms to contain male-determining loci essentially equivalent to two Y chromosomes. A patient possessing such an isochromosome would be expected to exhibit some of the clinical features common to XYY males. Above-average height and undescended testes in childhood are common features of the Y syndrome (Court Brown 1968), but a range of phenotypes is associated with the syndrome. Therefore an isochromosome for the short arms and a long-arm deletion cannot be distinguished on cytological evidence or the physical attributes of the patient alone. Absolute identification of a long-arm deletion of the Y chromosome is not possible at present.

A phenotypic male of average height with normal-sized testes, beard growth, and external genitalia was described by Meiser and Inhorn (1972). The male was surgically treated for undescended testes as a child. The Y chromosome was metacentric and did not exhibit the bright fluorescence characteristic of the distal portion of the long arm of the Y chromosome. Translocated Y-chromosome material was not detected. On the basis of the patient's average height, the abnormal Y chromosome was presumed to be one in which a long-arm deletion had taken place. The brother of this patient was normal and fertile and possessed the same morphologically abnormal Y chromosome.

Other cases of individuals with presumed long-arm deletions of the Y chromosome have been reported. A normal five-year-old boy of short stature was reported by Telfer and collaborators (1973). Muldal and Ockey (1962) have also described a family characterized by muscular dystrophy, hypospadias, and a Y chromosome in which a long-arm deletion had taken place. The males in this family were fertile. A tiny centric chromosome, the length of the short arms of the G-group autosomes, was reported by Nakagome and co-workers (1965) in a twenty-nine-month-old male. The male was normal sexually, and it was concluded that the centric fragment was composed by Y-chromosome material.

Long-arm deletions (or isochromosomes for the short arms) of the Y chromosome are associated with normal male development and fertility. Isochromosomes for the long arms of the Y chromosome are associated with hypoplastic female development, frequently without evidence of masculinization. The genes necessary for male development appear to be located primarily on the short arms of the Y chromosome. The presence of a small centric Y chromosome in a normal male (Nakagome et al. 1965) indicates that these loci are probably located close to the centromere. Nielsen and co-workers (1972) have reported a phenotypically normal male with testes of normal size in which the karyotype was 45,X/45,X ace+(?Yq−). Two cell lines were present, one which was 45,X and one which was normal except for an acentric fragment composed of only short-arm material. Gynecomastia was not present, beard growth and body hair were normal, and the patient did not exhibit features associated with Turner's syndrome aside from short stature. Testicular tissue was not karyotyped, and because of other factors it was not possible to determine if the male was fertile (J. Nielsen, personal communication, 1973). The short-arm, acentric fragment of the Y chromosome was present in sufficient proportions to cause testicular differentiation and male development, and was capable of producing a male phenotype in this individual. The above observations are a strong indication for the validity of the following hypothesis.

Hypothesis

We propose that male determination in man is controlled by a polygenic complex consisting of several gene loci that are tightly linked to each other and to the centromeric segment during nuclear divisions. This polygenic complex and the centromere act together as a single "supergene." Case studies indicate that male-determining loci specific for the male phenotype are located on the short-arm segment of this polygenic complex and that those loci necessary for normal male fertility are located on the long-arm segment of this complex. These loci can, for simplicity, be subdivided into a minimum of two subgroups, as shown in Figure 3, subgroup A being composed of those loci necessary for fetal development of the male gonad and subgroup B being composed of those loci necessary for normal spermatogenesis. More than two subgroups may exist; these two, however, may be considered basic and indispensable for normal male development and fertility. This proposed location of the male-determining gene loci provides a logical explanation of the observed abnormalities of sex differentiation associated with observed Y-chromosome aberrations.

Patients with isochromosomes for the long arm of the Y chromosome (46,XYqi) are phenotypically similar to those with the karyotype 45,X, indicating that the long arm of the Y chromosome is primarily inert and incapable of initiating testicular differentiation. These phenotypic females, however, are taller than the height normally associated with Turner's syndrome. This may be an indication that some specific growth factors, in addition to those factors necessary for male fertility, are located on the long arms. 45,X/46,XYqi mosaicism produces phenotypes similar to those associated with Turner's syndrome. The Turner phenotype, however, can be wholly attributed to the existence of the 45,X cell line. As a result, such cases of mosaicism are not entirely useful in determining the phenotypic effect of deleted or structurally abnormal Y chromosomes.

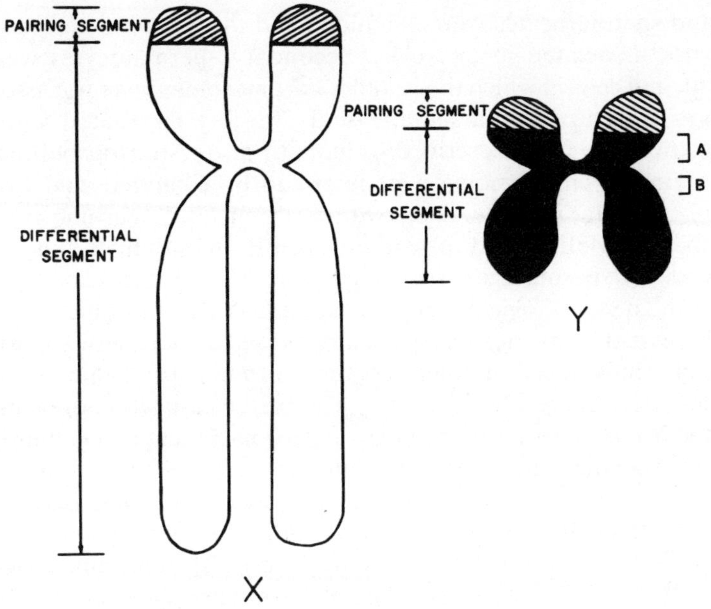

Figure 3. Diagrammatic representation of X and Y chromosomes in man showing homologous pairing segments. The small portion of the differential segment on both sides of the centromere of the Y chromosome is proposed to carry male determining gene loci: segment A responsible for fetal development of male gonads; segment B responsible for normal spermatogenesis and subsequently normal male fertility

Pericentric Inversions

Pericentric inversions of the Y chromosome are not reported frequently but are known to be inherited (Grace et al. 1972). Pericentric inversions of the Y chromosome characteristically produce normal fertile males with testes of normal size (Jacobs and Ross 1966), although fertility may be reduced as a result of such an inversion. Meiotic X–Y pairing may be affected, and individuals with pericentrically inverted Y chromosomes are sometimes prone either to production of a proportion of abnormal sperm or to breakdown of spermatogenesis at metaphase I (McIlree et al. 1966; Grace et al. 1972; Jacobs and Ross 1966). In an inversion of this type, which includes the centromere, those loci closely linked to the centromere remain intact.

The X and Y chromosomes in the male house mouse exhibit end-to-end association at diakinesis/metaphase of the first meiotic division, as do the X and Y chromosomes in man. Cattanach et al. (1968) have described a sterile mouse with complete absence of X–Y association in ninety-six meiotic metaphase-I cells examined. Spermatids and spermatozoa were absent. Beechey (1973) has also described a similar sterile mouse. Two hundred spermatocytes were examined and the X and Y chromosomes were not associated in any cells. Secondary spermatocytes were not present, and cells in which the X and Y chromosomes were not associated did not proceed beyond metaphase I. Sterility associated with X–Y dissociation has been described in humans also. An azoospermic male with a ring Y chromosome was described by Chandley and Edmond (1971), and an infertile male with a dicentric Y chromosome was described by McIlree and co-workers (1966). In both mice and humans, the block in spermatogenesis occurred at or soon after metaphase I. Therefore, it seems logical to conclude that X–Y chromosome association may be essential for normal spermatogenesis. Structurally abnormal X and Y chromosomes sometimes prevent normal pairing and, as a consequence, infertility results. Such may also be the case with pericentrically inverted Y chromosomes. Presence of a normal B segment (Figure 3), as well as association of the X and Y chromosomes, may be important for normal fertility. Some other prior disturbance other than dissociation, however, may be capable of producing the breakdown of spermatogenesis and the dissociation of the X and Y chromosomes (Beechey 1973). In the XYY mouse, XY and YY associations occur in a ratio of 2:1, indicating randomness of association. Therefore, X–Y association is not favored over the YY associations (Rathenberg and Muller 1973). The number of metaphase-II cells was decreased by more than half due to cell disintegration between diakinesis/metaphase I and metaphase II; the ratio of cells with XY and YY combinations, however, was still 2:1. In human XYY males, very few YY gametes are produced because of pairing behavior of the two Y chromosomes at metaphase I (Pearson 1972). Spermatogenesis, then, continues in the absence of X–Y association but is decreased. Since the pairing segments of the X and Y chromosomes are homologous, Y–Y associations may be capable of initiating further stages of spermatogenesis, but in a manner less efficient than normal X–Y associations.

Dicentric Y Chromosomes

Dicentric Y chromosomes are primarily composed of either short-arm or long-arm material. The formation of a dicentric depends upon breakage followed by reunion of the proximal ends of the chromatids of the Y chromosome during meiosis I in a normal male or by breakage in the two Y chromosomes of an XYY male followed by exchange (Cohen et al. 1973). The centric fragments will fuse together, resulting in a chromosome with two centromeres. The acentric fragments will be lost. As a result, dicentric chromosomes have varying duplications and deletions of Y-chromosome material. Breakage and reunion in the long arms lead to dicentrics for the short arms [dic(Yp)], while breakage and reunion of the short arms lead to dicentrics for the long arms [dic(Yq)]. Formation of a stable dicentric appears to depend on breakage and reunion close to the centromeric region. The more distal the break in either arm, the greater the distance between the two centromeres. In cases in which the distance between centromeres is significant, the chromosome may be unstable. During meiosis, the centromeres may move to opposite poles, form a bridge, and either break or become attenuated and be cut across by the cell wall. In mitosis, both centromeres would normally move to the same pole. In cases in which this does not occur, bridge formation or abnormal division will result (Cohen et al. 1973; Ying and Ives 1971; Angell et al. 1970).

Dicentrics for the long arms contain duplications of long-arm material and deletions of short-arm material, while dicentrics for the short arms contain short-arm duplications and long-arm deletions. As would be expected, the specific loci duplicated or deleted in dicentric Y chromosomes will be variable and, as a result, will show a variable phenotypic range of expression. A dicentric Y chromosome is usually present in association with mosaicism (Cohen et al. 1973). Ying and Ives (1971) have reported a child born with ambiguous genitalia and the karyotype 45,X/46,X dic (Yq). The right gonad was atrophic and fibrous, while the left gonad was an ovotestis. Cohen and co-workers (1973) have reported a similar case of an eleven-year-old girl. The karyotype was 45,X/46,X dic(Yq), the clitoris was enlarged, the vagina was of normal size, and the uterus was normally located. The right gonad was streak without evidence of testicular tissue, while the left gonad was composed of testicular tissue with a small segment of ovarian stroma. The Y chromosome in this patient was identified as a dicentric for the long arms. Banding revealed two brightly fluorescing regions at both ends of the chromosome. Two centromeres were distinctly visible. The similar phenotypes of these two patients would indicate that the dicentric (Yq) chromosome was composed of essentially the same short- and long-arm material in both individuals.

An individual with a dicentric (Yq) chromosome, not associated with chromosomal mosaicism, was reported by Armendares et al. (1972). The phenotype was female; the gonads were streak and composed of only ovarian stroma. The dicentric (Yq) in this nonmosaic individual was incapable of testicular differentiation, the small-arm material present in the patient apparently being insufficient. Angell and co-workers (1970) have reported two patients with the karyotype 45,X/46,X dic (Yq). In each case, features of Turner's syndrome were present. In one individual the gonad was streak, being composed entirely of the 45,X cell line.

Dicentric Y chromosomes for the long arm produce variable phenotypes; the amount of short-arm material present in the dicentric determines the degree of testicular differentiation. In cases accompanied by mosaicism, the proportion of Y-chromosome material in the gonads and the composition of the dicentric determine the presence or absence of testicular tissue. Dicentrics for the short arms of the Y chromosome should produce pronounced masculinization and a male phenotype, as the dicentric would be composed primarily of short-arm material. No case has been reported in which mosaicism was absent, however, and the cases reported of dicentrics for the short arms of the Y chromosome are questionable.

Starkman and Jaffe (1967) reported two cases in which the Y chromosome was thought to be a dicentric for the short arms. Both individuals were mosaic for a 45,X cell line. The individuals were phenotypically female and displayed some of the features of Turner's syndrome. The gonads in both were streak and composed of ovarian stroma. Leukocyte cultures from case 1 revealed a morphologically abnormal Y chromosome in 20 percent of the 46,XY cells, the Y chromosome being morphologically normal in the remaining 80 percent. A morphologically similar and abnormal Y chromosome was also observed in case 2. The morphology of the Y chromosome in both cases resembled that of a secondary constriction of the long arm reported by others (Muldal and Ockey 1960, 1962). The short arm appeared normal; the middle segment of the long arm ballooned out, however, and a close apposition was present at one point in the distal segment, so that the Y chromosome appeared dicentric. In some cells, a lightly staining region was observed in the middle of the long arm. A structurally similar Y chromosome was observed in the father of case 1. The secondary constriction of the long arm of the Y chromosome is more readily seen when coiling abnormalities are present (Muldal and Ockey 1962). The presumed dicentric Y chromosome was probably displaying a secondary constriction rather than two centromeres. Fluorescent banding could determine if the abnormal-appearing Y chromosome was a dicentric for the short arms, as the bright fluorescence of the long arm would be lacking (Pearson 1972). The individuals were mosaics possessing a 45,X cell line. It is, therefore, difficult to determine the potential effects of the Y chromosome. Gonadal tissue was karyotyped in case 2 and was found to be composed of only 45,X-containing cells.

Another case of a presumed dicentric (Yp) chromosome was reported (Angell et al. 1970) in which mosaicism for a 45,X cell line was present. The arms of the Y chromosome in the patient were distinctly shorter than the arms of the Y chromosome in her father. Morphologically normal Y chromosomes were also observed. Gonadal tissue was not karyotyped. The phenotypic sex was female and signs of Turner's syndrome were present. The abnormal-appearing chromosome may have been a dicentric, for the short arms indicated by size comparisons to the father's Y chromosome. The presence of a 45,X cell line in leukocyte cultures and the lack of information concerning the chromosomal constitution and histology of gonadal tissues make the phenotypic effect of this abnormal chromosome difficult to determine.

Dicentric (Yq) chromosomes are more frequently reported than dicentric (Yp) chromosomes. Dicentrics are usually accompanied by mosaicism, suggesting a postzygotic origin (Armendares et al. 1972). Female sex differentiation is usually common in cases of mosaicism in which one cell line contains a Y dicentric for the long arms. Some testicular differentiation does occur in these individuals, indicating some masculinizing effect of the dicentric. As we suggested previously, many reports of dicentric Y chromosomes for the short arm may be cases in which a secondary constriction of the long arms of a normal Y chromosome were observed.

POSSIBLE EXPLANATION OF THE XX-MALE AND XY-FEMALE PHENOMENA

Sex determination in man is normally precise; the XX-male and XY-female phenomena, however, appear to represent exceptions to this precise mechanism. Structural abnormalities of chromosomes, such as duplications and deletions of partial or whole chromosomes, are identifiable by karyotype analysis, but the XX male and XY female have apparently normal karyotypes.

The XX Male

Deletions, duplications, or translocations of chromosomal material, if of very small magnitude, may not be detectable by present methods of analysis. The quinacrine-fluorescence technique permits identification of each chromosome (Pearson 1972), but if a translocation very short in length took place it might escape detection (Breg et al. 1971). The Y chromosome, when stained by quinacrine, exhibits two very brightly fluorescing bands, located on the distal portion of the long arm (Schwinger and Pera 1972). The remainder of the Y chromosome exhibits two weakly fluorescing bands, one on each side of the centromere (Manolov et al. 1971). The presence of an essentially normal male phenotype in the XX male has led others to suggest a translocation of male-determining factors onto one of the X chromosomes (Ferguson-Smith 1966) or onto an autosome. Several attempts to detect Y fluorescent material in XX males have been undertaken (Andersson et al. 1972; Michalova et al. 1972; Caspersson et al. 1971; Breg et al. 1971; George and Polani 1970). Fluorescence was not detected. As mentioned previously, the translocation segment, if short, may not be detectable; and evidence points to the loci necessary for the male phenotype being located on the short arm of the Y chromosome, which fluoresces only weakly.

An opportunity for such a translocation of short-arm material would be present during paternal meiosis. Terminal associations via short homologous pairing segments have been reported in translocation trisomics of rye (Sybenga 1966b). In certain species such as rye and perhaps *Oenothera*, the region of pairing initiation (zygomere) is localized in a very short terminal segment (Sybenga 1966a), and chiasma formation is common to such short regions of rye chromosomes (Sybenga 1965). Furthermore, end segments may have the same probability of being involved in pairing irrespective of their length (Sybenga 1966b). In general, when pairing occurs in a small chromosomal segment in asynaptics and desynaptics, a chiasma usually follows (Abdel-Hameed 1972). Therefore, it is possible that the majority of observed X–Y end-to-end MI associations in man are genuine chiasmate associations, since they involve homologous telomeres and pairing segments. Cytological evidence for such chiasma is lacking, however; only a synaptinemal complex has been observed (Solari and Tres 1970). The synaptinemal complex may be required to facilitate genetic exchange at meiosis (de la Chapelle 1972; Moses 1968). The complex is found only in meiotic prophase (Ris 1961). In *Drosophila*, crossing-over does not occur in males, and synaptinemal complexes are found only in the oocytes of *Drosophila*. They are not found in spermatocytes (Meyer 1960). The function of the synaptinemal complex has not been definitely determined, but it is known to be a regular synaptic structure joining homologous bivalents along their lengths. Its primary function may be to pull homologous chromosomes together (Comings and Okada 1970) and increase the efficiency of recombination. Pairing consists of two processes: (1) chromosomal pairing, in which homologous chromosomes are brought in close association via the synaptinemal complex; and (2) molecular pairing, in which pairing of homologous DNA base pairs, recombination, and chiasmata take place (Comings and Okada 1970). The X and Y chromosomes must maintain synapsis until metaphase I in order for normal segregation to occur. The formation of the synaptinemal complex does not ensure continued association of two homologous chromosomes. Chiasma formation is necessary to prevent breakdown of bivalents until they align on the first meiotic metaphase plate. Chiasma or a chiasma-compensating mechanism has been proposed in certain *Clarkia* interspecific hybrids (Abdel-Hameed 1972) in which nonhomologous associations occur forming pseudobivalents apparently as a result of heterochromatin stickiness and pseudochiasma. Formation of translocation chromosomes requires nonhomologous pairing and reciprocal exchange. It is reasonable to assume that genetic crossing-over occasionally occurs between the X and Y chromosomes during meiosis. Such crossing-over between homologous pairing segments, however, may not result in recombination. Partial linkage between the X and Y chromosomes would be expected; this has, however, not been observed (Morton 1956).

Two possible pairing configurations of the X and Y chromosomes are represented in Figure 4. Crossing-over within the pairing segments in Figure 4a results in a normal X chromosome, a normal Y chromosome, a bridge, and a fragment. This configuration is unlikely. The configuration represented in Figure 4b, however, results in normal cross-over products, probably representative of actual pairing between the X and Y chromosomes. Nonhomologous regions of the short arms will be in close vicinity

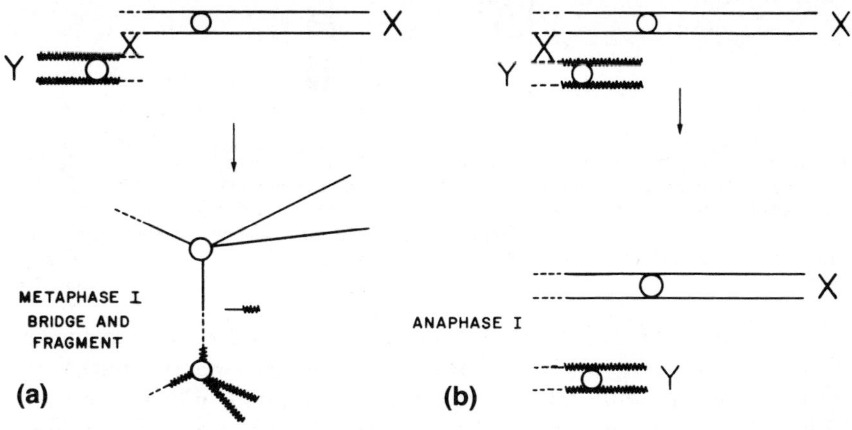

Figure 4. (a) Theoretical terminal association of X and Y chromosomes. Such association is unlikely, as it results in infertile gametes
(b) Proposed terminal association of X and Y chromosomes during first meiotic division. The products are normal, and X and Y chromosomes remain unchanged

during meiosis, and an opportunity exists for exchange outside, as well as within, pairing segments; reciprocal translocations of Y-chromosome loci onto the X chromosome and X-chromosome loci onto the Y chromosome would be expected. Of the four interchange sperm resulting, one contains a normal X chromosome, one contains a normal Y chromosome, and two sperm contain X–Y interchange chromosomes, represented by X^y and Y^x, as shown in Figure 5. The possible karyotype of the XX male might be 46,X^yX (de la Chapelle 1972; Ferguson-Smith 1966). The X^y chromosome, if composed of sufficient short-arm material, would be capable of producing testicular differentiation and a male phenotype.

The tissue cells of the XX male display sex chromatin (de la Chapelle 1972; Andersson et al. 1972; Lindsten et al. 1966; Boczkowski et al.

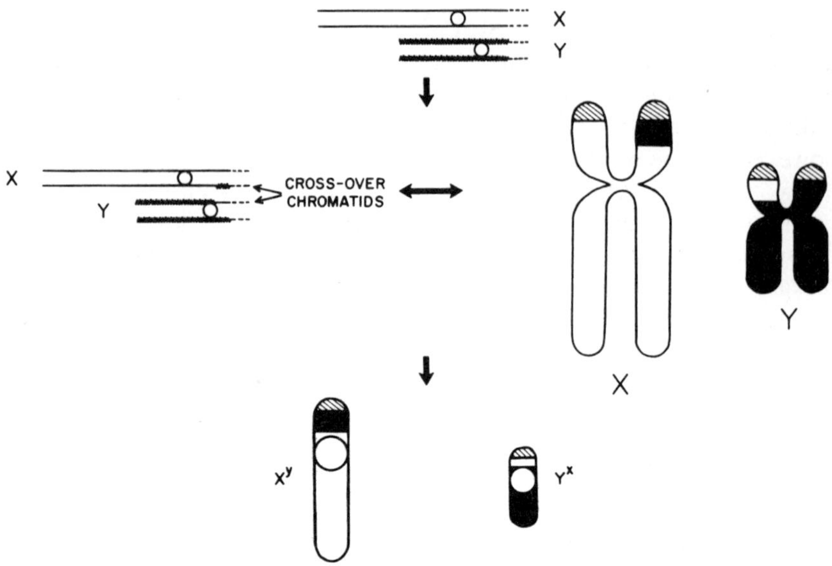

Figure 5. A diagrammatic representation of the X–Y interchange hypothesis. Genetic exchange proximal to the pairing segments produces translocated X^y and Y^x nonsister chromatids. The normal and translocated sister chromatids of X and Y will separate by anaphase II and will be included in separate gametes. Only the two interchange-type gametes are represented in the lower portion of the diagram

1968; Lisker et al. 1970; de la Chapelle et al. 1965, 1971; Michalova et al. 1972; Powers et al. 1970; Therkelsen 1964). The inactivation of one of the X chromosomes takes place normally, resulting in two genetically different cell lines. In a 46,X^yX individual, one cell line would contain an active normal X chromosome and the other cell line an active X chromosome containing some Y-chromosome material. Testicular tissue would arise in an X^yX individual only if a sufficient proportion of active X^y–containing cells was present in the gonads.

The sex chromosome complement XX^y was believed to explain the mode of origin of XX true hermaphrodites (Ferguson-Smith 1966; Jones et al. 1965). A true hermaphrodite is an individual in which both testicular and ovarian tissue is found (Barreiro et al. 1969). The external genitalia in an XO/XY mosaic can appear female, male, or intersexual. Ovarian follicles are rarely seen, although an XO/XY true hermaphrodite has been reported (Hirschhorn et al. 1960). In XX/Xy mosaicism, the phenotype can also range from normal-appearing female to normal-appearing male, similar to the phenotypic range of XO/XY mosaics (Miller 1964). Such a condition as XX/XY mosaicism, however, frequently results in a true hermaphrodite condition. The XX^y sex-chromosome complement being able to explain adequately XX true hermaphroditism is unlikely to account for the observed testicular-histology XX male. The testes of the XX male do not contain ovarian elements. Histology of the testes is frequently near normal, although degeneration of the seminiferous tubules occurs (de la Chapelle 1972), and the tubules are sparse and thickened (de la Chapelle 1972; de la Chapelle et al. 1965, 1971; Boczkowski et al. 1968; Lindsten et al. 1966; Lisker et al. 1970). The X inactivation may produce gonads that contain a greater proportion of normal active X chromosomes. The X^y chromosome resulting from an X–Y interchange may not always include sufficient male-determining loci to produce development of testicular tissue. Individuals with XX^y sex chromosomes containing sufficient Y male-determining genes would be likely to display an intersex condition, rather than that of a normal-appearing male.

We therefore believe that the XX-male condition most likely results from the chromosomal complement $46,X^yX^y$, shown in Figure 6. In this case two interchange events are necessary; X inactivation would result in all cell lines with a segment of active Y chromosome material. The

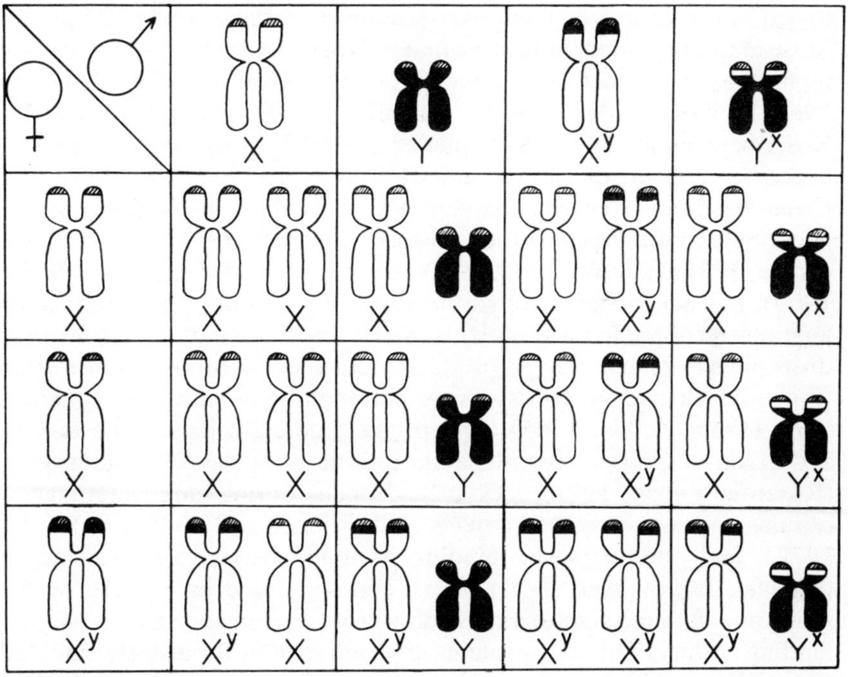

Figure 6. In addition to normal X and Y sperms produced in normal male spermatocytes, proposed X–Y interchange during first meiotic prophase would result in X^y and Y^x sperms in a low frequency during normal male spermatogenesis. See text for further explanation of XX-male and XY-female phenomena as well as other intersex phenotypes. The first line represents progeny resulting from fertilization of normal X ova; the second and third lines represent progeny resulting from X and X^y ova produced by carrier females

maternal parent must have the chromosomal complement $X^{y}X$. Either the other X chromosome can originate from spontaneous interchange in the paternal parent, or the paternal parent may have the sex-chromosome complement $X^{y}Y$. The XX male is not fertile and spermatogenesis is lacking (de la Chapelle 1972; de la Chapelle et al. 1965), but spermatogenic cells are sometimes seen. X–Y interchange would usually include only short-arm material proximal to the pairing segment and would not include portions of the B segment (Figures 3 and 5). The lack of spermatogenesis in the XX male may also result from the breakdown of spermatogenesis, as pairing between the X and Y chromosomes is impossible due to lack of a Y chromosome in these males.

The XY Female with Testicular Feminization

The feminized testes produce biologically active estrogens and androgens, but neither masculinization nor an anabolic response is produced in the complete syndrome. Administered androgens also do not cause masculinization, indicating that androgen insensitivity is not caused by biologically defective androgens (Griffiths et al. 1963; French et al. 1966; Morris and Mahesh 1963; Cardiff et al. 1972; Laurian et al. 1971; Rosenfield et al. 1971; Rivarola et al. 1967; Boczkowski 1968). A decreased conversion of testosterone to dihydrotestosterone in several tissues has been reported in patients with testicular feminization, and this decreased conversion was thought to be the primary defect in the syndrome (Wilson and Walker 1969; Wade et al. 1968; Strickland and Smith 1969), but administered dihydrotestosterone also does not produce an anabolic response in these patients (Strickland and Smith 1969). Antiandrogenic substances may be produced, but these substances must come from other than testicular tissues, as the condition remains after gonadectomy (French et al. 1965). The primary defect seems to be one of end-organ response. Hypothalamic–pituitary resistance to androgens (Rosenfield et al. 1971), lack of androgens during early fetal stages, cellular androgen-receptor defects or deficiencies (Bullock and Bardin 1972), and altered plasma-binding proteins specific for testosterone (Mauvais-Jarvis et al. 1970) have been suggested as a cause of the syndrome. Altered synthesis of binding proteins could produce the higher binding affinity of binding proteins in serum specific for testosterone that has been noted in the complete syndrome (Mauvais-Jarvis et al. 1970).

The incomplete form of testicular feminization (familial male pseudohermaphroditism with gynecomastia) may be a variant of the complete syndrome and, if androgen insensitivity of the type in the complete syndrome is present, it may be only partial (Saez et al. 1971). In man, two pathways of testosterone metabolism exist: oxidation of the 17β

hydroxy group via a 17β hydroxysteroid dehydrogenase and reduction by both a specific Δ^4-3-ketosteroid 5α and 5β reductases (Rose et al. 1973; Mauvais-Jarvis et al. 1968) and a specific 3β hydroxysteriod dehydrogenase (Mauvais-Jarvis et al. 1968). Studies of the incomplete syndrome indicate that the condition may result from a 17-ketosteriod reductase deficiency (Saez et al. 1971, 1972) or a 17β hydroxysteriod dehydrogenase deficiency (Goebelsmann et al. 1973). The incomplete syndrome may be produced as a result of enzyme defects or deficiencies directly influencing testosterone metabolism.

ANDROGEN RESPONSE. The transfer of a cytoplasmic androgen-receptor complex to the nuclear chromatin is thought to be an obligatory step in androgen response (Mainwaring et al. 1971; Mainwaring and Peterken 1971). This is also the case in estrogen response. The presence of estrogen-binding protcins in estrogen-sensitive cells has been established (Mueller et al. 1972), and the mechanism of estrogen action in target tissues is thought to be a two-step mechanism. The hormone enters the target cell and binds to an estrogen-binding protein (estrophile) in the cytoplasm. The 8S receptor complex formed in this association is relatively unstable, and in the presence of sodium or potassium chloride at concentrations of 0.2M or higher it is reversibly transformed into a 4S unit. The receptor complex undergoes a conformational change before entering the nucleus, detectable as the dissociated complex sediments at 5S (Jensen and DeSombre 1972). Entry into the nucleus is a temperature-dependent process (Mueller et al. 1972; Jensen and DeSombre 1972). The estrogen-receptor complex, after entry into the nucleus, becomes associated with the nuclear chromatin (Mueller et al. 1972). Increased RNA polymerase activity and acceleration of both RNA transcription and protein synthesis result (Mueller et al. 1972). Estrogen response appears to represent a general pattern for other classes of steroid hormones (Jensen and DeSombre 1972), and similarities exist in this receptor system and other types of steroid hormones. Certain catecholamine and peptide hormones also act via a two-step mechanism by (1) activating an adenyl cyclase system at the cell membrane which generates a cyclic nucleotide, which (2) in turn carries out a specific action in the cell (Sutherland and Robison 1966). Two general patterns of hormone control exist in mammalian cells (Jensen and DeSombre 1972), each consisting of two steps.

ANDROGEN RECEPTORS. The complete form of the syndrome of testicular feminization not only exists in man but is also present in rats, mice, and cattle (Bardin et al. 1970; Ohno and Lyon 1970; Short 1967). Bullock and Bardin (1972) have reported that mice and rats with testicular feminization lack detectable cytosol–androgen receptors. This receptor-complex deficiency was found in two species exhibiting the syndrome, suggesting that this may also be the primary defect in the syndrome in man. Since steroid-hormone response is dependent on the presence of such cytosol receptors, the lack of such receptors would prevent normal androgen response.

The hypothalamus and pituitary are normally responsive to steroid hormones. Receptor sites have been demonstrated in hypothalamic tissue (Kato and Villee 1967), and the hypothalamus can be considered a target tissue in itself. Rosenfield and co-workers (1971) administered dihydrotestosterone to a patient with complete testicular feminization. The plasma level of testosterone was not completely suppressed and the gonadotropin level was elevated, suggesting a hypothalamic–pituitary resistance to the effects of androgens. Provided the syndrome in man is caused by a lack of cytosol–androgen receptors, hypothalamic and pituitary resistance would be expected. Proper response of the hypothalamus and pituitary in a male pattern depends on these tissues being exposed to androgens in the proper amounts during fetal stages. Lack of cytosol–androgen receptors in early fetal stages or of target, pituitary, and/or hypothalamic tissues could cause androgen resistance later in development. Insufficient secretions of androgens in these critical fetal stages could also cause a female pattern of hypothalamic and pituitary response (Ohno et al. 1971). Testicular-feminized mice can be used to determine whether the hypothalamus and pituitary are reacting in a male or female pattern. Transplantation of the hypothalamus of a mouse with testicular feminization into a normal mouse would enable this determination. This test, however, would not enable us to distinguish between the absence of receptors and the lack of, or insufficiency of, androgens in fetal stages.

SERUM-BINDING PROTEINS. It is usually assumed that only unbound hormone is available for biological activity (Tavernetti et al. 1967). Increased binding affinity of serum proteins may be capable of preventing penetrance of androgens into target cells (Mauvais-Jarvis et al. 1970). A macromolecular binding factor for testosterone, other than albumin, was recently reported in human plasma (Pearlman and Crepy 1967; Rosner and Deakins 1968). This protein appears to be present in higher concentrations in females than in males and is increased in females during late pregnancy (Rosner and Deakins 1968). The binding protein is known to bind both estrogens and androgens (Mercier-Bodard et al. 1970), and the binding affinity for steroid hormones is higher in females than in males. However, the average concentration of testosterone in plasma is 6.6 ng/ml in men and 0.5 ng/ml in women (Smith 1960). The difference in binding of testosterone between females and males may be due to the differences in plasma testosterone levels. If the association is constant and the concentration of these binding proteins are the same in both sexes, the fraction of testosterone bound would depend on the plasma concentrations of testosterone (Rosner and Deakins 1968).

The relative binding affinity of serum protein for testosterone is significantly higher than in normal males in patients with testicular feminization, even though the plasma testosterone levels are within the range of those found in normal males (Mauvais-Jarvis et al. 1970). Increased affinity of specific plasma-binding proteins for testosterone would limit the unbound testosterone available in target cells for 5-α reduction. Hepatic 5-β reduction of testosterone is very low when the hormone is given intravenously but is normal when testosterone is administered orally in patients with testicular feminization. Administered orally, the hormone enters the liver before being bound to a specific protein (Yates 1967), which is in contrast to the results when it is injected into the peripheral circulation (Mauvais-Jarvis et al. 1970). Elevated binding affinity of serum proteins for testosterone may also be involved in the etiology of testicular feminization; it has not been proven, however, that the specific proteins binding testosterone in plasma are altered in the testicular-feminization syndrome. Elevated affinity of these proteins in the plasma, comparable to that observed in females in the first trimester of pregnancy, may be due to the inability of target tissues to accept the steroid hormone, thus preventing normal dissociation of the bound hormone as it enters the cytoplasm of target tissues. Androgen response in testicular feminization may be defective as a result of more than one metabolic or cellular defect. Further studies will be necessary before the exact defect(s) causing androgen insensitivity can be determined.

CHROMOSOMAL DEFECT. Individuals with XY testicular feminization seem to be capable of normal differentiation of their fetal testes; therefore, their Y chromosome is most probably normal. Two substances are believed to be produced by the fetal testes: (1) a nonsteroidal morphogenic substance which is concerned with suppression of the Mullerian ducts; and (2) a hormone, analogous to testosterone, concerned with masculinization of the Wolffian ducts and external genitalia (Laurian et al. 1971). Suppression of the Mullerian ducts apparently takes place, since uterus and tubes are absent. Stimulation of the Wolffian ducts, however, fails. If the hormone believed responsible for masculinization of the Wolffian ducts is similar in action to testosterone, a cytosol–androgen receptor defect or lack could account for the failure of masculinization in fetal stages.

Testicular feminization is a familial disorder transmitted via maternal carriers as an X-linked recessive or a male-limited autosomal dominant (Morris and Mahesh 1963; Laurian et al. 1971; Race and Sanger 1969). The carriers exhibit some penetrance of the condition, as decreased axillary and sexual hair are common (Morris and Mahesh 1963; Laurian et al. 1971; Boczkowski 1968). Localization of the genetic defect on the X chromosome has not been successful. Crossover studies indicate that it is not within measurable distance of either the X_g locus (Race and Sanger 1969) or the loci for color blindness and hemophilia (Morris and Mahesh 1963; Boczkowski 1968). The relative position of only a few X-linked loci are known (shown in Figure 2). Therefore, the possibility of an X-linked recessive mode of inheritance cannot be eliminated in testicular feminization. In the mouse, the loci for testicular feminization have been localized to the X chromosome (Lyon and Hawkes 1970). The fact that testicular feminization is X-linked in the mouse greatly increases the probability of X-linkage of corresponding loci in man and in the rat (Lyon and Hawkes 1970). Chromosomal deletions frequently act as lethal recessives (Swanson et al. 1967), and it may be possible that a deletion of certain loci on the X chromosome is genetically responsible for the syndrome. We propose that a deletion of certain loci on the X chromosome can explain the genetic defect and the recessive X-linked inheritance of testicular feminization. Due to X inactivation, the female carrier of the syndrome would exhibit variable penetrance of the deletion. This has been observed to be the case. Such a deletion on the X chromosome may have its origin through an X–Y interchange similar to that previously hypothesized for the XX male (see Figure 5). A genetic complex on the X chromosome specific for normal synthesis and/or action of cytosol–androgen receptors is proposed. Loci closely linked to this complex may be necessary for synthesis of enzymes involved in metabolism of androgens. The X^y chromosome would lack, in part or in whole, loci from this genetic complex, and would thus act as a genetic deletion within this complex. Deleterious mutations or the deletion of the X chromosome proposed above are similar in effect to that of lethals. Individuals with testicular feminization are sterile and therefore genetically dead (Ohno et al. 1971). The variations present between the complete and incomplete forms of the syndrome may be due to variations in the loci deleted from this polygenic complex.

OTHER PROPOSED MECHANISMS. Ohno et al. (1971) have proposed that, in a sense, the sex-determining genes may act in the same manner as does a regulator gene in an operon system. Two regulatory genes would be necessary: one to direct differentiation of the embryonic gonads, and the other to regulate manifestation of secondary sexual characteristics. The inducer of such a system would be the steroid hormone, maleness being the result of the induced state and femaleness being the result of the noninduced state of this regulatory system. Ohno and co-workers (1971) have further proposed that in X-linked testicular feminization a mutation of this regulatory locus has taken place. Such a mutation would render the regulatory system noninducible and female development would result. This is quite similar to the effect of the X-chromosome deletion that we have proposed above.

XY Gonadal Dysgenesis

The Y chromosome in the XY female with gonadal dysgenesis fails in differentiation of the embryonic gonads into testes. Female differentiation results. Mullerian-duct development is apparently not inhibited, and stimulation of the Wolffian ducts is lacking. Such a condition could result from lack of those loci necessary for male development on the Y chromosome (Figure 3). It has been proposed (Ferguson-Smith 1966) that such a deletion could originate through an X–Y interchange such as that proposed above for the XX male and the XY female with testicular feminization. "Pure" gonadal dysgenesis may result from fertilization of a normal X ovum by a Y^x-bearing sperm spontaneously produced by the paternal parent (Figure 6). Through interchange, the greater portion of the A segment (Figure 3) may be substituted. Degrees of masculinization are present in XY females with gonadal dysgenesis and are accountable by variations in the degree of A segment substituted.

Most cases of XY gonadal dysgenesis occur only in sibships of one generation without transmittance for two or more generations (Boczkowski et al. 1972; Chemke et al. 1970; Cohen and Shaw 1965; Espiner et al. 1970), but in some cases of the syndrome, the condition appears to be maternally transmitted. Sternberg et al. (1968) have reported a case in which three individuals were affected in three different sibships of the same family. A familial predisposition for meiotic X–Y-interchange events may be present in families exhibiting the syndrome, as illustrated by a similar case of a sibship consisting of two XY females with gonadal dysgenesis and a dysgenetic male pseudohermaphrodite reported by Chemke and co-workers (1970). The three sibs were the last of seven children; the paternal parent had worked in a uranium mine from one year before the conception of the sibs until death. Penetrating ionizing radiation from X-rays and gamma rays and that from radium, other radioactive atoms, and neutrons from nuclear reactors are well known mutagens (Stern 1973). X-rays can produce chromosome breaks followed by random reunions, resulting in chromosomal aberrations (Sankaranarayanan 1971; Brown and Bertke 1969). The father of these sibs may then have been prone to chromosome breakage and translocation between the X and Y chromosomes. The degree of A segment translocated and lacking from the Y chromosome most likely varies from one incident of translocation to another. The XY gonadal-dysgenetic females are probably examples of X–Y interchanges with most of the A segment deleted, while the XY dysgenetic male pseudohermaphrodite of the same sibship likely resulted from an X–Y-interchange translocation including a lesser deletion of the A segment.

Incidence of XX Males and XY Females

The frequency of the XX-male and XY-female phenomena is very low, the XX male occurring in approximately 0.004 percent (de la Chapelle 1972) to 0.006 percent (Abdel-Hameed and Gluesing 1979) of male births and the XY female with testicular feminization occurring in approximately 0.002 percent to 0.006 percent of newborn females (Stenchever et al. 1969). Both conditions probably occur more frequently, as most cases are detected by chance (Sternberg and Barclay 1967; Kaplan et al. 1969; de la Chapelle 1972). Interchange events between the X and Y chromosomes would occur more frequently in certain genotypes and we would expect familial differences for such interchanges. Employing the incidence of XX male births (0.004 percent), the frequency of X–Y interchange would be approximately 1.0 percent; i.e., on the average, interchange will occur in 1.0 percent of spermatocytes undergoing spermatogenesis. It should be noted that this frequency represents only those cases of interchange that include sufficient male-determining loci and consequently can be detected in the XX male. The incidence of X–Y interchange is then much greater than the estimate of 10^{-4} to 10^{-6} for point mutations (Cavalli-Sforza and Bodmer 1971). In comparison to point mutations, X–Y-interchange events are relatively frequent. Due to familial predisposition and difficulties in estimating the actual incidence of interchange, X–Y interchange probably occurs at a greater frequency than 1.0 percent.

SUMMARY AND CONCLUSIONS

The male phenotype, in mammals and humans, is dependent on the presence of Y-chromosome material capable of initiating testicular differentiation. Evolution has maximized the genetic differences between the X and Y chromosomes and minimized the homology between them (Cavalli-Sforza and Bodmer 1971), but in most instances pairing is adequate to prevent nondisjunction during meiosis. However, the small homologous segments between the X and Y chromosomes predispose them to meiotic nondisjunction and can in part account for the relatively high incidence of sex-chromosome aneuploidy.

Rare recombination events between the X and Y chromosomes may break the tight linkage of male-determining loci on the Y chromosome.

Such translocations would allow genetic combinations that can adequately account for the XX male, the XY female, and other intersex phenomena. Both X–Y translocation and nondisjunctional events can cause a wide range of sex-chromosome abnormalities. Sex-chromosome aneuploidy can produce abnormal sex differentiation deviating from that of normal male and female. Female differentiation usually occurs in the absence of a Y chromosome. Sex chromosome aneuploidies producing female differentiation range from 45,X Turner's syndrome to 48,XXXX. In contrast, male differentiation usually occurs in the presence of a Y chromosome. The range of abnormal sex-chromosome karyotypes producing male development include 47,XXY to 49,XXXXY Klinefelter's syndrome and 47,XYY. Recently cases of individuals with the sex-chromosome complement 48,XYYY have also been reported (Hunter and Quaife 1973; Schoepflin and Centerwall 1972). Sex abnormalities associated with apparently normal karyotypes, such as XX true hermaphroditism, XX males, XY females with testicular feminization, XY females with gonadal dysgenesis, and XY dysgenetic male pseudohermaphrodites, may result through X–Y-interchange events. The frequency of X–Y-interchange events can be estimated from the incidence of XX males: X–Y-interchange events probably occur at a greater frequency than 1.0 percent, as this figure represents only those translocations involving a greater portion of the proposed A segment necessary for male differentiation.

Meiosis is a highly specialized and ordered developmental process and as such must be controlled by an elaborate genetic system. X–Y dissociation during paternal meiosis may either cause breakdown of spermatogenesis at metaphase I or decrease the efficiency of gamete formation, thereby producing complete to partial azoospermia. The genetic system controlling meiosis may then be intimately bound with the behavior of the X and Y chromosomes.

The XY female with testicular feminization may occur as a result of chromosomal deletions of X-linked loci essential to normal synthesis of protein androgen receptors in target tissues, while the XY female with gonadal dysgenesis may result from deletion of Y-chromosome material necessary for male differentiation. The XX male may occur through X–Y-interchange events resulting in translocation of male-determining loci on the X chromosome. The karyotype of the XX male may be 46,$X^{y}X$ (Ferguson-Smith 1966). Since there is inactivation of one X chromosome in the XX male nuclei, we therefore propose 46,$X^{y}X^{y}$ as the most likely karyotype of the XX male. The XX-male and XY-female phenomena may result from the same type of chromosomal abnormality (Figure 6), and through X–Y-interchange events a variety of intersex conditions may be produced. Nondisjunctional events producing sex-chromosome aneuploidy, centromeric misdivision resulting in isochromosomes, and breakage and loss of portions of chromosomes also result in a variety of intersex states. In one group, the duplication, translocation, or deletion of sex-chromosome material involves whole or major portions of the X or Y chromosomes, while in the second group, deletions or duplications resulting from X–Y interchange are small and involve few gene loci. While the phenomena of XX males and XY females appear to represent exceptions to the normal process of sex differentiation in man, these phenomena may, in fact, represent intersex states produced by chromosome abnormalities not unlike, except on a smaller scale, those states resulting from sex-chromosome aneuploidies.

REFERENCES

ABDEL-HAMEED, F.
1972 Chromosomal abnormalities in F_1 — diploid hybrids between *Clarkia lassenensis* and *C. amoena* ssp. *huntiana*. *Genetica* 43:3–24.

ABDEL-HAMEED, F., L. M. GLUESING
1979 "The XX male and XY female in man," in *Physiological and morphological adaptation and evolution.* Edited by William Stini. World Anthropology. The Hague: Mouton.

ANDERSSON, L., S. BERGMAN, J. REITALU, S. ANSEHN
1972 A case of XX male: cytogenetic findings by autoradiography and fluorescence microscopy. *Hereditas* 70: 311–314.

ANGELL, R., F. GIANELLI, P. E. POLANI
1970 Three dicentric Y chromosomes. *Annals of Human Genetics* 34:39–50.

ARMENDARES, S., L. BUENTELLO, F. SALAMANCA, J. CANTU-GARZA
1972 A dicentric Y chromosome without evidence of sex chromosomal mosaicism, 46,XY q dic, in a patient with features of Turner's syndrome. *Journal of Medical Genetics* 9:96–100.

BACHMAN, R., F. F. DE LA CRUZ, M. AL-AISH, F. SANTELL
1971 Short arm deletion of an X-chromosome in a 19 year old girl. *American Journal of Mental Deficiency* 75:435–441.

BARDIN, C. W., L. BULLOCK, G. SCHNEIDER, J. E. ALLISON, A. J. STANLEY
1970 Pseudohermaphrodite rat: end organ insensitivity to testosterone. *Science* 167:1,136–1,137.

BARREIRO, E., A. PERALTA, J. MONEREO, J. A. USANDIZAGA, F. CONTRERAS, A. SWARTZ
1969 Two new cases of true hermaphroditism. *Journal of Medical Genetics* 6:426–428.

BEECHEY, C. V.
1973 X–Y chromosome dissociation and sterility in the mouse. *Cytogenetics* 12:60–67.

BOBROW, M., P. L. PEARSON, M. C. PIKE, O. S. EL-ALFI
1971 Length variation in the quinacrine-binding segment of human Y chromosomes of different sizes. *Cytogenetics* 10:190–198.

BOCZKOWSKI, K.
1968 Genetical studies in testicular feminization syndrome. *Journal of Medical Genetics* 5:181–188.

BOCZKOWSKI, K., Z. JANCZEWSKI, J. PHILLIP, M. MIKKELSEN
1968 A clinical and cytological study of XX males. *Hereditas* 62:285–292.

BOCZKOWSKI, K., J. TETER
1969 Intersexuality characterized by female external genitals and intra-abdominal rudimentary testes. *American Journal of Obstetrics and Gynecology* 103:1,167–1,168.

BOCZKOWSKI, K., J. TETER, J. STEINNADEL
1972 Sibship occurrence of XY gonadal dysgenesis with disgerminoma. *American Journal of Obstetrics and Gynecology* 113:952–955.

BREG, W. R., G. M. DILLARD, E. M. SHORT, P. W. ALLDERDICE, D. A. MILLER, O. J. MILLER
1971 Fluorescence and Y translocation in XX males. *Lancet* 1:1,241–1,242.

BROWN, W. V., E. M. BERTKE
1969 *Textbook of cytology.* Saint Louis: C. V. Mosbey.

BULLOCK, L. P., C. W. BARDIN
1972 Androgen receptors in testicular feminization. *Journal of Clinical Endocrinology and Metabolism* 35:935–937.

CARDIFF, C., V. UDEN, D. W. KILLINGER
1972 Steroid biosynthesis in gonadal tissue from patients with testicular feminization. *Canadian Journal of Biochemistry* 50:849–855.

CASPERSSON, T., A. DE LA CHAPELLE, J. LINDSTEN
1971 Absence of brightly fluorescent Y material in XX men. *Annals of Genetics* 14:173–176.

CATTANACH, B. M., C. E. POLLARD, J. H. ISSAACSON
1968 Ethyl methanesulfonate-induced chromosome breakage in the mouse. *Mutation Research* 6:297–307.

CAVALLI-SFORZA, L. L., W. F. BODMER
1971 *The genetics of human populations*. San Francisco: W. H. Freeman.

CHANDLEY, A. C., P. EDMOND
1971 Meiotic studies on a subfertile patient with a ring Y chromosome. *Cytogenetics* 10:295–304.

CHEMKE, J., R. CARMICHAEL, J. M. STEWART, R. H. GEER, A. ROBINSON
1970 Familial XY gonadal dysgenesis. *Journal of Medical Genetics* 7:105–111.

COHEN, M. M., M. H. MACGILLIVRAY, V. J. CAPRARO, T. A. ACETO
1973 Human dicentric Y chromosomes. Case report and review of the literature. *Journal of Medical Genetics* 10:74–79.

COHEN, M. M., M. W. SHAW
1965 Two XY siblings with gonadal dysgenesis and a female phenotype. *New England Journal of Medicine* 272:1,083–1,088.

COMINGS, D. E., T. A. OKADA
1970 Whole mount electron microscopy of meiotic chromosomes and the synaptonemal complex. *Chromosoma* 30:269–286.

COURT BROWN, W. M.
1968 Males with an XYY sex chromosome complement. *Journal of Medical Genetics* 5:341–359.

DE LA CHAPELLE, A.
1972 Nature and origin of males with XX sex chromosomes. *American Journal of Human Genetics* 24:71–105.

DE LA CHAPELLE, A., H. HORTLING, J. WENNSTROM, M. NIEMI, C. J. JOHANSSON
1965 Two males with female chromosomes. *Acta Endocrinologica Supplement* 100:90.

DE LA CHAPELLE, A., S. SIMILA, M. LANNING, M. KONTTURI, C. JOHANSSON
1971 Two further males with female karyotype. *Humangenetik* 11:286–294.

DEWHURST, C. J.
1971 The XY female. *American Journal of Obstetrics and Gynecology* 109:564–588.

EGGEN, R. R.
1969 "Cytogenetics," in *Clinical diagnosis by laboratory methods*. Edited by I. Davidson and J. B. Henry, 1,208–1,240. Philadelphia: W. B. Saunders.

ESPINER, E. A., A. M. O. VEALE, V. E. SANDS, P. H. FITZGERALD
1970 Familial syndrome of streak gonads and normal male karyotype in five phenotypic females. *New England Journal of Medicine* 283:6–11.

FERGUSON-SMITH, M. A.
1966 X–Y chromosomal interchange in the aetiology of true hermaphroditism and of XX Klinefelter's syndrome. *Lancet* 2:475–476.

FERGUSON-SMITH, M. A., E. BOYD, M. E. FERGUSON-SMITH, J. G. PRITCHARD, *et al.*
1969 Isochromosome for the long arm of the Y chromosome in patient with Turner's syndrome and sex chromosome mosaicism (45,X/46,XYqi). *Journal of Medical Genetics* 6:422–425.

FRENCH, F. S., B. BAGGETT, J. J. VAN WYK, L. M. TALBERT, W. R. HUBBARD, *et al.*
1965 Testicular feminization: clinical, morphological and biochemical studies. *Journal of Clinical Endocrinology* 25:661–677.

FRENCH, F. S., J. J. VAN WYK, B. BAGGETT, W. E. EASTERLING, L. M. TALBERT, F. R. JOHNSTON
1966 Further evidence of a target organ defect in the syndrome of testicular feminization. *Journal of Clinical Endocrinology* 26:493–503.

GEORGE, K. P., P. E. POLANI
1970 Heterochromatin and XX males. *Nature* 228:1,215–1,216.

GERMAN, J.
1970 Studying human chromosomes today. *American Scientist* 58:182–201.

GOEBELSMANN, V., R. HORTON, J. H. MESTMAN, J. J. ARCE, Y. NAGATA, *et al.*
1973 Male pseudohermaphroditism due to testicular 17-β hydroxysteroid dehydrogenase deficiency. *Journal of Clinical Endocrinology and Metabolism* 36:867–879.

GRACE, H. J., F. E. ALLY, M. A. PARUK
1972 46,X inv (Yp+q−) in four generations of an Indian family. *Journal of Medical Genetics* 9:293–297.

GRIFFITHS, K., J. K. GRANT, W. G. WHYTE
1963 Steroid biosynthesis *in vitro* by cryptorchid testes from a case of testicular feminization. *Journal of Clinical Endocrinology* 23:1,044–1,055.

HAFEZ, E. S. E.
1968 "Intersexuality," in *Reproduction in farm animals,* second edition: 333–336. Philadelphia: Lea and Febiger.

HIRSCHHORN, K., W. H. DECKER, H. L. COOPER
1960 Human intersex with chromosome mosaicism of type XY/XO. Report of a case. *New England Journal of Medicine* 263:1,044–1,048.

HOFFENBERG, R., W. P. U. JACKSON
1957 Gonadal dysgenesis in normal looking females. A genetic theory to explain the variability of the syndrome. *British Medical Journal* 1:1,281–1,284.

HUNTER, H., R. QUAIFE
1973 A 48,XYYY male: a somatic and psychiatric description. *Journal of Medical Genetics* 10:80–96.

ISMAIL, A. A. A., D. W. DAVIDSON, A. R. SONKA, J. A. LORAINE, G. L. FOSS
1972 Hormonal studies in a patient with testicular feminization syndrome. *Acta Endocrinologica* 71:409–416.

JACOBS, P. A., A. ROSS
1966 Structural abnormalities of the Y chromosome in man. *Nature* 210:352–354.

JENSEN, E. V., E. R. DeSOMBRE
1972 Mechanism of action of the female sex hormones. *Annual Review of Biochemistry* 41:203–230.

JONES, H. W., M. A. FERGUSON-SMITH, R. H. HELLER
1965 Pathologic and cytogenetic findings in true hermaphroditism. *Journal of Obstetrics and Gynecology* 25:435–447.

JOST, A.
1947 Recherches sur la différenciation sexuelle de l'embryon de lapin III. Rôle des gonades foetales dans la différenciation sexuelle somatique. *Archives d'Anatomie Microscopique et de Morphologie Expérimentale* 36:271.

KAPLAN, S. A., W. H. SNYDER, S. LITTLE
1969 Inguinal hernias in females and the testicular feminization syndrome. *American Journal of Diseases of Children* 117:243–251.

KATO, J., C. A. VILLEE
1967 Preferential uptake of estradiol by the anterior hypothalamus of the rat. *Endocrinology* 80:567–575.

LAURIAN, L., J. O. FINKELSTEIN, N. SOFERMAN, M. LOEWENTHAL, L. DELEGDISCH
1971 Testicular feminization syndrome. *Israel Journal of Medical Science* 7:251–254.

LEE, C. R., W. D. MacDIARMID, G. E. CARTWRIGHT, M. M. WINTROBE
1968 Hereditary X-linked, sideroachrestic anemia. The isolation of two erythrocyte populations differing in X_g^a blood type and porphyrin content. *Blood* 32:59–70.

LINDSTEN, J., C. G. BERGSTRAND, K. G. TILLINSER, H. G. SCHWARZASHER, L. TREPOLO, *et al.*
1966 A clinical and cytogenetical study of three patients with male phenotype and apparent XX sex chromosome constitution. *Acta Endocrinologica* 52:91–112.

LISKER, R., F. FLORES, A. COBO, F. G. ROJAS
1970 A case of XX male syndrome. *Journal of Medical Genetics* 7:394–398.

LYON, M. F.
1962 Sex chromatin and gene action in the mammalian X chromosome. *American Journal of Human Genetics* 14:135–148.

LYON, M. F., S. G. HAWKES
1970 X-linked gene for testicular feminization in the mouse. *Nature* 227:1,217–1,219.

MacDIARMID, W. D., G. R. LEE, G. E. CARTWRIGHT, M. M. WINTROBE
1967 X-inactivation in an unusual X-linked anemia and the X_g^a blood group. *Clinical Research* 15:132.

MAINWARING, W. I. P., F. R. MANGAN, B. M. PETERKEN
1971 Studies on the solubilized ribonucleic acid polymerase from rat ventral prostate gland. *Biochemical Journal* 123:619–628.

MAINWARING, W. I. P., B. M. PETERKEN
1971 A reconstituted cell-free system for the specific transfer of steroid receptor complexes into nuclear chromatin isolated from rat ventral prostate gland. *Biochemical Journal* 125:285–295.

MANOLOV, G., Y. MANOLOVA, G. FISKESJO, A. LEVAN
1971 The complexity of the fluorescent pattern of the human Y chromosome. *Hereditas* 68:328–331.

MAUVAIS-JARVIS, P., J. P. BERCOVICI, O. CREPY, F. GAUTHIER
1970 Studies on testosterone metabolism in subjects with testicular feminization syndrome. *Journal of Clinical Investigation* 49:31–40.

MAUVAIS-JARVIS, P., H. H. FLOCH, J. P. BERCOVICI
1968 Studies on testosterone metabolism in human subjects with normal and pathological sexual differentiation. *Journal of Clinical Endocrinology and Metabolism* 28:460–471.

McILREE, M. E., W. H. PRICE, W. M. COURT BROWN, W. S. TULLOCH, *et al.*
1966 Chromosome studies on testicular cells from 50 subfertile men. *Lancet* 2:69–71.

MEISER, L. F., S. L. INHORN
1972 Normal male development with Y chromosome long arm deletion (Yq−). *Journal of Medical Genetics* 9:373–377.

MERCIER-BODARD, C., A. ALFSEN, E. E. BAULIEU
1970 "Sex steroid binding plasma protein (SBP). Karolinska Symposia on Research Methods in Reproductive Endocrinology," in *Acta Endocrinologica* (supplement). Edited by E. Diczfalusy, 204–224. Copenhagen: Bogtrykkeriet Forum.

MEYER, G. F.
1960 "The fine structure of spermatocyte nuclei of *Drosophila melanogaster*," in *Proceedings of the European Regional Congress of Electronmicroscopy, Delft, 1960*, volume two: 951–954. Delft: De Nederlandse Vereniging voor Electronenmicroscopie.

MICHALOVA, K., J. NEUWIRTH, J. RABOCH
1972 Evidence of the absence of Y fluorescent material in XX men. *Humangenetik* 14:251–254.

MILLER, O. J.
1964 Sex chromosome anomalies. *American Journal of Obstetrics and Gynecology* 90:1,078–1,139.

MONEY, J., A. A. EHRHARDT
1973 *Man and woman, boy and girl*. Baltimore: Johns Hopkins University Press.

MORRIS, J. M.
1953 The syndrome of testicular feminization in male pseudohermaphrodites. *American Journal of Obstetrics and Gynecology* 65:1,192–1,211.

MORRIS, J. M., V. B. MAHESH
1963 Further observations on the syndrome, testicular feminization. *American Journal of Obstetrics and Gynecology* 87:731–748.

MORTON, N. E.
1956 Sequential tests for the detection of linkage. *American Journal of Human Genetics* 7:277–318.

MOSES, M. J.
1968 Synaptinemal complex. *Annual Revue of Genetics* 2:363–412.

MUELLER, G. C., B. VONDERHAAR, U. H. KIM, M. L. MAHIEU
1972 Estrogen action: an inroad to cell biology. *Hormone Research* 28:1–49.

MULDAL, S., C. H. OCKEY
1960 The "double male": a new chromosome constitution in Klinefelter's syndrome. *Lancet* 2:492–493.
1962 Deletion of Y chromosome in a family with muscular dystrophy and hypospadias. *British Medical Journal* 1:291–294.

NAKAGOME, Y., M. SASAKI, I. MATSUI, M. KAWAZURA, Y. FUKUYAMA
1965 A mentally retarded boy with minute Y chromosome. *Journal of Pediatrics* 67:1,163–1,167.

NIELSEN, J., V. FRIEDRICH, A. CHRISTENSEN, H. H. GODT, L. S. STROMGREA
1972 A phenotypic male with karyotype 45,X/45,X,ace+(?Yq−). *Humangenetik* 15:319–326.

OHNO, S., M. F. LYON
1970 X-linked testicular feminization in the mouse as a non-inducible regulatory mutation of the Jacob–Monod type. *Clinical Genetics* 1:121–127.

OHNO, S., U. TETTENBORN, R. DOFUKU
1971 Molecular biology of sex differentiation. *Hereditas* 69:107–124.

Original Article Series
1972 Standardization in human cytogenetics. *Birth Defects: Original Article Series* 8:7. The National Foundation, March of Dimes, 1275 Mamaroneck Avenue, White Plains, New York.

PEARLMAN, W. H., O. CREPY
1967 Steroid protein interactions with particular reference to testosterone binding by human serum. *Journal of Biological Chemistry* 242:182–189.

PEARSON, P.
1972 The use of new staining techniques for human chromosome identification. *Journal of Medical Genetics* 9:264–275.

PEARSON, P. L., M. BOBROW
1970 Definitive evidence for the short arm of the Y chromosome associating with the X chromosome during meiosis in the human male. *Nature* 226:959–961.

POLANI, P. E., R. ANGELL, F. GIANNELLI
1964 Evidence that the X_g locus is inactivated in structurally abnormal X chromosomes. *Nature* 207:613–616.

POWERS, H. O., R. L. NEU, H. SMULYAN, L. I. GARDNER
1970 An adult phenotypic male with 46,XX chromosome complement. *Journal of Clinical Endocrinology and Metabolism* 31:576–579.

RACE, R. R., R. SANGER
1969 X_g and sex chromosome abnormalities. *British Medical Bulletin* 25:99–103.

RATHENBERG, R., D. MULLER
1973 X and Y chromosome pairing and disjunction in a male mouse with an XYY sex-chromosome constitution. *Cytogenetics* 12:87–92.

RIS, H.
1961 The annual invitation lecture: ultrastructure and molecular organization of genetic systems. *Canadian Journal of Genetics and Cytology* 3:95–120.

RIVAROLA, M. A., J. M. SAEZ, W. J. MEYER, F. M. KENNY, C. J. MEGLON
1967 Studies of androgens in the syndrome of male pseudohermaphroditism with testicular feminization. *Journal of Clinical Endocrinology* 27:371–378.

ROSE, L. I., R. H. UNDERWOOD, D. E. PARKER, G. H. WILLIAMS
1973 Δ^4−3-ketosteroid 5α-reductase in the adult human testis. *Journal of Clinical Endocrinology and Metabolism* 36:995–998.

ROSENFIELD, A. L., A. M. LAWRENCE, S. LIAO, R. L. LANDAU
1971 Androgens and androgen responsiveness in the feminized testes syn-

drome. Comparison of complete and incomplete forms. *Journal of Clinical Endocrinology* 32:625–632.

ROSNER, W., S. M. DEAKINS
1968 Testosterone-binding globulins in human plasma: studies on sex distribution and specificity. *Journal of Clinical Investigation* 47:2,109–2,116.

RUSSELL, L. B.
1963 Mammalian X chromosome action: inactivation limited in spread and in region of origin. *Science* 140:976–978.

SAEZ, J. M., E. DE PERETTI, A. M. MORERA, M. DAVID, J. BERTRAND
1971 Familial male pseudohermaphroditism with gynecomastia due to testicular 17-ketosteroid reductase defect. I. Studies *in vivo*. *Journal of Clinical Endocrinology and Metabolism* 32:604–610.

SAEZ, J. M., A. M. MORERA, E. DE PERETTI, J. BERTRAND
1972 Further *in vivo* studies in male pseudohermaphroditism with gynecomastia due to a testicular 17-ketosteroid defect (compared to a case of testicular feminization). *Journal of Clinical Endocrinology and Metabolism* 34:598–600.

SANKARANARAYANAN, K.
1971 Recent advances in mammalian radiation genetics and their relevance to the problem of genetic risk estimates in man. *International Journal of Environmental Studies* 1:187–193.

SCHOEPFLIN, G. S., W. R. CENTERWALL
1972 48,XYYY: a new syndrome? *Journal of Medical Genetics* 9:356–380.

SCHWINGER, E., F. PERA
1972 On splitting of the Y fluorescent body in man. *Humangenetik* 14:107–111.

SHORT, R. V.
1967 Reproduction. *Annual Revue of Physiology* 29:373–400.

SMITH, I.
1960 *Chromatographic and electrophoretic techniques*, second edition. New York: Interscience.

SMITH, M., W. J. PEPLER
1969 Female phenotype associated with XY karyotype: report of a case with gonadal dysgenesis. *South African Medical Journal* 43:1,414–1,415.

SOLARI, A. J., L. L. TRES
1970 The three dimensional reconstruction of the XY chromosomal pair in human spermatocytes. *Journal of Cell Biology* 45:43–53.

SOUTHERN, A. L., S. TOCHIMOTO, N. C. CARMODY, K. ISURUI
1965 Plasma production rates of testosterone in normal adult men and women and in patients with the syndrome of feminized testes. *Journal of Clinical Endocrinology* 25:1,441–1,450.

SPERLING, K., I. LACKMAN
1971 Large human Y chromosome with two fluorescent bands. *Clinical Genetics* 2:352–355.

STARKMAN, M. N., R. B. JAFFE
1967 Chromosome aberrations in XO/XY mosaic individuals and their fathers. *American Journal of Obstetrics and Gynecology* 99:1,056–1,066.

STENCHEVER, M. A., A. B. P. NG, G. K. JONES, J. A. JARVIS
1969 Testicular feminization syndrome. Chromosomal, histologic, and genetic studies in a large kindred. *Obstetrics and Gynecology* 33:649–657.

STERN, C.
1973 *Principles of human genetics*: San Francisco. W. H. Freeman.
STERNBERG, W. H., D. L. BARCLAY
1967 Women with male sex chromosomes: the syndromes of testicular feminization and XY gonadal dysgenesis. *Birth Defects: Reprint Series 141.* The National Foundation, March of Dimes, 1275 Mamaroneck Avenue, White Plains, New York.
STERNBERG, W. H., D. L. BARCLAY, H. W. KLOEPFER
1968 Familial XY gonadal dysgenesis. *New England Journal of Medicine* 278:695–700.
STRICKLAND, A. L., F. S. SMITH
1969 Absence of response to dihydrotestosterone in the syndrome of testicular feminization. *Journal of Clinical Endocrinology and Metabolism* 29:1,284–1,286.
SUTHERLAND, E. W., G. A. ROBISON
1966 The role of cyclic-3′5′-AMP in responses to catecholamines and other hormones. *Pharmacology Revue* 18:145–161.
SWANSON, C. P., T. MERZ, W. J. YOUNG
1967 *Cytogenetics.* New Jersey: Prentice-Hall.
SYBENGA, J.
1965 The quantitative analysis of chromosome pairing and chiasma formation based on the relative frequencies of MI configurations. III. Telocentric trisomics. *Genetica* 36:351–361.
1966a The zygomere: a hypothetical unit of chromosome pairing initiation. *Genetica* 37:186–198.
1966b The quantitative analysis of chromosome pairing and chiasma formation based on the relative frequencies of MI configurations. V. Interchange trisomics. *Genetica* 37:481–510.
TAVERNETTI, R. R., W. ROSENBAUM, W. G. KELLY, N. P. CHRISTY, M. S. ROGINSKY
1967 Evidence for the presence in human plasma of an estrogen-binding factor other than albumin: abnormal binding of estradiol in men with hepatic cirrhosis. *Journal of Clinical Endocrinology and Metabolism* 27:920–926.
TELFER, M., D. BAKER, I. ROLLIN
1973 Probable long arm deletion of Y chromosome in boy of short stature. *Lancet* 1:608.
TETSUJI, K., K. OHAMA, T. SOFUNI, T. IKEDA
1971 An XY human female with ovaries. *American Journal of Obstetrics and Gynecology* 110:98–102.
THERKELSEN, A. J.
1964 Sterile male with the chromosome constitution 46,XX. *Cytogenetics* 3:207–218.
WADE, A. D., G. S. WILKINSON, J. C. DAVIS, T. N. A. JEFFCOATE
1968 The metabolism of testosterone, androstenedione, and oestrone by testes from a case of testicular feminization. *Journal of Endocrinology* 42:391–403.
WILSON, J. D., J. D. WALKER
1969 The conversion of testosterone to 5α- and androstan-17β-ol-3-one (dihydrotestosterone) by skin slices of man. *Journal of Clinical Investigation* 48:371–379.
YATES, F. E.
1967 The liver and the adrenal cortex. *Gastroenterology* 53:477–482.

YING, K. L., E. J. IVES
1971 Mitotic behavior of a human dicentric Y chromosome. *Cytogenetics* 10:208–218.

ADDENDUM: NOTE OF PROOF

The recent detection of the histocompatibility antigen determined by the Y chromosome (H–Y antigen) in the heterogametic sex of vertebrate animals (McKusick and Ruddle 1977) is considered to be one of the most important discoveries made in the area of gonadal tissue differentiation. As a rule laboratory animals of highly inbred strains usually accept tissue transplants from other animals of that strain. The major exception to this rule is the fact that females of many inbred strains of mice will reject tissue from males of the same strain since the females lack the H–Y antigen. Furthermore, it seems that the gene locus for this H–Y male antigen has been conserved on the Y chromosome through a long period of vertebrate evolution during the last several hundred million years (McKusick and Ruddle 1977). Studies of various aberrations of the Y chromosome indicate that the H–Y locus is located on the short arm close to the centromere of the Y chromosome in man, and this locus and the one that determines the differentiation of embryonic testes may indeed be one and the same gene locus (McKusick and Ruddle 1977). On the other hand, the presence of one or more genes on the long arm of the Y chromosome that are essential to normal spermatogenesis and spermiogenesis has been indicated by the abnormalities found in patients with Y deletions (McKusick and Ruddle 1977).

Comparison of the mean height of individuals with XYY, XXY, XY, and XX karyotypes may suggest that the human Y chromosome carries one or more genes for stature. But the Y influence on stature may be mediated by a mechanism other than androgen, since XY females with gonadal dysgenesis are characteristically taller than and yet very similar, especially with regard to the differentiation of their internal and external genitalia, to XO Turner females. Convincing evidence for the presence of other genes, including the "hairy ear rims," is lacking; thus the H–Y antigen represents the only gene locus that has been unequivocally located so far on the human Y chromosome (McKusick and Ruddle 1977).

McKUSICK, V. A., F. H. RUDDLE
1977 The status of the gene map of the human chromosomes. *Science* 196:390–405.

Critical Approach to the Interpretative Problem of the DNA Nuclear Content Variations in Primates

M. G. MANFREDI ROMANINI and G. F. DeSTEFANO

We have considered the DNA nuclear content of a cell in G1 period (e.g. the peripheral blood lymphocyte) as the standard genome size characteristic of the species. Our data on various specimens of mammalian families have revealed that within this class the DNA nuclear content values show remarkable variations (Manfredi Romanini, Minazza, et al. 1971; Capanna and Manfredi Romanini 1971a, 1971b, 1973), and exact taxonomic values have emerged from the comparison between the data on microchiropters and those on several species of the order of primates. On the other hand, the fact that large variations exist within mammalians appeared conspicuously from the work of Benirschke et al. (1970), who assigned to the *Orycteropus* a "genome size" of 167 percent compared to man's, while our data on microchiropters show very small values: for *Minioterus schreibersi*, 50 percent less than man's (Capanna and Manfredi Romanini 1971a).

We have previously emphasized the importance that the dosage may have, in terms of the DNA nuclear content: (1) within groups, for phylogenetic interpretation in conjunction with the morphological description of the karyotype or with descriptive data on the kinds of DNAs in the nucleus, even though the ratio between reassociation speeds of the various DNA fractions has not yet been given a clear interpretation in functional terms (Kohne et al. 1972); and (2) in the comparison among close species with known karyotypes, in order to throw light on the possible mechanisms of species formation.

With regard to primates particularly, our data have been applied to both these areas of investigation, within and between species. From the phylogenetic point of view, the remarkable constancy of the DNA nuclear content in the platyrrhines has been reported (Manfredi Romanini 1968; Fontana 1969, 1971). In addition, the existence of a separation in the Pongidae, based on DNA nuclear content, between *Pongo* on the one hand and chimpanzee, gorilla, and man on the other, is correlated with the karyotype of these species (Chiarelli 1962; Klinger et al. 1963). This has recently been confirmed with methods very different from the protein-sequence methods of Goodman et al. (1972).

With regard to the analysis of species formation phenomena, the most interesting result we have obtained from data on DNA nuclear content dosage is a clarification of the relationship between *Galago crassicaudatus* (2n=62) and *Galago senegalensis* (2n=38). We have shown that there is no statistically significant difference between these two species in DNA content (Manfredi Romanini et al. 1972). The previous hypothesis, that species formation occurred by endoreduplication, was not corroborated by this experimental test. Our conclusions are that the difference between the two karyotypes is the result of centric fusion or fission mechanisms which occurred in either the 24 metacentrics of *G. senegalensis* or the 48 acrocentrics of *G. crassicaudatus.*

Unfortunately, this work on primates, which has now been carried on for several years, has met with substantial technical difficulties which must be duly considered, since there may be some criticism of the data we are gradually obtaining.[1] These difficulties concern the possibility of obtaining several specimens of the same population whether it lives free in nature or has become acclimatized and reproductive in captivity. Recently De Boer (1973) in a cytogenetic study on the Loridídae has pointed out the existence of individuals with a different number of chromosomes within an apparently homogeneous population and a continuing polymorphism in other primate populations which are apparently still evolving. In this sense we can well include the cytogenetic data which were obtained by Jones et al. (1973) on different specimens of *Saimiri sciureus* and the latest data of Formenti (1975). The latter author, in a study of the DNA nuclear content of species of the genus *Cercopithecus,* found large significant differences within the subspecies *C. aethiops aethiops* among individuals from different zoos, whereas excellent

[1] The data on DNA nuclear content of lymphocytes were obtained by means of photometry at 550μ. The smears of peripheral blood lymphocytes, dried in the air, were fixed with 10 percent formalin and then colored through Feulgen reaction according to Itikawa and Ogura (1954). Publications on analytical techniques contain descriptions of the types of photometric equipment used, essentially a Zeiss histophotometer, a Deeley's integrator microdensitometer and, more recently, an integrator microdensitometer and interferometer M 86 Vickers. The optical complex used on every occasion was a 10× eyepiece and a 100× lens. The possibility of comparison between the data obtained with the different kinds of photometric equipment is guaranteed, in relative terms, by constant reference to control smears of the same individual man. Our data are always given with reference to the fact that they concern the DNA fraction which results in Feulgen positive staining under the abovementioned technical conditions. The problem of the ratio between this value and the total DNA present has not yet been solved. It is part of a program of research into the real quantitative value of the Feulgen reaction, which has been carried on in our institute for several years.

agreement was found among individuals of the same populations, as shown in Figure 1. Unfortunately, he did not have the chromosomal plates at his disposal, but it is clear that when the DNA nuclear content varies from 8.21 pg., for *C. aethiops tantalus,* to 9.50 pg., for *C. griseoviridis* (a difference of 15 percent), among individuals which are classified by zoologists as belonging to the same subspecies, *C. aethiops aethiops*, we must suppose that a course of differentiation must have occurred which was not expressed in the karyotype itself. Our institute's research from 1966 to the present has always been done by comparing data obtained with the same lot of Feulgen stain and calculated in comparison with a control human blood smear which in recent years was not only always a male of the *Homo sapiens* species but also always the same human male. The existence of this constant control entitles us to consider valid, in a relative sense, our quantitative ratios for the different groups of primates. For the same reason, we are compelled to be critical of values showing differences among individuals which were obtained previous to the use of such specific controls. Thus, recently, data on nuclear DNA content of an individual of *Hylobates lar,* obtained with photometry in 1973 by one of us (DeStefano, unpublished data) with the same control as those of Porcelli (1971a), revealed a DNA nuclear content slightly less

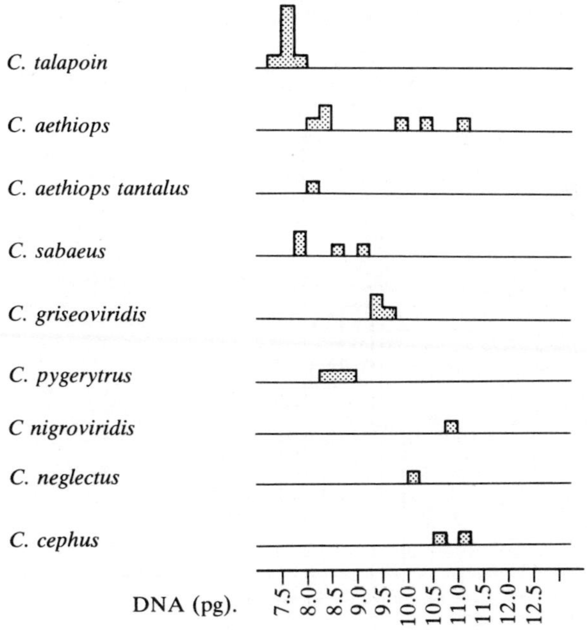

Figure 1. Distribution of DNA (Feulgen positive nuclear content) in various species of genus *Cercopithecus* (from Formenti 1975)

than man's (8 percent less), in contrast to those of Porcelli, in three specimens of which the nuclear DNA is 26.7 percent less than the human DNA. It should be noted that the specimens studied by Porcelli came from the zoo of Rome, while ours, supplied by Professor B. Chiarelli, came from the Primatology Center of Turin.

PRESENT STATE OF KNOWLEDGE ABOUT THE "GENOME SIZE" OF PRIMATES

Even considering the above-mentioned technical criticisms, we want to emphasize the fact that the data on the DNA nuclear content are valid when a great number of specimens for the same species are available. One can then use mean values for a species, which include the usually small variations among individuals of the same population (DeStephano and Scalmana, photometric study of human lymphocytes, in progress). The final variabilities which are bound to differences among populations can then be ascertained. In the outline described by one of us (Manfredi Romanini 1972) and shown in Figure 2, although the more lateral data are questionable, this procedure is exemplified. (The questionable data are for *Tarsius* [Porcelli 1971b], of which we had only one specimen, and for *Hylobates*, which the recent data of DeStefano would shift to the right side of the histogram.) If we include the recent data of Formenti, the

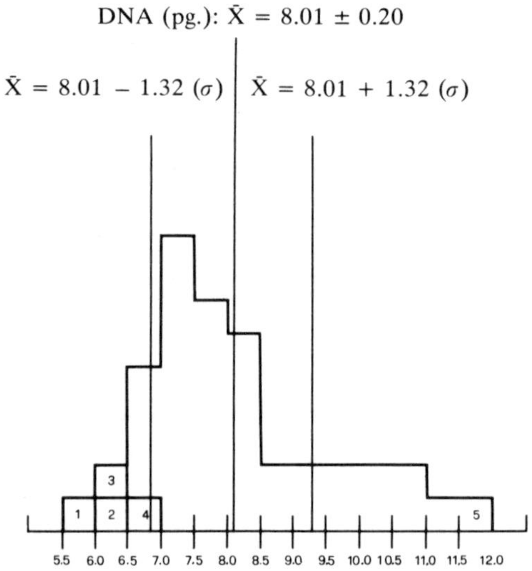

Figure 2. DNA nuclear content (pg.) in primate lymphocytes: (1) *Hylobates agilis;* (2) *Hylobates lar* (Porcelli 1971a); (3) *Symphalangus syndactylus*; (4) *Hylobates lar* (DeStefano, unpublished data); (5) *Tarsius syrichta*

distribution is limited to a narrow area around a mean value of 7.97 pg. This value has been recalculated in pg., assigning a value of 7.3 pg. to man, according to Bachmann (1972). From a general view, it appears that the variability of DNA content in primates varies from a minimum of 6.10 in *H. lar* to a maximum of 11.18 pg. in *T. syrichta* (a 50 percent increase in the latter over the former, or a 2:1 ratio). This variation is remarkably lower than that of the chromosome numbers, which goes from 20 for *Callicebus torquatus* to 80 for *Tarsius syrichta* or a 4:1 ratio. Kleinfeld and Chu (1958) had already reported the noncorrelation between chromosome number and DNA nuclear content, while the lack of correlation between DNA content and Matthey's fundamental number was pointed out by Manfredi Romanini and Fontana (1968). The average data for the separate groups of primates tend to be superimposed between some groups (e.g. Cebidae and Cercopithecidae), whereas in others they reveal a difference that, in the present state of knowledge, could have a very definite taxonomic significance, e.g. between Lemuridae and Galagidae. We have included in this our data on *Perodicticus potto*, of which we have studied only two specimens, but we are in the process of proving this out by studying new specimens of several species of lemur, as shown in Figure 3. These data will have significance from a cytogenetic point of view as well (see the discussions concerning the mechanisms of species formation in Chu and Swomley 1961 and Egozcue 1969), especially with reference to the Lemuridae, which are the only primates who have microchromosomes.

Regarding the distribution of values within Cercopithecoidea, it is well known that this subfamily includes groups of populations whose habitats are very different. The substantial constancy of the chromosome number in the whole group, with the exception of the genus *Cercopithecus*, does not have an exact correspondence with the DNA content. In addition, comparison of species with an equal chromosome number (2n=42) in which the DNA content varies by 40 percent (e.g. *Macaca* and *Cercocebus*) emphasizes the importance of the "genome size" quantitative dosage for an acceptable interpretation of cytogenetic data.

A particular case within the group Cercopithecoidea is the genus *Cercopithecus*. Chiarelli (1964) discussed the variability of the chromosome number, which he expressed in terms of increase; the latest data of Formenti (1975) give support to this idea. A generic correlation between the DNA nuclear content and the genomic length, as it was measured by Chiarelli and Vaccarino (1964), appears to exist. Unfortunately the comparison is at present limited to *Cercopithecus talapoin* (2n=54), whereas the most recent taxonomy refers to the species *C. aethiops* (2n=66). In these the DNA nuclear content is about 7.65 pg. in *C. talapoin*, about 8.83 pg. for the group of *C. aethiops aethiops* subspecies, and about 10.87 for *C. cephus*, as shown in Table 1. From Formenti's data

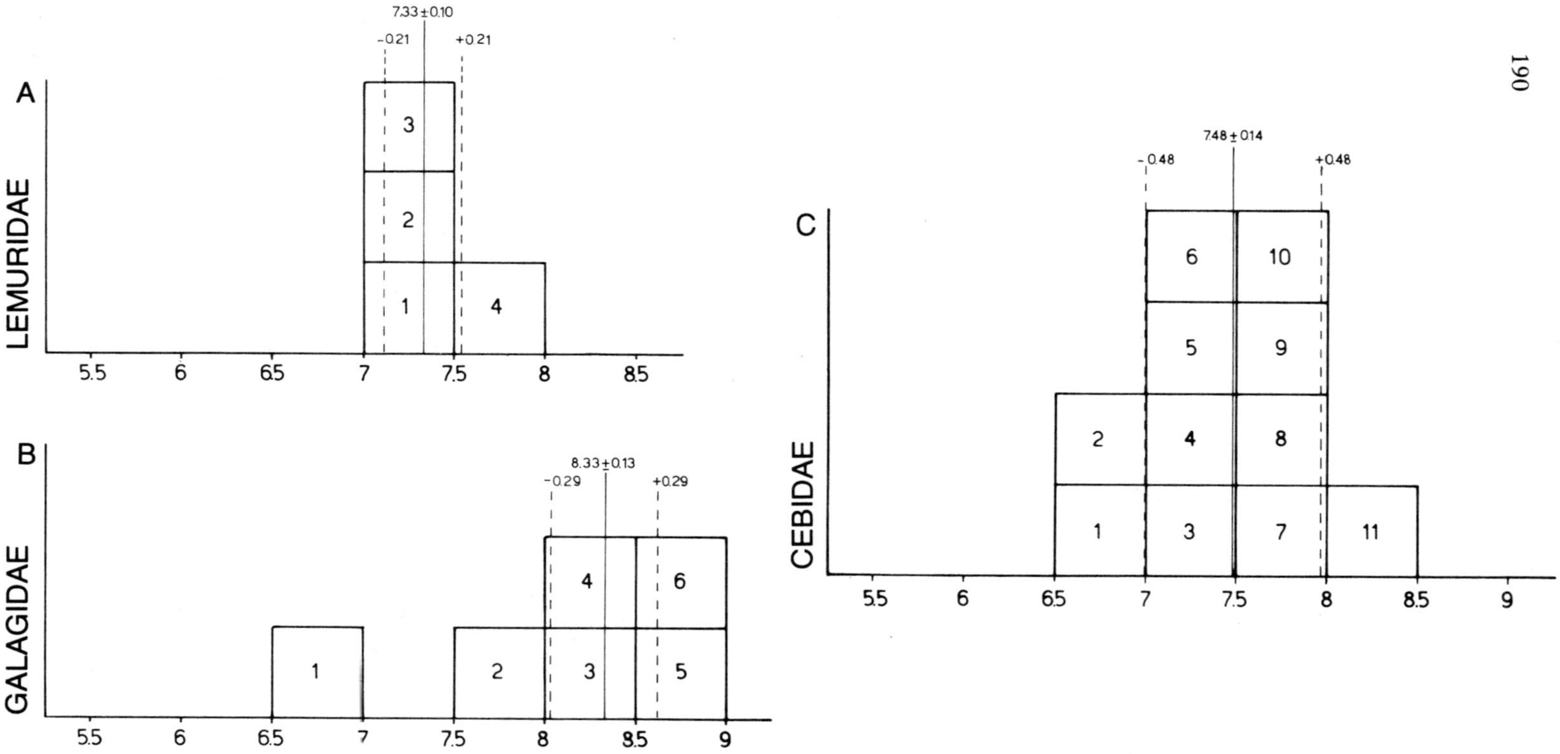

Figure 3. DNA nuclear content (pg.): (A) Lemuridae: (1) *Lemur mongoz*: (2) *L. catta*; (3) *L. macaco*; (4) *L. fulvus*; (B) Galagidae and Lorisidae: (1) and (6) *Perodicticus potto*; (2) and (4) *Galago crassicaudatus*; (3) and (5) *Galago senegalensis*; (C) Cebidae: (1) *Ateles geoffroyi*; (2) *Saimiri sciureus*; (3) *Cebus nigrivittatus*; (4) *Cebuella pygmaea*; (5) *Ateles paniscus*; (6) *Cebus albifrons*; (7) *Alouatta palliata*; (8) *Ateles belzebuth*; (9) *Cebus apella*; (10) *Cebus capucinus*; (11) *Alouatta caraya*:

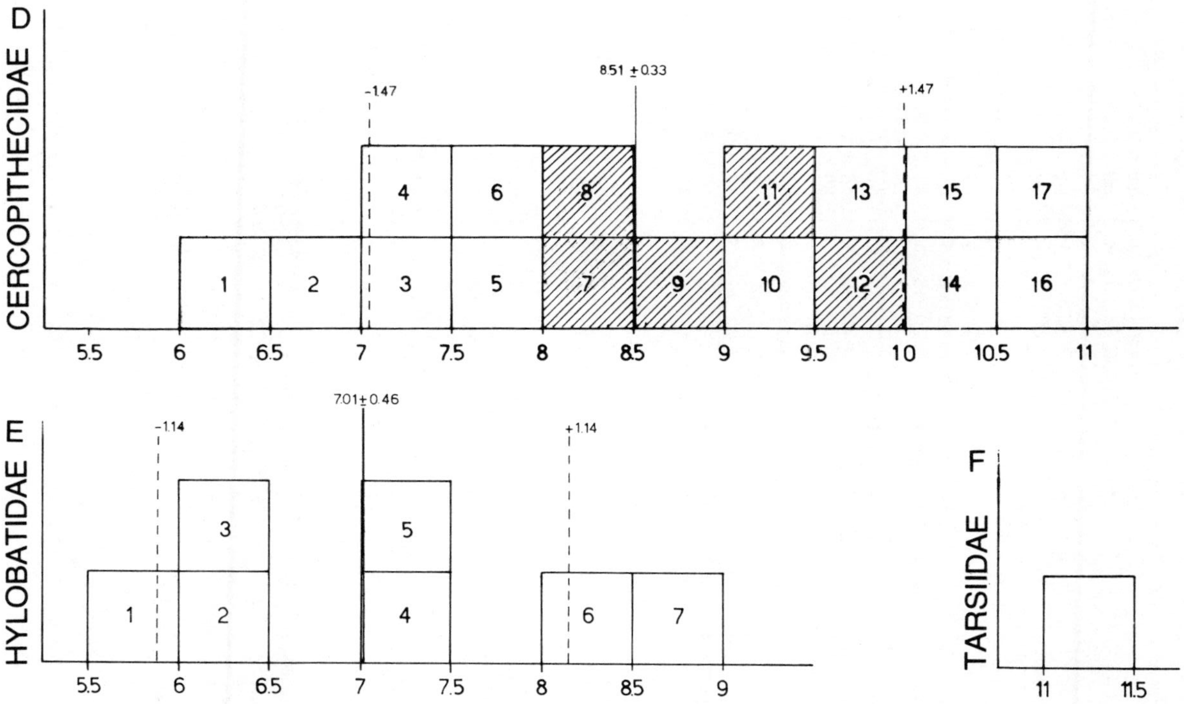

(D) Cercopithecidae: (1) *Macaca silenus*; (2) *Macaca mulatta*; (3) *Cercopithecus patas*; (4) *Papio hamadryas*; (5) *Colobus satanas*; (6) *Cercopithecus talapoin*; (7) *Cercopithecus aethiops tantalus*; (8) *Cercopithecus sabaeus*; (9) *Cercopithecus pygerythrus*; (10) *Nasalis larvatus*; (11) *Cercopithecus aethiops*; *(12) Cercopithecus griseoviridis*; (13) *Cercocebus galeritus*; (14) *Cercopithecus neglectus*; (15) *Cercocebus torquatus*; (16) *Cercopithecus nigroviridis*; (17) *Cercopithecus cephus*; (E) Hylobatidae: (1) *Hylobates agilis*; (2) *Hylobates lar*; (3) *Symphalangus syndactylus*; (4) *Hylobates lar* (DeStefano, unpublished data); (5) *Gorilla gorilla*; (6) *Pan troglodytes*; (7) *Pongo pygamaeus*; (F) Tarsiidae: *Tarsius syrichta*

Table 1. Chromosomal formula, values of DNA nuclear content, and nuclear area

Species	2n	M	S	A	nFa	X	Y	DNA ± s.d. (arbitrary units)	DNA (pg.)
Lemur catta	56	–	10	44	64	A	A	12.18 ± 0.11	7.19
Lemur fulvus	48–60	–	16	30	62	A	A	12.80 ± 0.16	7.55
Lemur macaco	44	–	20	22	62	A	A	12.68 ± 0.16	7.48
Lemur mongoz	58	–	4	52	60	A	A	12.07 ± 0.10	7.11
Galago senegalensis	38	–	24	12	60	S	S	13.97 ± 0.09	8.24
Tarsius syrichta	80	–	–	–	–	–	–	18.95 ± 0.77	11.18
Cebuella pygmaea	44	4	28	10	–	S	A	12.31 ± 0.12	7.26
Cebus albifrons	54	4	16	32	72	S	A	14.07 ± 0.13	8.31
Cebus apella	54	4	20	28	76	A	A	13.01 ± 4.19	7.71
Cebus capucinus	52	4	14	34	70	S	A	13.45 ± 4.27	7.93
Cebus nigrivittatus	–	–	–	–	–	–	–	12.02 ± 0.19	7.09
Saimiri sciureus	44	4	26	12	72	S	A	11.66 ± 0.13	6.87
Alouatta caraya	52	4	16	30	70	S	A	13.59 ± 3.05	8.01
Alouatta palliata	–	–	–	–	–	–	–	12.82 ± 0.70	7.56
Ateles belzebuth	34	–	30	2	62	S	A	12.82 ± 0.15	7.56
Ateles geoffroyi	34	–	30	2	62	SA–S		11.48 ± 0.12	6.77
Ateles paniscus	34	–	30	2	62	SA–S		12.26 ± 0.20	7.23
Cercopithecus aethiops	60	–	36	22	94	SA–S		15.34 ± 0.11	9.35
Cercopithecus aethiops tantalus	60	–	–	–	–	S	S	13.89 ± 0.13	8.21
Cercopithecus cephus	66	–	48	16	112	S	–	18.40 ± 0.08	10.87
Cercopithecus griseoviridis	60	–	–	–	–	–	–	16.09 ± 0.07	9.50
Cercopithecus neglectus	58–60–62	–	–	–	–	–	–	17.18 ± 0.09	10.14
Cercopithecus nigroviridis	60	–	36	22	–	S	A	18.32 ± 0.09	10.82
Cercopithecus patas	54	–	38	14	90	SA–S		12.43 ± 0.24	7.33
Cercopithecus pygerythrus	58 ?	–	–	–	–	–	–	14.78 ± 0.06	8.73
Cercopithecus sabaeus	60	–	36	22	–	S	S	14.19 ± 0.07	8.38
Cercopithecus talapoin	54	–	38	14	–	S	A	12.95 ± 0.02	7.65
Cercocebus galeritus	42	–	40	–	80	S	S	16.72 ± 0.29	9.86
Cercocebus torquatus	42	–	40	–	80	S	–	17.31 ± 0.29	10.22
Macaca mulatta	42	–	40	–	80	S	S	11.11 ± 0.26	6.55
Macaca silenus	42	–	40	–	80	S	S	11.24 ± 0.18	6.63
Papio hamadryas	42	–	40	–	80	S	S	12.48 ± 0.48	7.36
Colobus satanas	44	–	42	–	84	S	–	12.72 ± 0.13	7.51
Nasalis larvatus	48	–	46	–	92	S	–	15.27 ± 0.19	9.01
Hylobates agilis	44	–	42	–	84	S	–	10.11 ± 0.11	5.78
Hylobates lar	44	–	42	–	84	S	S	10.35 ± 0.16	6.10
Symphalangus syndactylus	50	–	46	2	94	S	S	10.54 ± 0.22	6.21
Pongo pygmaeus	48	–	26	20	72	S	S	14.50 ± 8.40	8.55
Pan troglodytes	48	–	34	12	80	S	A	13.60 ± 4.60	8.03
Gorilla gorilla	48	–	30	16	76	S	S	12.62 ± 7.30	7.44
Homo sapiens	46	–	34	10	78	S	A	12.36 ± 0.11	7.30

an important element emerges supporting the new taxonomic rearrangement of the genus *Cercopithecus:* the variability among the subspecies of *aethiops* which modern taxonomy includes in the genus *Cercopithecus aethiops (C. aethiops tantalus, C. griseoviridis, C. sabaeus, C. pygerythrus)* does not exceed the variability among populations of the same subspecies *aethiops aethiops*, considering that the average value of this group is midway between *C. talapoin* and *C. cephus*.

CONCLUSIONS

At present, the value of the DNA nuclear content in primates is clearly set around a value of 7.97 pg. ± 0.20; the range is 51.61 percent. The DNA content is not correlated with variations of chromosome numbers or with Matthey's fundamental number. This value may have an exact taxonomic significance at different levels. This is clear in the comparison between Lemuridae and Galagidae and in confirming the inclusion in the species *aethiops* the numerous subspecies which until now were considered separate from it. On the other hand, it emphasizes differences which are still cytogenetically unproven, e.g. in the comparison among the various subfamilies of the family Cercopithecidae, all with 42 chromosomes and formed with submetacentric one (Egozcue 1969). No new data is added to those which are already present in the literature for the group on Pongidae; the fact that Hylobatidae is separate from Pongidae according to their DNA content is in contrast to cytogenetic results. Finally, we think it is important to conclude this critical revision concerning DNA nuclear content in primates by making it clear that from now on the research on DNA nuclear content in this group needs to be scheduled in much wider terms than it has been previously. It will be necessary to do several blood smears of several individuals coming from natural habitats or from different zoos, at least for the more common species. In addition, it would be desirable if, parallel with our research on circulating lymphocytes, we could have metaphasic plates at our disposal. It is true that the DNA content we measured as positive Feulgen material is only one of the instruments for the interpretation of the genome characteristics, and it cannot be discussed apart from all the other qualitative and quantitative information on the subject.

REFERENCES

BACHMANN, KONRAD

1972 Genome size in mammals. *Chromosoma* (Berlin) 37:85–93.

BENIRSCHKE, KURT, DORIS H. WURSTER, R. J. LOW

1970 The chromosome complement of the Aerdvark, *Orycteropus afer. Chromosoma* (Berlin) 31: 68–78.

CAPANNA, ERNESTO, MARIA GABRIELLA MANFREDI ROMANINI

1971a Nuclear DNA content and morphology of the karyotype in certain palearctic Microchiroptera. *Caryologia* 24(4):471–482.

1971b Contenuto in DNA nei nuclei postcinetici in due popolazioni chromosomicamente differenti di *Rattus rattus* L. *Rendiconti Accademia Nazionale dei Lincei* 51:105–110.

1973 Contenu en ADN des noyaux postkinétiques et évolution du caryotype chez les chiroptères. *Periodicum Biologorum* 75:55–60.

CHIARELLI, BRUNETTO

1962 Comparative morphometric analysis of the primate chromosomes. I: The chromosomes of the anthropoid apes and man. *Caryologia* 15:99–121.

1964 Incremento genomico nel genere "*Cercopithecus*." *Archivio per l'Antropologia e la Etnologia* 94.

1968 Chromosome polymorphism in the genus *Cercopithecus*. *Cytologia* 33:1–16.

CHIARELLI, BRUNETTO, C. VACCARINO

1964 Cariologia ed evoluzione del genere *Cercopithecus*. *Atti Associazione Genetica Italiana* 9:329–339.

CHU, ERNEST H. Y., BARBARA A. SWOMLEY

1961 Chromosomes of lemurine lemurs. *Science* 133:1,925–1,926.

DE BOER, LOEBERT E. M.

1973 Cytotaxonomy of the Lorisoidea (primates: Prosimii) 1: Chromosome studies and karyological relationships in the Galagidae. *Genetica* 44:407–421.

EGOZCUE, J.

1969 "Primates," in *Comparative mammalian cytogenetics*. Edited by K. Benirschke, 357–389. Berlin: Springer.

FONTANA, FRANCESCO

1969 Contenuto in DNA nucleare in proteine totali ed aree nucleari nei linfociti di alcune specie di Ceboidea. *Rivista di Antropologia* 56:219–224.

1971 "The nuclear DNA content, the nuclear area and the amounts of total proteins in the lymphocytes from some species of South American monkeys," in *Proceedings of the Third International Congress of Primatology, Zurich, 1970*, volume two: 153–157. Basel: S. Karger.

FORMENTI, DANIELE

1975 Nuclear DNA content in some species of the genus *Cercopithecus* (primates: Cercopithecidae). *Genetica* 45:307–313.

GOODMAN, MORRIS, JOHN BARNABAS, WILLIAM G. MOORE

1972 Man, the conservative and revolutionary mammal. Molecular findings on this paradox. *Journal of Human Evolution* 1:663–686.

ITIKAWA, O., Y. OGURA

1954 The Feulgen reaction after hydrolysis at room temperature. *Stain Technology* 19:13–15.

JONES, T. C., R. W. THORINGTON, M. M. HU, E. ADAMS, R. W. COOPER

1973 Karyotypes of squirrel monkeys (*Saimiri sciureus*) from different geographic regions. *American Journal of Physical Anthropology* 38:269–278.

KLEINFELD, R., E. H. Y. CHU

1958 DNA determination of kidney cell cultures of three species of monkeys. *Cytologia* 23:452–459.

KLINGER, HAROLD P., JOHN L. HAMERTON, DAVID MUTTON, ERNST M. LANG

1963 "The chromosomes of the hominoidea," in *Classification and human evolution*. Edited by S. L. Washburn. Chicago: Aldine.

KOHNE, DAVID E., JOHN A. CHISCON, BILL H. HOYER

1972 Evolution of primate DNA sequences. *Journal of Human Evolution* 1:627–644.

MANFREDI ROMANINI, MARIA GABRIELLA

1967 Deoxyribonucleic acid amount in the lymphocytes of anthropoid apes

(First Congress of the International Primatological Society, Frankfurt, 1966). *Neue Ergebnisse der Primatologie* 8(183):167–170.

1968 Dati sul contenuto in DNA, in proteine e sul volume dei linfociti di alcune specie di scimmie platirrine (Atti VIII Congresso Nazionale della Società Italiana di Istochimica, Siena, 1967). *Rivista di Istochimica Normale e Patologica* 14:2.

1972 Nuclear DNA content and area of primate lymphocytes as a cytotaxonomical tool. *Journal of Human Evolution* 1:23–40.

MANFREDI ROMANINI, MARIA GABRIELLA, LOEBERT E. M. DE BOER, BRUNETTO CHIARELLI, SUSANNA TINOZZI MASSARI

1972 DNA nuclear content in the cytotaxonomy of *Galago senegalensis* and *Galago crassicaudatus*. *Journal of Human Evolution* 1:437–476.

MANFREDI ROMANINI, MARIA GABRIELLA, FRANCESCO FONTANA

1968 Nuclear DNA content (a.u.) in some species of Lemuroidea. *Rivista di Antropologia* 55:85–94.

MANFREDI ROMANINI, MARIA GABRIELLA, MARIELLA MINAZZA, ERNESTO CAPANNA

1971 DNA nuclear content in lymphocytes from *Mus musculus* L. and *Mus poschiavinus* Fatio. *Bollettino di Zoologia* 38:321–326.

PORCELLI, FRANCA

1971a "Nuclear areas and DNA content of lymphocytes in Hylobatinae" in *Proceedings of the Third International Congress of Primatology, Zurich, 1970,* volume two: 149–152. Basel: S. Karger.

1971b "Nuclear area and DNA content of lymphocytes in *Tarsius syrichta* L," in *Proceedings of the Third International Congress of Primatology, Zurich, 1970*, volume two: 158–160. Basel: S. Karger.

Bands or Chromomeres: A Functional Interpretation of Banding and Speculation About a Model of Nucleus and Chromosome Organization in the Eukaryots

B. CHIARELLI

Since 1970, our attention in chromosome banding has been focused on its functional and ultrastructural interpretation rather than on its benefits to the study of human cytogenetics and cytotaxonomic problems (Chiarelli and Sarti Chiarelli 1972). Only under pressure from other researchers did we develop a comparison of human and chimpanzee chromosomes (Lin et al. 1973).

As discussed in previous papers (Chiarelli et al. 1973), I interpreted the bands as the results of the denaturation process of the proteins constituting the chromosomes, which can eventually reveal the chromomeric structure of the early mitotic or leptotene chromomeres.

On the basis of this interpretation and on observations of optical and electron microscopy available in the literature (DuPraw 1970; Chiarelli 1970, 1974; Englehardt and Pusa 1972; Hoskins 1968; Ardito 1973), I decided to propose a model of chromosome organization which presumes to explain the whole structure of the interphase nucleus and the chromosome organization process from interphase to metaphase for any eukaryot.

The model is broken down into eleven points, most of which are currently only partially supported by experimental evidence. In a following and more extensive paper, supporting evidence from the literature and from personal observation will be provided to substantiate each of these points.

1. Each metaphase chromosome is the result of the condensation of a 200 Å chromatin filament which is composed of the double helix of DNA surrounded by a proteinaceous matrix.

The author is indebted to David Shafer for the stimulating discussions and for his revision of the English text.

2. The filament forms not just a single chromosome but the whole haploid cellular set of chromosomes (genome), which is ring-shaped.

3. Thus in each eukaryot's nucleus would exist two chromosomal structures formed by only two circular chromatin filaments, one of maternal and the other of paternal origin.

4. These filaments are not free in the nucleoplasm but are attached to several points connected with the annuli of the internal surface of the nuclear envelope.

5. The two filaments (maternal and paternal) would have attachments in homologous regions to the inner annuli of the same or an adjacent pore.

6. The starting signal for reduplication is mediated by the nuclear envelope and could originate from the centriole.

7. The reduplication of the chromatin filament would start therefore at these several attachment points near the pores of the nuclear envelope and continue along the filament which extends into the nuclear sap.

8. As reduplication goes on, the newly formed filament would develop a folded structure and produce entangled aggregations in the single filament. These aggregations would grow progressively larger as more of the filament was condensed together. When reduplication is over each section of the chromatin filament included between two attachment points on different annuli would have obtained two joined chromatin complexes. With the optical and electron microscope these chromatin aggregations could be identified as chromomeres.

9. Not all the points of attachment with the nuclear envelope would have the same significance. Most of them would be relatively weak and would detach as soon as reduplication was over. A few, however, would be stronger and would last until the disappearance of the nuclear envelope. In the metaphasic chromosomal structure these latter attachment points could be assigned to centromeric and telomeric regions of each chromosome.

10. At metaphase a number of dense masses of each haploid filament appear as chromosomes.

11. The meiotic process is reductive. Therefore the processes described in points 7 and 8 could take place in reverse order.

This doubtless oversimplified model is supported, as stated, by indirect information drawn from scientific literature and by personal observations, both of which will be expounded in a more detailed paper.

To mention just one of the consequences this model of superchromosomal organization could have in reinterpreting some cytogenetic phenomena, it is worth mentioning the change of chromosome morphology as due not to mutation mechanisms but simply to changes in the significance of one of the filament's attachment points to the nuclear envelope (from normal to telomeric or centromeric), bringing a reinterpretation of such mechanisms as centric fusion, tandem fusion, centric fission, and simple translocation.

If the model is verified, moreover, the interpretation of the organization of the genome, such as chromosome behavior and function, structural mutation, and meiotic process, would be significantly affected.

REFERENCES

ARDITO, G.
1973 Trypsin and trypsin-versene banding on chromosomes. *Bollettino di Zoologia*. 41(2):127–128.

CHIARELLI, B.
1970 Struttura e ultrastruttura dei cromosomi nei mammiferi. *Aggiornamenti di Radiobiologia* 3:9 (IX Congr. Naz. Ass. It. di Radiobiologia Medica, Varese, 1968).

CHIARELLI, B.
1974 Bands or chromomeres: a functional interpretation of banding and speculation about a model of nucleous and chromosome organization in the Eucaryots. *Bollettino di Zoologia* 41:123–125.

CHIARELLI, B.
1976 A circular model of the superchromosomal organization in the Eukaryota. *The Nucleus* 19:71–74.

CHIARELLI B., C. C. LIN
1972 Comparison of fluorescence patterns in human and chimpanzee chromosomes. *Genen en Phaenen* 5:103–106.

CHIARELLI B., M. SARTI-CHIARELLI
1972 Bandeggiamento dei cromosomi con tripsina. *Rivista di Antropologia* 57 (1970–1971) 269–272.

CHIARELLI B., M. SARTI-CHIARELLI, G. ARDITO
1973 Hypothetical mechanism of action involved in producing bands with trypsin on chromosomes. *Bollettino di Zoologia* 40:209–211.

DU PRAW, E. J.
1970 *DNA and chromosomes*. New York: Holt, Rinehart and Winston.

ENGELHARDT P., K. PUSA
1972 Nuclear pore complexes: "Press-stud" elements of chromosomes in pairing and control. *Nature* 200:163.

HOSKINS, G. C.
1968 Sensitivity of microsurgically removed chromosomal spindle fibre to enzyme disruption. *Nature* 217:748.

LIN C. C., B. CHIARELLI, L. E. M. DE BOER, M. N. COHEN
1973 A comparison of the fluorescent karyotypes of the chimpanzee (*Pan troglodytes*) and man. *Journal of Human Evolution* 2:311–321.

SECTION FIVE

Specialized Techniques for Chromosome Studies

Technical Notes on the Workshop on Chromosome Banding

E. RABINO MASSA and G. ARDITO

The recent development of new techniques for chromosome studies reopened the interest in karyological research. The introduction of the banding techniques originated by Caspersson (Caspersson et al. 1968) has certainly contributed a great deal to congenital pathology by more clearly assessing abnormal chromosomal rearrangements in pathological conditions.

Comparison of the chromosomes of different species by means of these new techniques is expected to contribute new emphasis to evolutionary cytotaxonomy.

With such techniques it is possible not only to detect with certainty homologous pairs in a metaphasic plate, but possibly to establish homology among the chromosomes of different species, so that it is possible to reconstruct the history of the changes which occurred among related species and refer them to inversion, translocation, and other mutations which are known to play important roles in the evolution of karyotypes.

A general review of different techniques and preliminary results obtained to date has recently been published (Caspersson and Zech, editors 1973).

The technical procedures used during the workshop follow. The procedure used for each technique includes the original reference for the technique. The techniques used during the workshop, however, did not necessarily follow exactly the methods described in the references.

Q-banding (Lin et al. 1973)
Treat slide in 0.5 percent (w/v) quinacrine dihydrochloride for ten to fifteen minutes at pH 4.3 to 4.5.

The cooperation of Dr. C. C. Lin, Dr. B. Lomholt, and Mr. K. Bahng was indispensable for the success of the workshop.

Wash the stained slide in distilled H_2O (pH 4.3 to 4.5, adjusted with 0.1N HCl) for five to ten minutes (three changes).
Mount the slide in pH 4.3 to 4.5 distilled H_2O and seal the cover slip with wax.
Examine the chromosomes through a fluorescent microscope equipped with darkfield illumination.

G-banding: NaOH (Arrighi and Hsu 1971)
Fix metaphase chromosomes in methanol and acetic acid (3 : 1).
Immerse in 0.07N NaOH for ninety seconds or in 0.04N NaOH for two to three minutes.
Wash slides in several changes of ethanol 70 percent, 95 percent, and 100 percent, then air dry.
Incubate in Sorenseen buffer for twenty-four hours at pH 6.8 at 55°C.
Rinse again in distilled water, dry at 37°C., and stain with Giemsa.

G-banding: trypsin (Seabright 1971; Chiarelli et al. 1972)
Prepare slides with the usual system.
Dip them in a trypsin solution (0.25 g. of Difco trypsin 1:250 in 100 cc. of Hanks solution; pH adjusted to 6.8 to 7.2 with $NaHCO_3$ 0.1M) for ten seconds to one minute at 25°C.
Wash in distilled water.
Dye with Giemsa buffered with Sorenseen buffer (1:4) for five minutes.
Rinse in phosphate buffer, allow to dry.

T-banding (Dutrillaux 1973)
Dip the slides for ten minutes at 87°C. in the following solution:
  94 ml. distilled water +
  3 ml. phosphate buffer, pH 6.7, +
  3 ml. Giemsa (to be added just before use).
Destain according to the following procedure:
  alcohol 90 percent for five minutes;
  alcohol 70 percent for five minutes;
  alcohol 50 percent for thirty minutes;
  distilled water for fifteen minutes.
Stain in acridine orange (5 mg./100 ml.) for twenty minutes.
Rinse in phosphate buffer and mount in buffer.

Q-banding: quinacrine mustard (Caspersson et al. 1971)
Treat slides as follows:
  alcohol 70 percent for five minutes;
  alcohol 50 percent for five minutes;
  distilled water for five minutes (twice);
  McIlvaine buffer, pH 7.0, for five minutes:

McIlvaine buffer: 20 ml. citric acid (monohydrate) 0.1M (21.0g/l) + 80 ml. disodium phosphate (dihydrate) 0.2M (35.6g/l) per 100 ml. of buffer;
quinacrine mustard for twenty minutes:
quinacrine mustard: dilute stock solution in distilled water and keep at 0°C in darkness. For staining, dilute in McIlvaine buffer (pH 7.0), final concentration 50 μg./ml.;
McIlvaine buffer for five minutes (three times); mount in buffer.

R-banding (Dutrillaux and Lejeune 1971)
Treat slide at 95°C. in 2M NaCl buffered with tris-HCl 0.01M, pH 7.0, with additional $MgSO_4$, 0.01M for thirty minutes.
Rinse in phosphate buffer (pH 7.0).
Stain in buffered Giemsa.
Wash and mount in buffer.

C-banding (Arrighi and Hsu 1971)
Treat slides as follows:
hydrolyze at room temperature in HCl 0.2N for $1\frac{1}{2}$ hours;
rinse in distilled water and dry;
wash in 2 × SSC, alcohol 70 percent, and alcohol 95 percent:
2 × SSC: NaCl 0.3M (8.7 g.) + tribasic sodium citrate 0.03M (4.41 g.) in 500 ml. of distilled water;
immerse quickly in NaOH 0.07N;
wash in alcohol 70 percent and 95 percent, then air dry;
place in 2×SSC for five hours at 65°C.;
wash in alcohol 70 percent and 95 percent;
stain in buffered Giemsa (in phosphate buffer, pH 6.8 to 7.20) for three to four minutes;
wash in distilled water.

G-banding: heat (Lomholt and Mohr 1971)
Prepare buffer composed of 0.9M NaCl and 0.1M sodium citrate (adjusted to pH 7.1 by 0.1M citric acid).
Dilute the buffer in three times its volume of redistilled water (the final pH 7.43).
Incubate the prepared slides in the diluted buffer for three hours at 69° to 70°C.
Leave the slides in the diluted buffer to cool to about 30°C.
Stain for twenty minutes in Giemsa (Merck Giemsa solution) diluted 1/10 with 1/15M phosphate buffer (pH 6.95).
Wash twice with distilled water and air dry.

G-banding: tris buffer (Kato and Moriwaki 1972)
Incubate slides in tris buffer at pH 7.0 for twenty-four hours at 55°C.

Rinse in distilled water.
Air dry and stain with Giemsa, pH 7.0.

G-banding (Sumner et al. 1971)
Place slides in 2 × SSC at 65°C. for 1½ hours:
2 × SSC: NaCl 0.3M (8.7 g.) + tribasic sodium citrate 0.03M (4.41 g.) in 500 ml. of distilled water.
Wash in distilled water.
Stain with buffered Giemsa (in phosphate buffer, pH 6.8 to 7.2) for three to four minutes.
Wash in distilled water.

G-banding: pepsin (Rabino Massa, unpublished data)
Dissolve 0.12 g. of pepsin in 50 cc. of KH_2PO_4 (pH 5.6), then incubate at 37°C. for one to two hours.
Rinse in distilled water.
Wash in alcohol 100 percent and stain with Giemsa adjusted to pH 7.0 with Sorenseen.
Rinse and air dry (or dry in alcohol 100 percent).
Mount with DPX.

G-banding: sodium salts (diethylbarbiturate or arsenate) (Rabino Massa, unpublished data)
Keep slides for twenty-four hours in sodium salts solution (2 percent) in Sorenseen buffer at 55°C.
Then rinse in distilled water, dry, and stain with Giemsa.

G-banding: veronal buffer (Rabino Massa, unpublished data)
Keep slides in veronal buffer at pH 7.0 for twenty-four hours at room temperature.
Rinse in distilled water.
Stain with Giemsa (pH 7.0).

REFERENCES

ARRIGHI, F. E., T. C. HSU
1971 Localization of heterochromatin in human chromosomes. *Cytogenetics* 10:81–86.

CASPERSSON, T., S. FARBER, G. E. FOLEY, J. KUDYNOWSKI, E. J. MODEST, *et al.*
1968 Chemical differentiation along metaphase chromosomes. *Experimental Cell Research* 49:219–222.

CASPERSSON, T., G. LOMAKKA, L. ZECH
1971 The 24 fluorescence patterns of the human metaphase chromosomes: distinguishing characters and variability. *Hereditas* 67:89–102.

CASPERSSON, T., L. ZECH, *editors*
1973 *Chromosome identification — technique and application in biology and medicine.* Proceedings of the twenty-third Nobel Symposium, 1972. New York: Academic Press.

CHIARELLI, B., M. SARTI-CHIARELLI
1972 Bandeggiamento dei cromosomi con tripsina. *Rivista di Antropologia* 57:269–272.

CHIARELLI, B., M. SARTI-CHIARELLI, D. A. SHAFER
1972 Chromosome banding with trypsin. *Genetica* 43:190–194.

DUTRILLAUX, B.
1973 Nouveau système de marquage chromosomique: les bandes T. *Chromosoma* 41:395–402.

DUTRILLAUX, B., J. LEJEUNE
1971 Sur une nouvelle technique d'analyse du cariotype humaine. *Comptes Rendues de l'Académie des Sciences* 272:2,638–2,640.

KATO, H., K. MORIWAKI
1972 Factors involved in the production of banded structures in mammalian chromosomes. *Chromosoma* 38:105–120.

LIN, C. C., I. A. UCHIDA, P. F. KRUSSE, M. K. PATTERSON
1973 "Fluorescent banding of chromosomes (Q bands)," in *Methods and applications of tissue cultures*. Edited by Paul F. Kruse, Jr., and M. K. Pattersson Jr., 778–779. New York: Academic.

LOMHOLT, B., J. MOHR
1971 Human karyotyping by heat-Giemsa staining and comparison with fluorochrome techniques. *Nature New Biology* 234(47):109–110.

SEABRIGHT, M.
1971 A rapid banding technique for human chromosomes. *Lancet* 2:971–972.

SUMNER, A. T., H. J. EVANS, R. A. BUCKLAND
1971 New technique for distinguishing between human chromosomes. *Nature New Biology* 232:31–32.

Meiotic Chromosome Preparations

A. FALEK

1. Obtain fresh testicular tissue from biopsy or autopsy.
2. Cut the outer membraneous capsule (albuginea) away from the seminiferous tubules and discard.
3. Place small portions of the tissue in a 1 percent sodium citrate solution and tease out the tubules with teasing needles until all the tissue is used.
4. Decant the sodium citrate solution into a centrifuge tube, taking care not to get the tubules, and centrifuge at 1,000 rpm for ten minutes.
5. Remove the supernatant and add freshly prepared fixative (3:1 methanol and acetic acid) dropwise.
6. Refrigerate for thirty minutes and resuspend the cell button.
7. Centrifuge, remove the supernatant, and add fixative, resuspending the cell button.
8. Repeat step 7 until the supernatant is clear.
9. Drop about three drops of the cell suspension on cold slides and spread by the use of an alcohol flame.
10. Let slides dry and stain the preparations with Giemsa stain.

SECTION SIX

Primate Chromosome Atlas

Introduction

A collection of primate karyotypes is presented in this appendix. The karyotypes are reproduced exactly as they were published in the different journals, books etc. without any attempt to unify both the nomenclature and the chromosome disposition.

We decided to include only normally stained karyotypes in the atlas, even if many karyotypes with G-, Q-, R-, and C-bands appear in the literature. However, as a number of these karyotypes are still largely incomplete (of about 120 published primate karyotypes no more than forty or fifty are presented with bands), we plan to postpone their publication until more data becomes available. The quality of some of the photographs is poor: some of the karyotypes, in fact, were published more than fifteen years ago and it was not possible to obtain the originals. In these cases the karyotype has been photographed directly from the journal or book.

For each karyotype the following information is given:

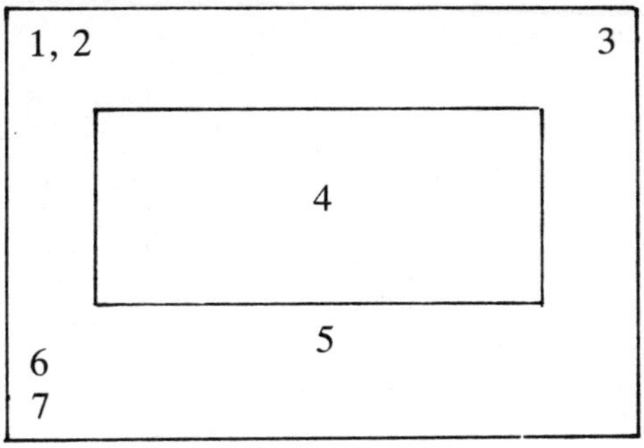

1. Code number of the species
2. Name of the species
3. Diploid number
4. Karyotype
5. Reference of the karyotype presented
6. The author who published the first data on the species
7. The author who published the first photograph of the species.

Obviously 6 and 7 often coincide.

I would like to thank the authors, editors, and publishers for their kind permission to reproduce the previously published photographs.

Table 1. Prosimii

	2n	S–M	A	X	Y
Tupaiidae					
1.1.1.1. *Tupaia glis*	60–62	12–14	44–48	M–S	A
1.1.1.2. *Tupaia montana*	52–68	6	60	M	A
1.1.1.5. *Tupaia minor**	66	–	–	–	–
1.1.3.1. *Urogale everetti*	44	36	8	–	–
Lorisidae					
1.2.1.1. *Loris tardigradus*	62	34–38	26–22	S	S–A
1.2.2.1. *Nycticebus coucang*	50–52	48	–	S	S
1.2.2.2. *Nycticebus pygmaeus**	50	–	–	–	–
1.2.3.1. *Arctocebus calabarensis*	52	50	–	S	S
1.2.4.1. *Perodicticus potto*	62	24	36	S	A
Galagidae					
1.3.1.1. *Galago senegalensis*	36–38	22–24–30	14–12–6	S	S–A
1.3.1.2. *Galago crassicaudatus*	62	6–30	54–30	S	A–S
1.3.1.3. *Galago alleni*	40	20–22	16–18	S	A–S
1.3.1.4. *Galago demidovii*	58	6	50	S	A
Lemuridae					
1.4.1.1. *Microcebus murinus*	66	–	64	A	A
1.4.2.1. *Cheirogaleus major*	66	–	64	A	A
1.4.2.2. *Cheirogaleus medius*	66	–	64	S–M	?
1.4.3.1. *Phaner furcifer*	48	16	30	S	A?
1.4.4.1. *Hapalemur griseus*	54–58	6–10	50–42	A	A
1.4.4.2. *Hapalemur simus*	60	4	54	M	A
1.4.5.1. *Lemur catta*	56	8–10–14	46–44–50	S–A	A
1.4.5.2. *Lemur variegatus*	46	18	26	S	A
1.4.5.3. *Lemur macaco*	44–48–52–58–60	20–16–4	22–30–52 54	A	A
1.4.5.4. *Lemur rubriventer*	50	14	34	A	?
1.4.5.5. *Lemur mongoz*	58–60	4	52–54	A	A
1.4.5.6. *Lemur fulvus*	48–50–60	14	34	A	A
1.4.6.1. *Lepilemur mustelinus*	20–22–26–34–38	2–6–18	26–32	M	A
Indridae					
1.5.1.1. *Propithecus diadema*	42–48	32	8	M	A
1.5.1.2. *Propithecus verreauxi*	48	30–28	16–18	M	A
1.5.2.1. *Avahi laniger**	64	2	60	M	A
1.5.3.1. *Indri indri*	40–44	32	6	M	A
Daubentoniidae					
1.6.1.1. *Daubentonia madagascariensis**	30	–	–	–	–
Tarsiidae					
1.7.1.2. *Tarsius syrichta*	80	12	66	S	A?
1.7.1.3. *Tarsius bancanus*	80	14	66	–	–

The code numbers for the species are the ones used in Chiarelli, B. (1972). *Taxonomic atlas of living primates*. London: Academic Press.

Key:

* No published karyotypes for this species.

2n = Diploid number.

S–M: Number of submetacentric and metacentric chromosomes.

A: Number of acrocentric chromosomes.

X and Y: Morphology of X and Y chromosomes (M = metacentric; S = submetacentric; and A = acrocentric).

1.1.1.1. *Tupaia glis* 2n = 60–62

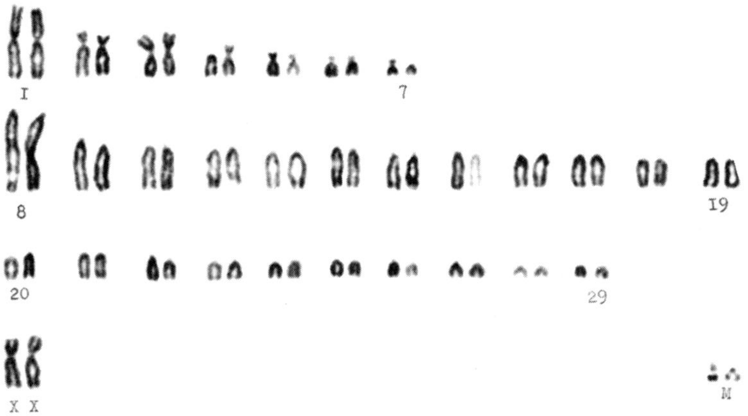

Egozcue, J., et al. (1968). *Folia Primatologica* 8: 152.

Chu, E.H.Y., and Bender, M.A. (1961). Cytogenetics and evolution of primates. *Annals of the New York Academy of Sciences* 102: 253–266.
Klinger, H.P. (1963). The somatic chromosomes of some primates (*Tupaia glis, Nycticebus coucang, Tarsius bancanus, Cercocebus aterrimus, Symphalangus syndactylus*). *Cytogenetics* 2: 140–151.

1.1.1.2. *Tupaia montana* 2n = 52–68

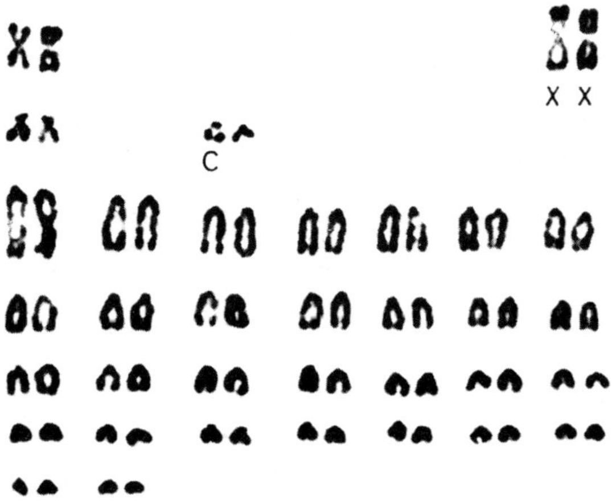

Arrighi, F.E., et al. (1969). *Cytogenetics* 8: 203.

Arrighi, F.E., Sorenson, M.W., and Shirley, L.R. (1969). Chromosomes of the tree shrews (Tupaiidae). *Cytogenetics* 8: 199–208.

1.1.3.1. *Urogale everetti* 2n = 44

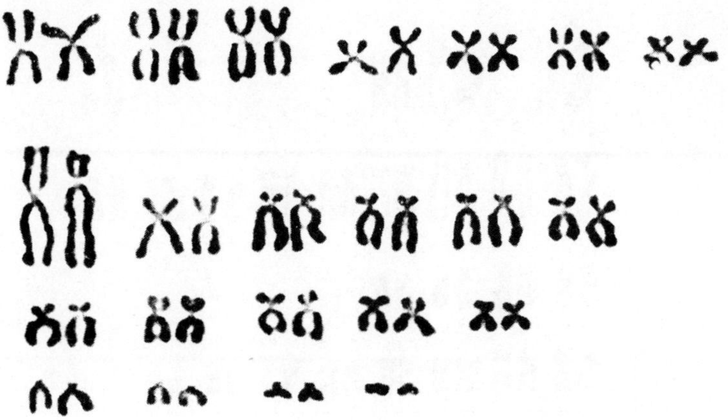

Arrighi, F.E., et al. (1969). *Cytogenetics* 8: 204.

Bender, M.A., and Chu, E.H.Y. (1963). "The chromosomes of primates," in *Evolutionary and genetic biology of the primates*, volume one. Edited by J. Buettner-Janusch, 261–310. New York, London: Academic Press.
Arrighi, F.E., Sorenson, M.W., and Shirley, L.R. (1969). Chromosomes of the tree shrews (Tupaiidae). *Cytogenetics* 8: 199–208.

1.2.1.1. *Loris tardigradus* 2n = 62

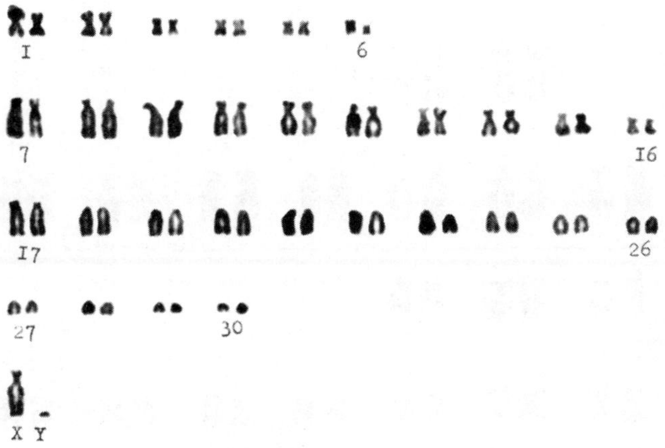

Egozcue, J., et al. (1966). *Mammalian Chromosomes Newsletter* 22: 204.

Egozcue, J. Ushijima, R.N., and Vilarasau de Egozcue, M. (1966). The chromosomes of the slender loris (*Loris tardigradus*). *Mammalian Chromosomes Newsletter* 22: 204.

1.2.2.1. *Nycticebus coucang* 2n = 50–52

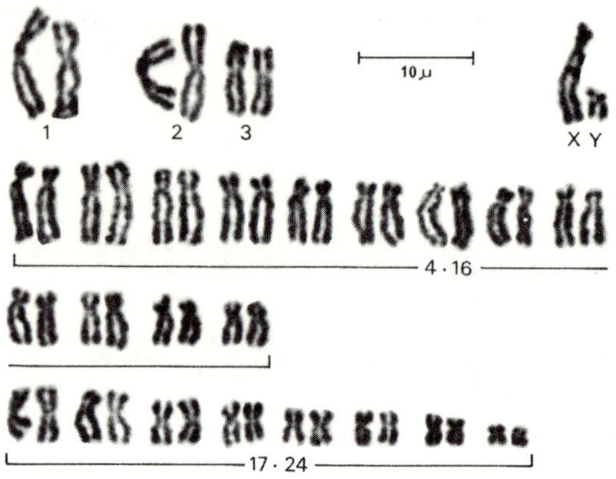

De Boer, L.E.M. (1973). *Genetica* 44: 342.

Bender, M.A., and Mettler, L.E. (1958). Chromosome studies of primates. *Science* 128: 186–190.

Klinger, H.P. (1963). The somatic chromosomes of some primates (*Tupaia glis, Nycticebus coucang, Tarsius bancanus, Cercocebus aterrimus, Symphalangus syndactylus*). *Cytogenetics* 2: 140–151.

1.2.3.1. *Arctocebus calabarensis* 2n = 52

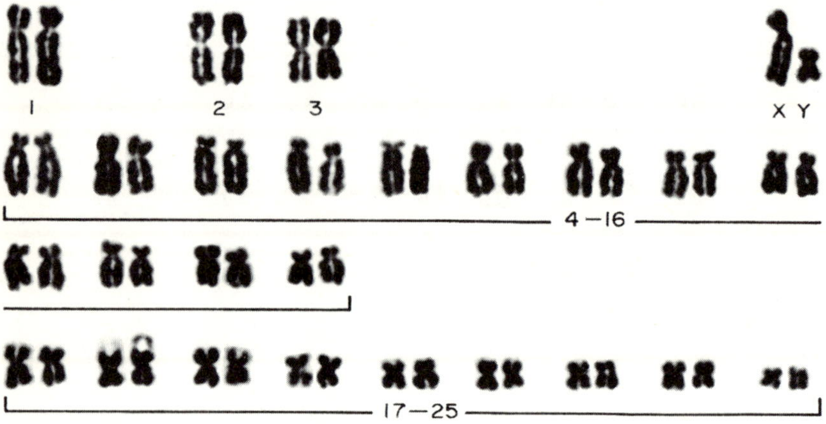

De Boer, L.E.M. (1973). *Journal of Human Evolution* 2(4): 275

Egozcue, J., and Vilarasau de Egozcue, M. (1966). The chromosomes of the angwantibo (*Arctocebus calabarensis*). *Mammalian Chromosomes Newsletter* 20: 53–54.

1.2.4.1. *Perodicticus potto* 2n = 62

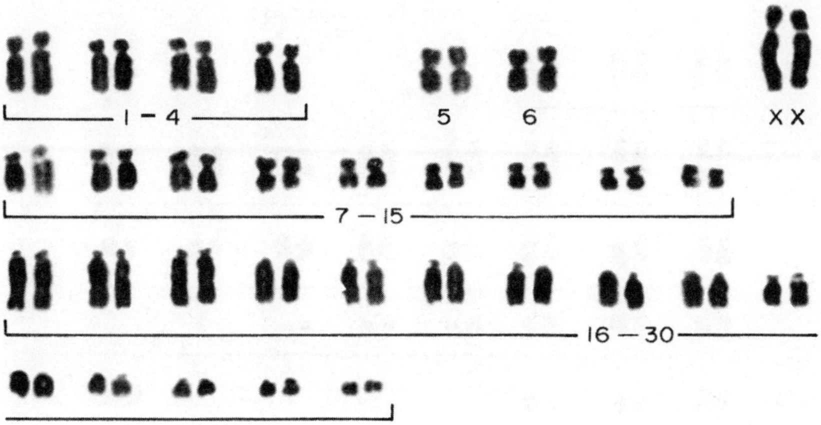

De Boer, L.E.M. (1973). *Journal of Human Evolution* 2(4): 274.

Chu, E.H.Y., and Bender, M.A. (1961). Chromosome cytology and evolution in primates. *Science* 133: 1,399–1,405.

Bender, M.A., and Chu, E.H.Y. (1963). "The chromosomes of primates," in *Evolutionary and genetic biology of the primates*, volume one. Edited by J. Buettner-Janusch, 261–310. New York, London: Academic Press.

1.3.1.1. *Galago senegalensis* 2n = 36–38

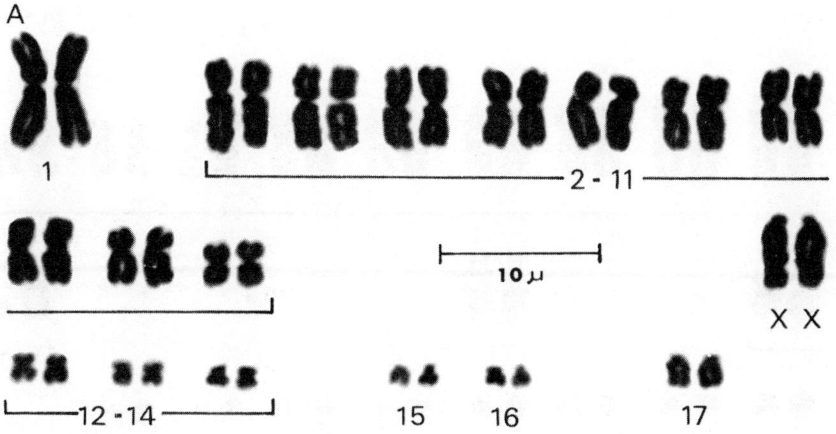

De Boer, L.E.M. (1973). *Genetica* 44: 186.

Matthey, R. (1955). Les chromosomes de *Galago senegalensis* Geoffroy (Prosimii–Lorisidae–Galaginae). *Revue Suisse de Zoologie* (supplement) 62: 190–197.

Chu, E.H.Y., and Bender, M.A. (1961). Chromosome cytology and evolution in primates. *Science* 133: 1,399–1,405.

1.3.1.2. *Galago crassicaudatus* 2n = 62

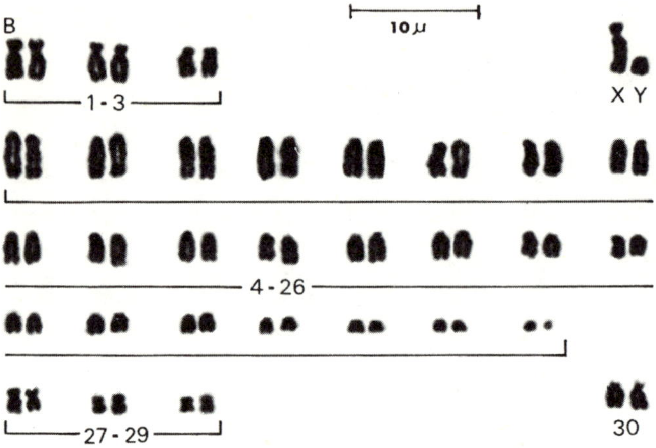

De Boer, L.E.M. (1973). *Genetica* 44: 186.

Chu, E.H.Y., and Bender, M.A. (1961). Chromosome cytology and evolution in primates. *Science* 133: 1,399–1,405.

Bender, M.A. and Chu, E.H.Y. (1963). "The chromosomes of primates," in *Evolutionary and genetic biology of the primates*. Edited by J. Buettner-Janusch, 261–310. London, New York: Academic Press.

1.3.1.3. *Galago alleni* 2n = 40

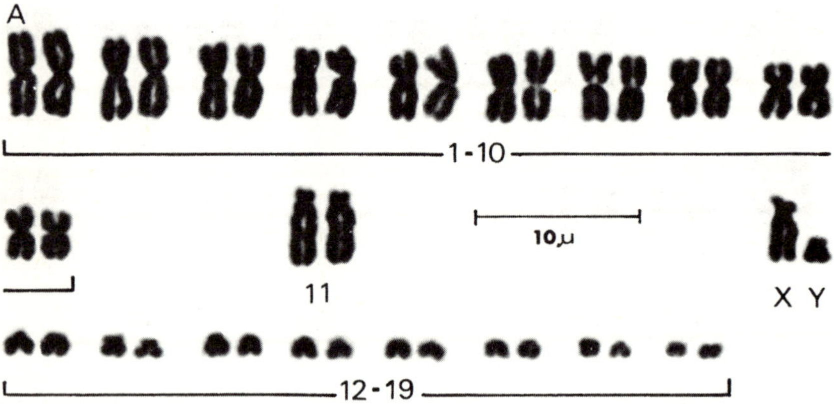

De Boer, L.E.M. (1973). *Genetica* 44: 186.

De Boer, L.E.M. (1972). The karyotype of *Galago alleni* Waterhouse 1837 (2n = 40), compared with that of *Galago senegalensis* Geoffroy 1796 (2n = 38) (primates, Prosimii: Galagidae). *Genetica* 43: 183–189.

1.3.1.4. *Galago demidovii* 2n = 58

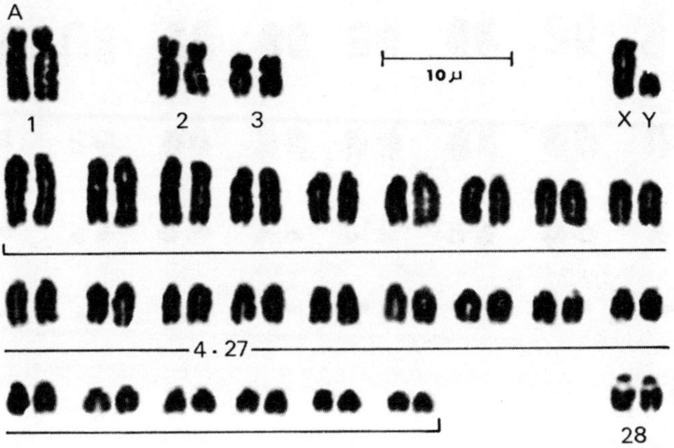

De Boer, L.E.M. (1973). *Genetica* 44: 186.

De Boer, L.E.M. (1972). The karyotype of *Galago demidovii* Fisher 1808 (2n = 58) (primates, Prosimii: Galagidae). *Genen en Phaenen* 15: 19–22.

1.4.1.1. *Microcebus murinus* 2n = 66

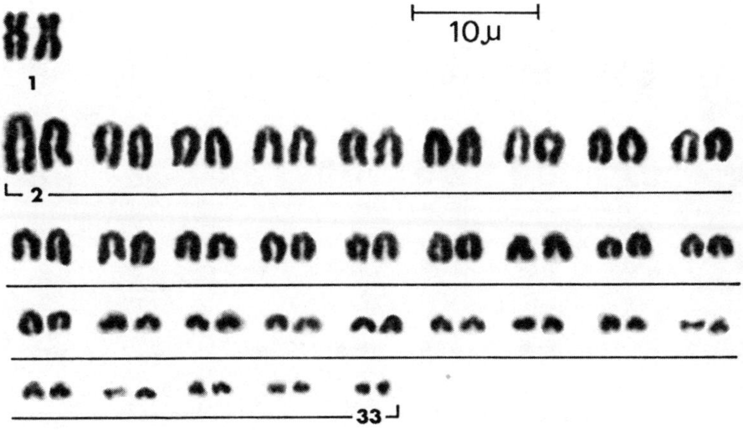

De Boer, L.E.M. – Original.

Chu, E.H.Y., and Swomley, B.A. (1961). Chromosomes of lemurine lemurs. *Science* 133: 1,925–1,926.

De Boer, L.E.M. This volume.

1.4.2.1. *Cheirogaleus major* — 2n = 66

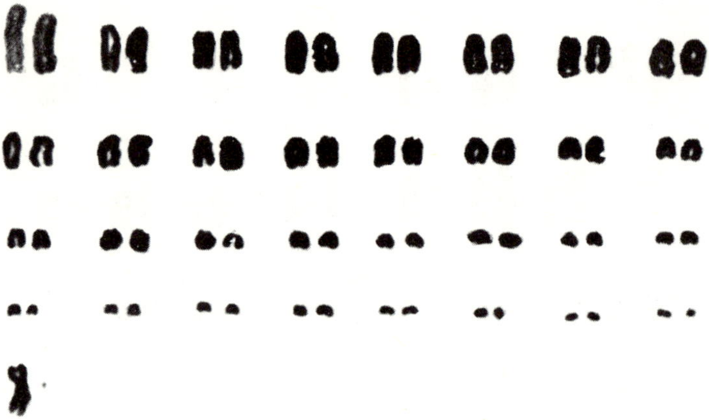

Rumpler, Y., and Albignac, R. (1973). *American Journal of Physical Anthropology* 38(2): 262.

Rumpler, Y., and Albignac, R. (1973). Cytogenetic study of the endemic Malagasy lemurs subfamily Cheirogaleinae Gregory 1915. *American Journal of Physical Anthropology* 38: 261–264.

1.4.2.2. *Cheirogaleus medius* — 2n = 66

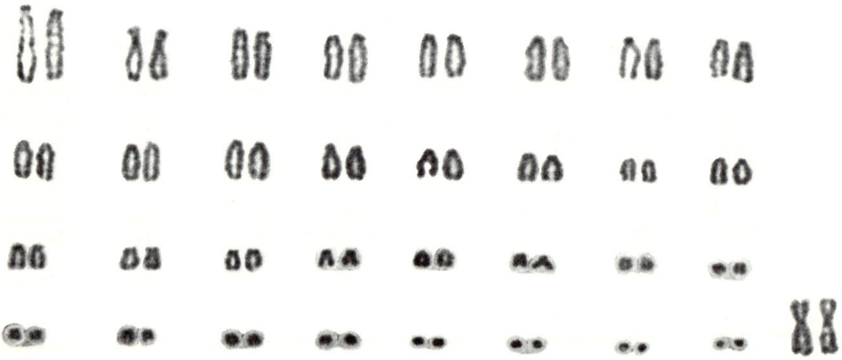

Wurster-Hill, D.H. (1973). *Journal of Human Evolution* 2(4): 261.

Rumpler, Y., and Albignac, R. (1973). Cytogenetic study of the endemic Malagasy lemurs subfamily Cheirogaleinae Gregory 1915. *American Journal of Physical Anthropology* 38: 261–264.

Wurster-Hill, D.H. (1973). Chromosomes of the Lemuridae. *Journal of Human Evolution* 2(4): 259–264.

1.4.3.1. *Phaner furcifer* 2n = 48

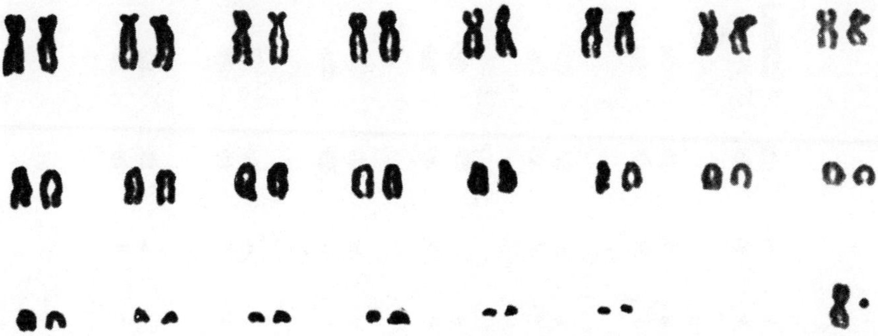

Rumpler, Y., and Albignac, R. (1973). *American Journal of Physical Anthropology* 38(2): 263.

Rumpler, Y., and Albignac, R. (1973). Cytogenetic study of the endemic Malagasy lemurs subfamily Cheirogaleinae Gregory 1915. *American Journal of Physical Anthropology* 38: 261–264.

1.4.4.1. *Hapalemur griseus* 2n = 54–58

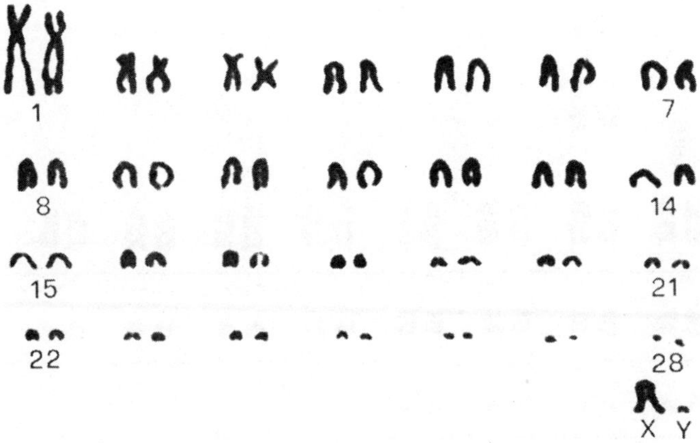

Rumpler, Y., and Albignac, R. (1973). *Journal of Human Evolution* 2(4): 270.

Chu, E.H.Y., and Swomley, B.A. (1961). Chromosomes of lemurine lemurs. *Science* 133: 1,925–1,926.

Bender, M.A., and Chu, E.H.Y. (1963). "The chromosomes of primates," in *Evolutionary and genetic biology of the primates*, volume one. Edited by J. Buettner-Janusch, 261–310. London, New York: Academic Press.

1.4.4.2. *Hapalemur simus* 2n = 60

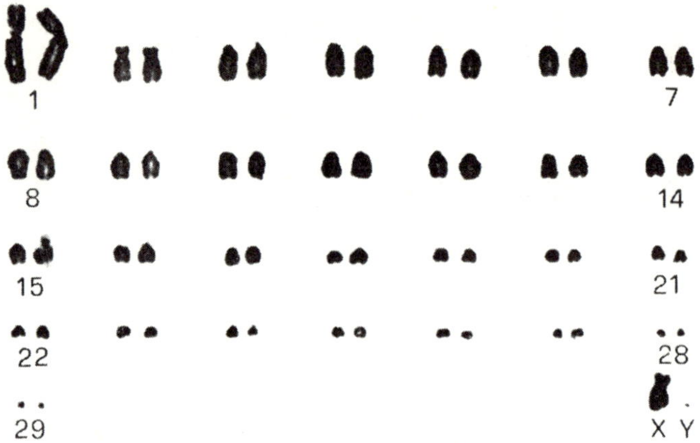

Rumpler, Y., and Albignac, R. (1973). *Journal of Human Evolution* 2(4): 270.

Rumpler, Y., and Albignac, R. (1973). Cytogenetic study of the endemic Malagasy lemur: *Hapalemur* I. Geoffroy, 1851. *Journal of Human Evolution* 2(4): 267–270.

1.4.5.1. *Lemur catta* 2n = 56

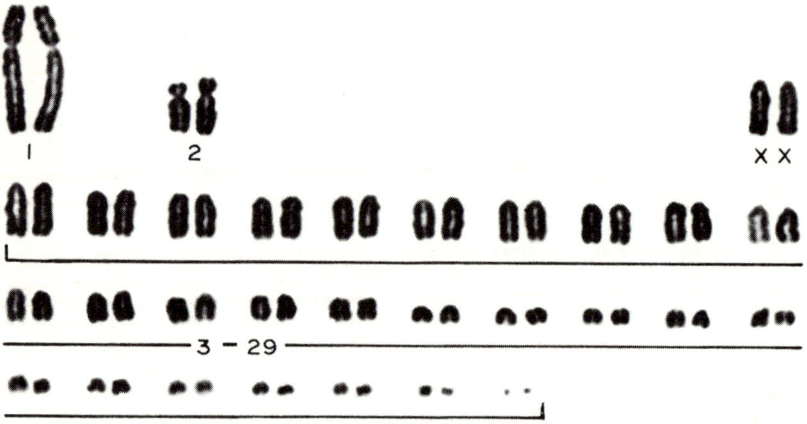

De Boer, L.E.M. (1973). *Journal of Human Evolution* 2(4): 274.

Chu, E.H.Y., and Swomley B.A. (1961). Chromosomes of lemurine lemurs. *Science* 133: 1,925–1,926.

Bender, M.A., and Chu, E.H.Y. (1963). "The chromosomes of primates," in *Evolutionary and genetic biology of the primates.* Edited by J. Buettner-Janusch, 261–310. London, New York: Academic Press.

1.4.5.2. *Lemur variegatus* 2n = 46

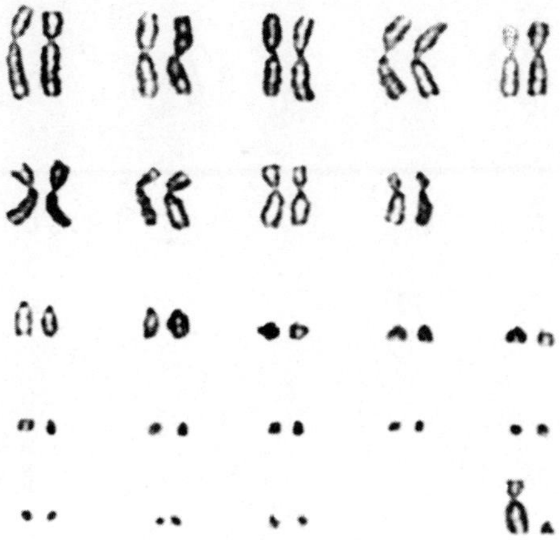

Wurster-Hill, D.H. (1973). *Journal of Human Evolution* 2(4): 262.

Chu, E.H.Y., and Swomley, B.A. (1961). Chromosomes of lemurine lemurs. *Science* 133: 1,925–1,926.

Wurster-Hill, D.H. (1973). Chromosomes of the Lemuridae. *Journal of Human Evolution* 2(4): 259–264.

1.4.5.3. *Lemur macaco* 2n = 44–48–52–58–60

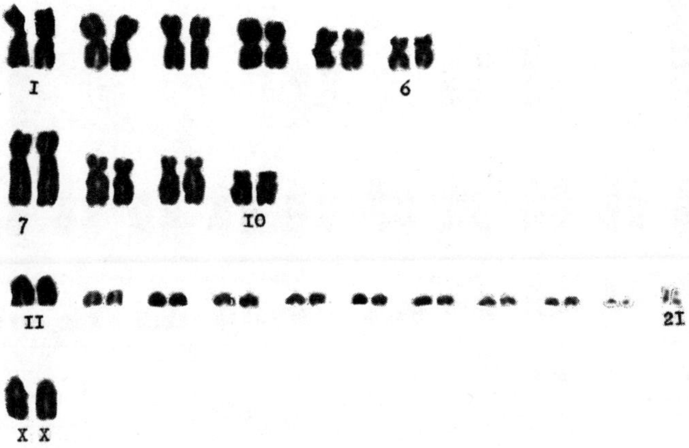

Egozcue, J. (1967). *American Journal of Physical Anthropology* 26(3): 343.

Chu, E.H.Y., and Swomley, B.A. (1961). Chromosomes of lemurine lemurs. *Science* 133: 1,925–1,926.

Chu, E.H.Y., and Bender, M.A. (1961). Chromosome cytology and evolution in primates. *Science* 133: 1,399–1,405.

1.4.5.4. *Lemur rubriventer* 2n = 50

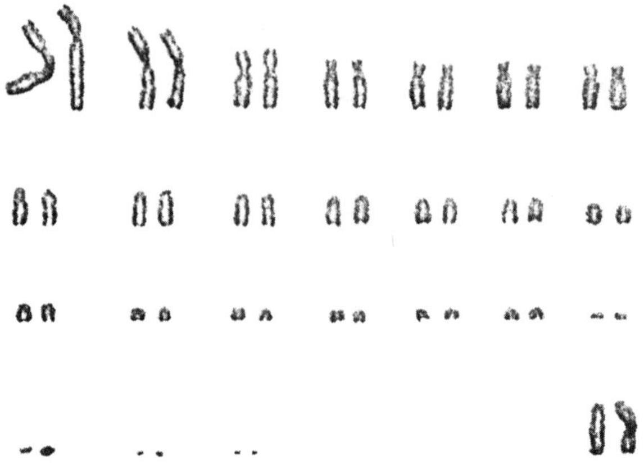

Rumpler, Y., and Albignac, R. (1971). *Comptes Rendus des Séances de la Société de Biologie* 165(3): 746.

Rumpler, Y., and Albignac, R. (1971). Etude cytogénétique de *Varecia variegata* et de *Lemur rubriventer. Comptes Rendus des Séances de la Société de Biologie.* 165(3): 741–747.

1.4.5.5. *Lemur mongoz* 2n = 58–60

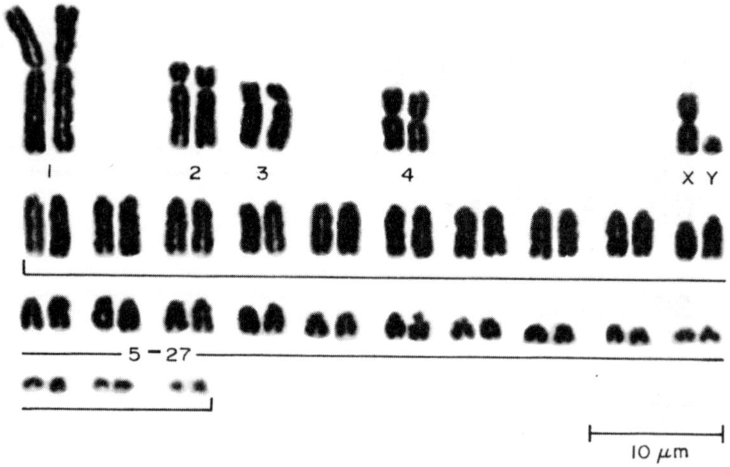

De Boer, L.E.M. (1973). *Journal of Human Evolution* 2(4): 274.

Chu, E.H.Y., and Swomley B.A. (1961). Chromosomes of lemurine lemurs. *Science* 133: 1,925–1,926.

Egozcue, J. (1967). Chromosome variability in the Lemuridae. *American Journal of Physical Anthropology* 26(3): 341–348.

1.4.5.6. *Lemur fulvus* 2n = 48–50–60

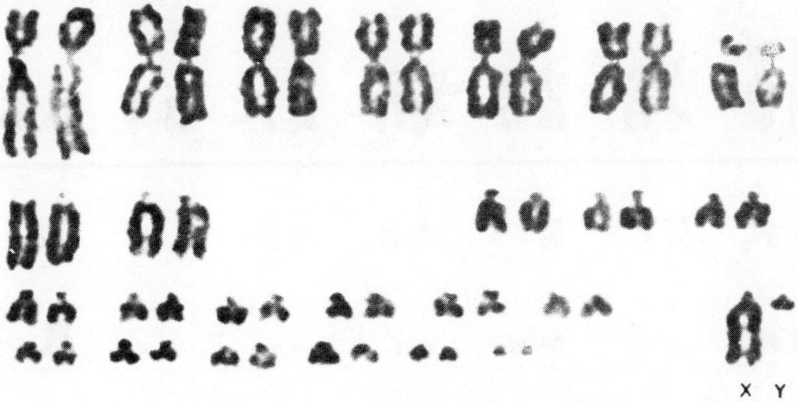

Hamilton, A.E., Buettner-Janusch, J., and Chu, E.H.Y. (1977). *American Journal of Physical Anthropology* 46(3): 399.

Chu, E.H.Y., and Swomley, B.A. (1961). Chromosomes of lemurine lemurs. *Science* 133: 1,925–1,926.

1.4.6.1. *Lepilemur mustelinus* 2n = 20–22–26–34–38

Rumpler, Y. – Original.

Rumpler, Y. (1972). Cytogenetic contributions to a new classification of lemurs. *Research Seminar on Prosimian Biology*. London.

Buettner-Janusch, J., Hamilton, A.E., and Bergeron, J.A. (1973). Chromosomes of lemuriformes. A chromosome complement of *Lepilemur mustelinus* (I. Geoffroy 1851). *American Journal of Physical Anthropology* 39: 1–6.

1.5.1.1. *Propithecus diadema*

2n = 42–48

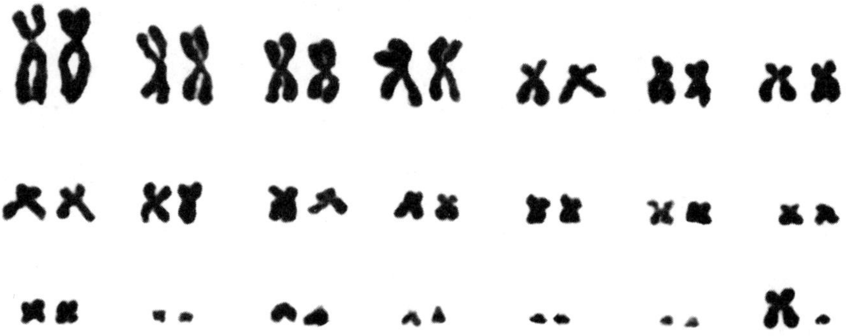

Rumpler, Y., and Albignac, R. This volume.

Rumpler, Y., and Albignac, R. This volume.

1.5.1.2. *Propithecus verreauxi*

2n = 48

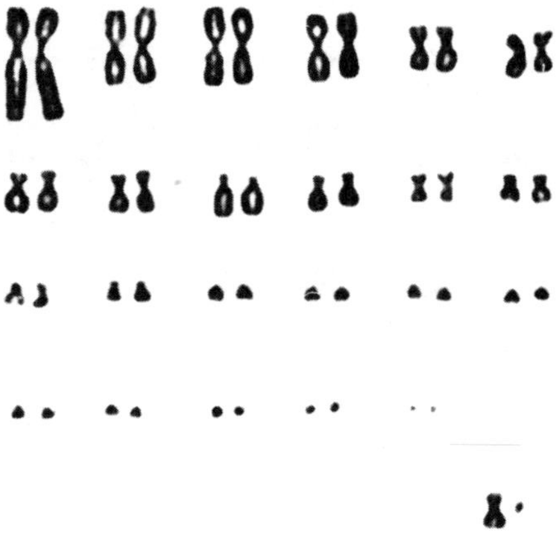

Rumpler, Y. – Original.

Chu, E.H.Y., and Bender, M.A. (1961). Cytogenetics and evolution of primates. *Annals of the New York Academy of Sciences* 102: 253–266.
Rumpler, Y., and Albignac, R. This volume.

1.5.3.1. *Indri indri* 2n = 40–44

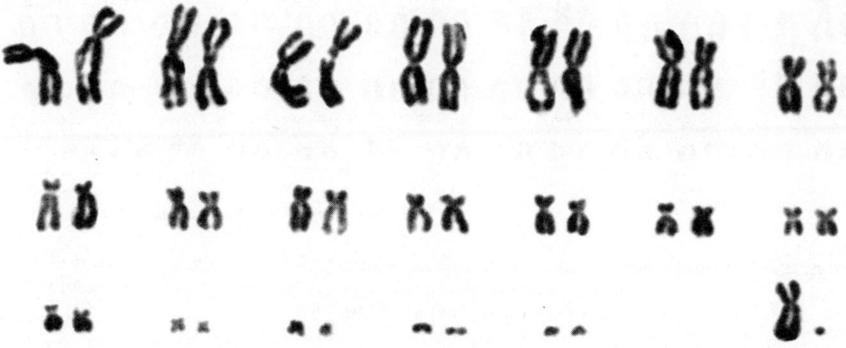

Rumpler, Y., and Albignac, R. This volume.

Rumpler, Y., and Albignac, R. This volume.

1.7.1.2. *Tarsius syrichta* 2n = 80

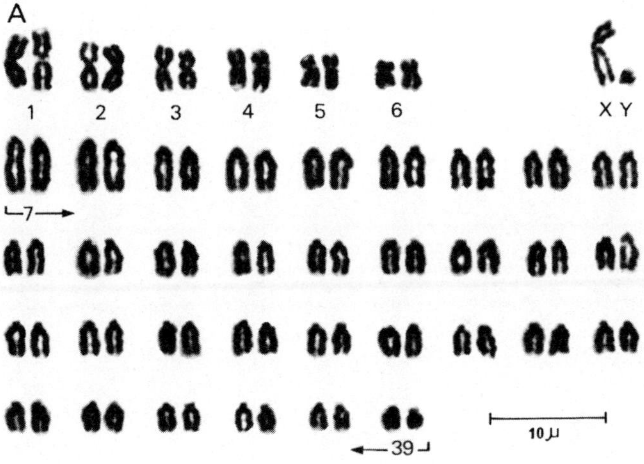

De Boer, L.E.M. and de Boer-van der Vlist, J. (1973). *Genen en Phaenen* 16(3): 67.

De Boer, L.E.M., and de Boer-van der Vlist, J. (1973). The somatic chromosomes and the idiogram of *Tarsius syrichta carbonarius* Heude, 1898 (primates, Tarsioidea). *Genen en Phaenen* 16(3): 65–71.

1.7.1.3. *Tarsius bancanus* 2n = 80

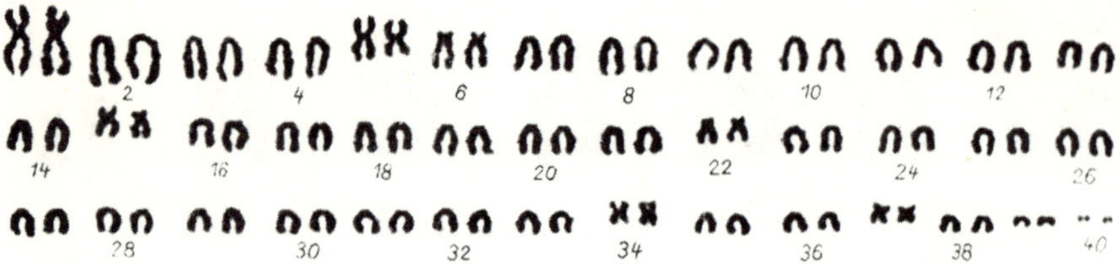

Klinger, H.P. (1963). *Cytogenetics* 2: 145.

Klinger, H.P. (1963). The somatic chromosomes of some primates *(Tupaia glis, Nycticebus coucang, Tarsius bancanus, Cercocebus aterrimus, Symphalangus syndactylus)*. *Cytogenetics* 2: 140–151.

Table 2. Simiae platyrrhinae

	2n	S–M	A	X	Y
Callithrix					
2.1.1.1. *Callithrix argentata*	44	28–32	10–14	S	M–A
2.1.1.2. *Callithrix humeralifer*	44–46	30–32–34	10–14	S	S–M–A
2.1.1.3. *Callithrix jacchus*	44–46	32–34	10–12	S	A
2.1.1.4. *Callithrix chrysoleucos*	46	30	14	S	?
Cebuella					
2.1.2.1. *Cebuella pygmaea*	44	32	10	S	A
Leontideus					
2.1.3.1. *Leontideus rosalia*	46	32–34	10–12	S	M
Saguinus					
2.1.4.1. *Saguinus tamarin**	46	–	–	–	–
2.1.4.2. *Saguinus midas*	46	30	14	S	?
2.1.4.3. *Saguinus fuscicollis*	46	30–34	10–14	S	M
2.1.4.5. *Saguinus nigricollis*	46	30–32–34	10–12–14	S	M
2.1.4.7. *Saguinus mystax*	46	30	14	S	M
2.1.4.10 *Saguinus oedipus*	46	30–32–34	10–12–14	S–M	M–A
2.1.4.11. *Saguinus leucopus*	46	30	14	S	S
Callimico					
2.2.1.1. *Callimico goeldii*	48	28–32	16–18	S	A
Aotes					
2.3.1.1. *Aotes trivirgatus*	49–50–51–52–53–54	18–20–21–22–30–32–34	18–20–22–25–28–30–32	S–M	S–M–A
Callicebus					
2.3.2.1. *Callicebus cupreus*	46–50?	20–22	24–26	S	S–M–A
2.3.2.2. *Callicebus torquatus*	20	10	10	–	–
Pithecia					
2.3.3.1. *Pithecia pithecia*	46–48	18	28	S	S
Chiropotes					
2.3.4.1. *Chiropotes satanas*	54	24	30	–	–
Cacajao					
2.3.5.2. *Cacajao rubicundus*	45–46	20–22	24–25	–	–
Alouatta					
2.3.6.1. *Alouatta villosa*	53–54	22	30	S	?
2.3.6.2. *Alouatta fusca*	50	20	28	S	A
2.3.6.3. *Alouatta seniculus*	44	12–16	26–30	A	S–M
2.3.6.5. *Alouatta caraya*	52	20	30	S	A
Saimiri					
2.3.7.1. *Saimiri sciurea*	44	28–30–32	10–12–14	S	A
Cebus					
2.3.8.1. *Cebus apella*	54	26	26	S–A	A
2.3.8.2. *Cebus capucinus*	52–54	18–20–24–26	26–28–30–34	S–A	A
2.3.8.3. *Cebus albifrons*	52–54	18–20	32–34	S	A
Ateles					
2.3.9.1. *Ateles paniscus*	34	30	2	S–M	A
2.3.9.2. *Ateles fusciceps*	34	30	2	S	S

Table 2. Simiae platyrrhinae (*continued*)

	2n	S–M	A	X	Y
2.3.9.3. *Ateles belzebuth*	34	30	2	M	A
2.3.9.4. *Ateles geoffroyi*	34	30	2	S–M	S–M–A
Brachyteles					
2.3.10.1. *Brachyteles arachnoides*	62	24	36	S	?
Lagothrix					
2.3.11.1. *Lagothrix lagothricha*	62	30	30	S	A

The code numbers for the species are the ones used in Chiarelli, B. (1972). *Taxonomic atlas of living primates*. London: Academic Press.

Key:

* No published karyotypes for this species.

2n = Diploid number.

S–M: Number of submetacentric and metacentric chromosomes.

A: Number of acrocentric chromosomes.

X and Y: Morphology of X and Y chromosomes (M = metacentric; S = submetacentric; and A = acrocentric).

2.1.1.1. *Callithrix argentata* 2n = 44

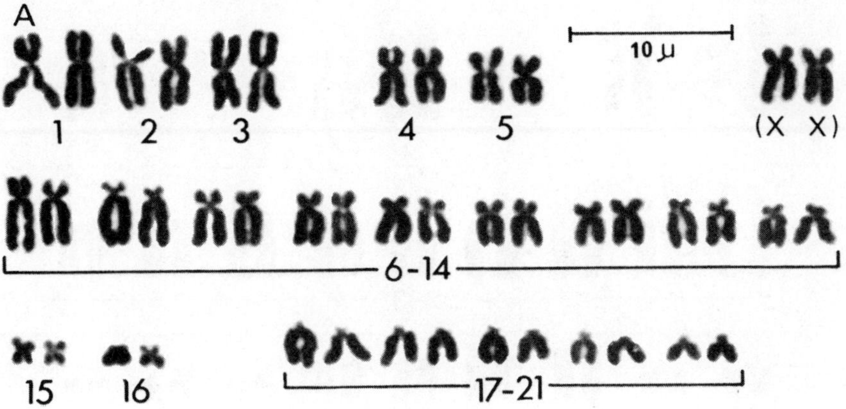

De Boer, L.E.M. (1974). *Genen en Phaenen* 17(1–2): 15.

Egozcue, J., Perkins, E.M., and Hagemenas, F. (1968). Chromosomal evolution in marmosets, tamarins and pinchés. *Folia Primatologica* 9: 81–94.

2.1.1.2. *Callithrix humeralifer* 2n = 44–46

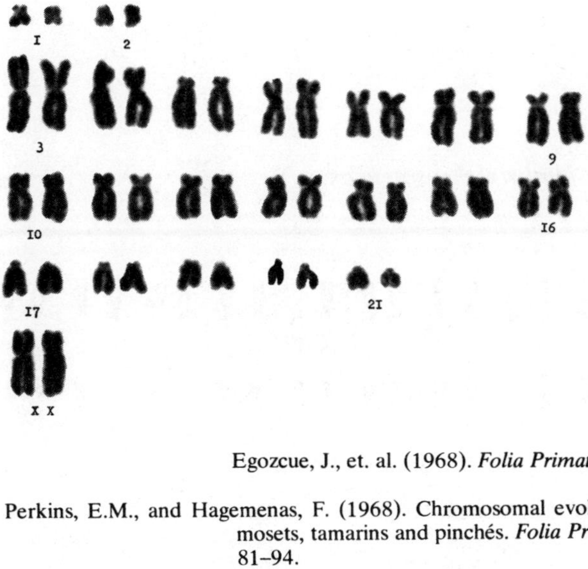

Egozcue, J., et. al. (1968). *Folia Primatologica* 9: 88.

Egozcue, J., Perkins, E.M., and Hagemenas, F. (1968). Chromosomal evolution in marmosets, tamarins and pinchés. *Folia Primatologica* 9: 81–94.

2.1.1.3. *Callithrix jacchus* 2n = 44–66

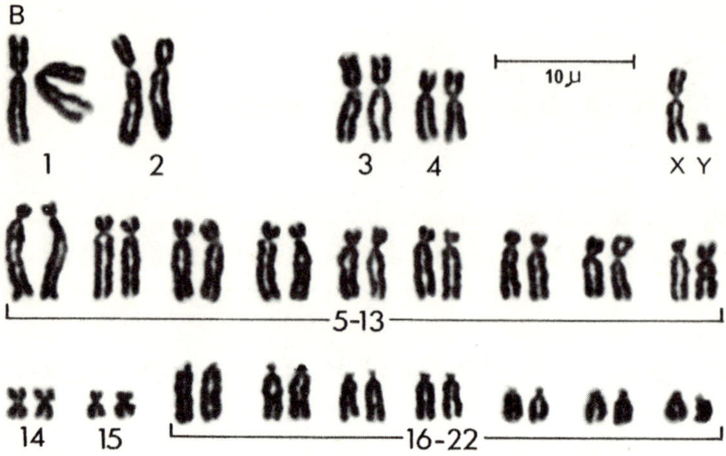

De Boer, L.E.M. (1974). *Genen en Phaenen* 17(1–2): 15.

Chiarelli, B. (1961). Cited by Bender, M.A., and Chu, E.H.Y. (1963). "The chromosomes of primates," in *Evolutionary and genetic biology of the primates*. Edited by J. Buettner-Janusch, London, New York: Academic Press.

Benirschke, K., Anderson J.M., and Brownhill, L.E. (1962). Marrow chimerism in marmosets. *Science* 138: 513–515.

2.1.1.4. *Callithrix chrysoleucos* 2n = 46

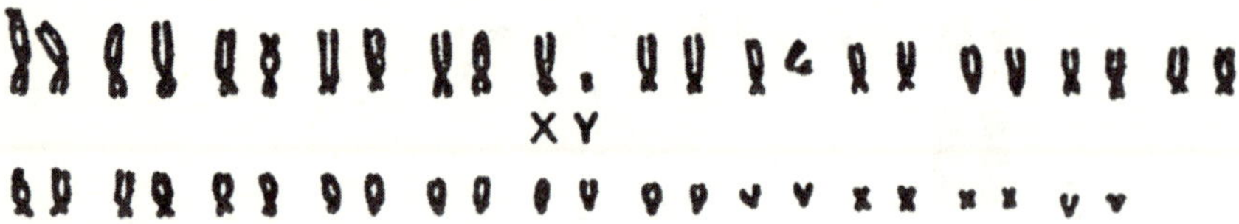

Bender, M.A., and Mettler, L.E. (1960). *Cytologia* 25: 402.

Bender, M.A., and Mettler, L.E. (1960). Chromosome studies of primates. II. *Callithrix*, *Leontocebus* and *Callimico*. *Cytologia* 25: 400–404.

2.1.2.1. *Cebuella pygmaea* 2n = 44

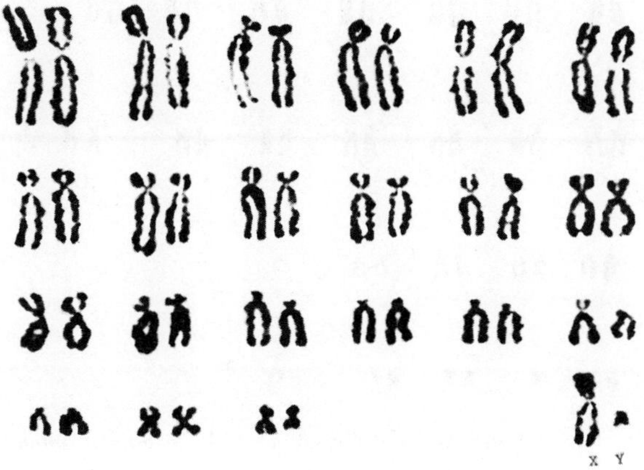

Hsu, T.C., and Benirschke, K. (1967). *An atlas of mammalian chromosomes*, volume one: 47.

Benirschke, K., and Brownhill, L.E. (1962). Further observations on marrow chimerism in marmosets. *Cytogenetics* 1: 245–257.

2.1.3.1. *Leontideus rosalia* 2n = 46

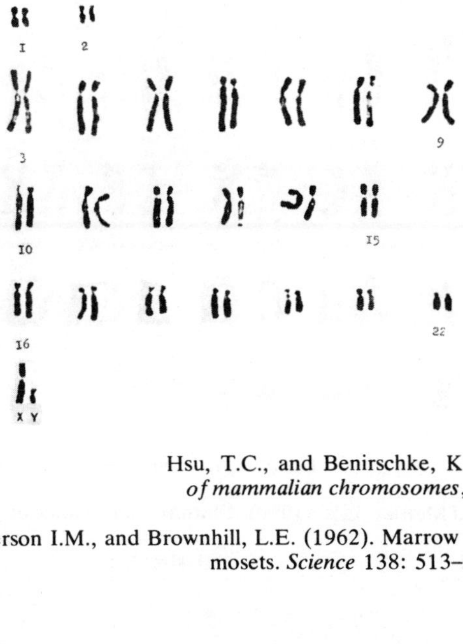

Hsu, T.C., and Benirschke, K. (1971). *An atlas of mammalian chromosomes*, volume five: 245.

Benirschke, K., Anderson I.M., and Brownhill, L.E. (1962). Marrow chimerism in marmosets. *Science* 138: 513–515.

2.1.4.2. *Saguinus midas* 2n = 46

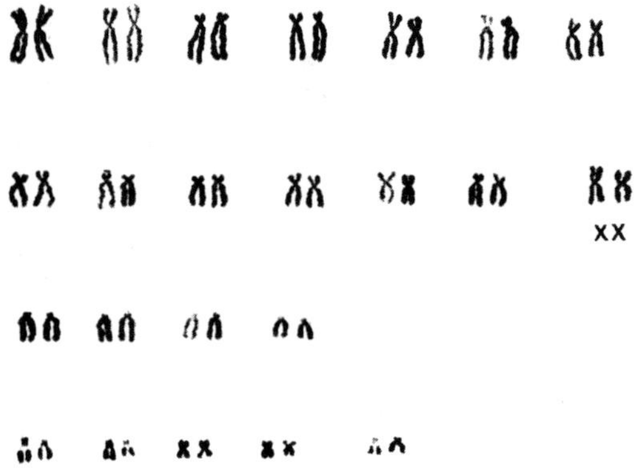

Schmid, W., and Glaser, D. (1977). *Folia Primatologica* 28: 155

Benirschke, K., and Brownhill, L.E. (1963). Heterosexual cells in testes of chimeric marmoset monkeys. *Cytogenetics* 2: 231–241.

Schmid, W., and Glaser, D. (1977). The banded karyotype of *Saguinus midas tamarin* Link, 1795 (Callithricidae, primates). *Folia Primatologica* 28: 154–158.

2.1.4.3. *Saguinus fuscicollis* 2n = 46

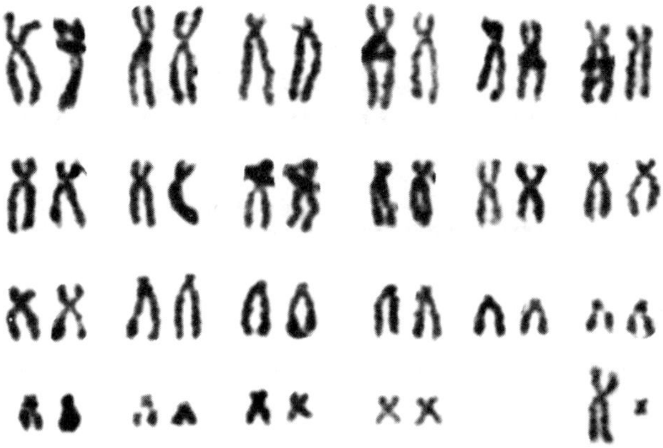

Egozcue, J., et al. (1969). *Folia Primatologica* 10: 155.

Bender, M.A., and Mettler, L.E. (1960). Chromosome studies of primates. II. *Callithrix*, *Leontocebus* and *Callimico*. *Cytologia* 25: 400–404.

2.1.4.5. *Saguinus nigricollis* 2n = 46

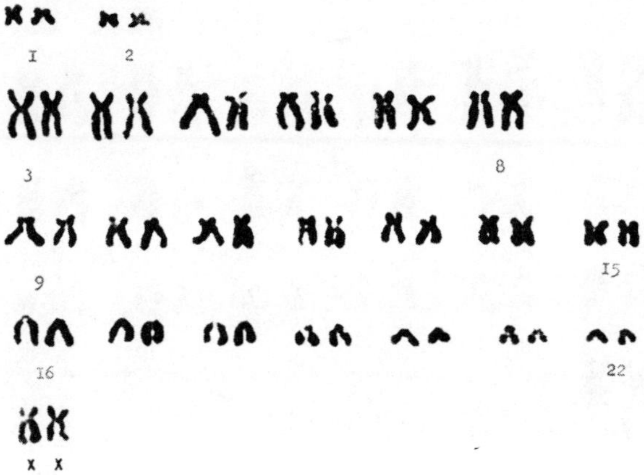

Hsu, T.C., and Benirschke, K. (1968). *An atlas of mammalian chromosomes*, volume two: 100.

Benirschke, K., and Brownhill, L.E. (1962). Further observations on marrow chimerism in marmosets. *Cytogenetics* 1: 245–257.

2.1.4.7. *Saguinus mystax* 2n = 46

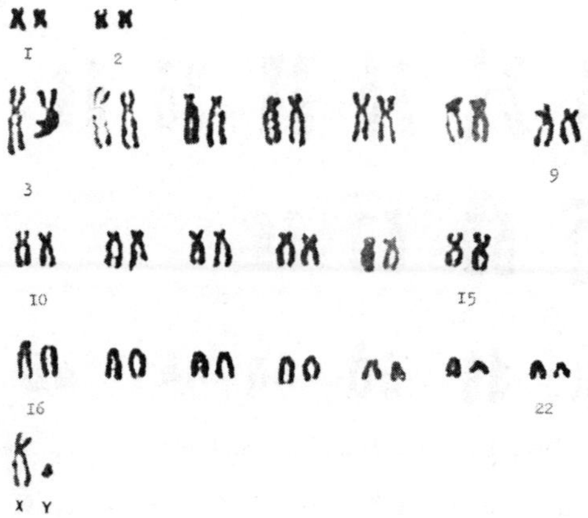

Hsu, T.C., and Benirschke, K. (1968). *An atlas of mammalian chromosomes*, volume two: 99.

Benirschke, K., and Brownhill, L.E. (1963). Heterosexual cells in testes of chimeric marmoset monkeys. *Cytogenetics* 2: 231–241.

2.1.4.10. *Saguinus oedipus* 2n = 46

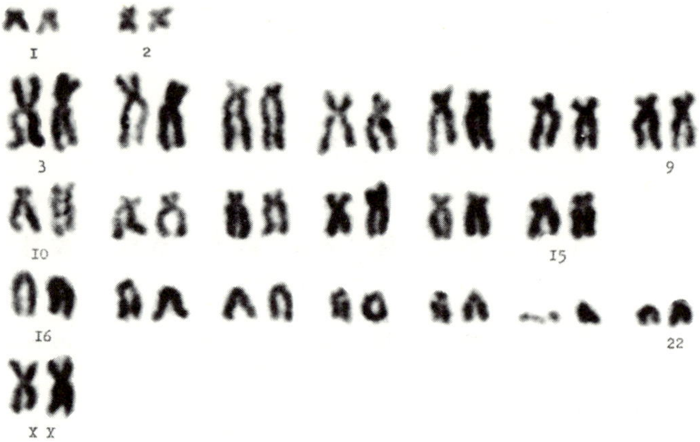

Egozcue, J. – Original.

Chiarelli, B. (1961). Some chromosome numbers in primates. *Mammalian Chromosomes Newsletter* 6: 3.
Benirschke, K., and Brownhill, L.E. (1962). Further observations on marrow chimerism in marmosets. *Cytogenetics* 1: 245–257.

2.1.4.11. *Saguinus leucopus* 2n = 46

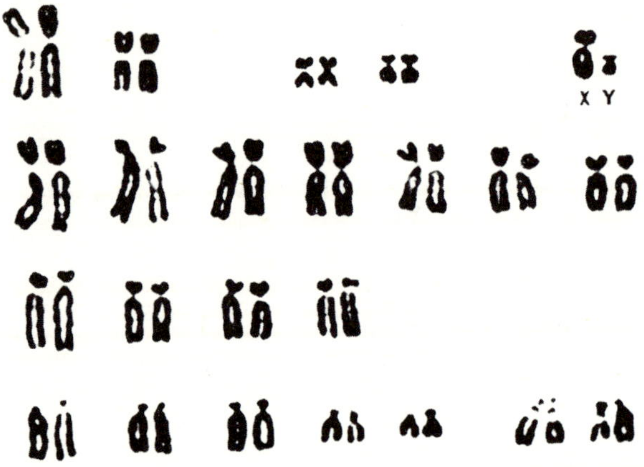

Hsu, T.C., and Hampton, S.H. (1970). *Folia Primatologica* 13: 186.

Benirschke, K., and Brownhill, L.E. (1963). Heterosexual cells in testes of chimeric marmoset monkeys. *Cytogenetics* 2: 231–241.
Hsu, T.C., and Hampton, S.H. (1970). Chromosomes of Callithricidae with special reference to an XX/"XO" sex chromosome system in Goeldi's marmoset (*Callimico goeldii* Thomas 1904). *Folia Primatologica* 13: 183–195.

2.2.1.1. *Callimico goeldii* 2n = 48

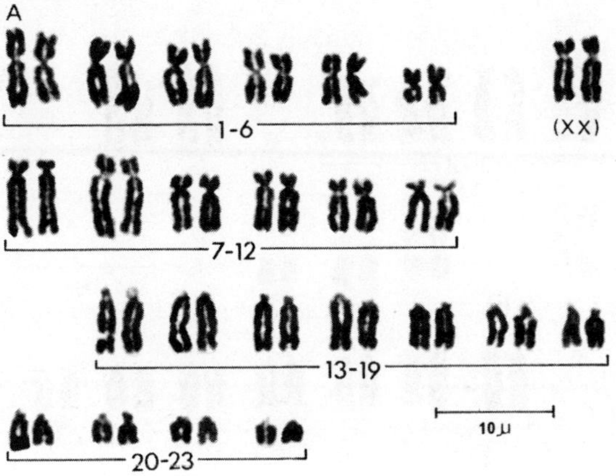

De Boer, L.E.M. (1974). *Genen en Phaenen* 17(1–2): 26.

Bender, M.A., and Mettler, L.E. (1960). Chromosome studies of primates. II. *Callithrix*, *Leontocebus* and *Callimico*. *Cytologia* 25: 400–404.

2.3.1.1. *Aotes trivirgatus* 2n = 49–50–51–52–53–54

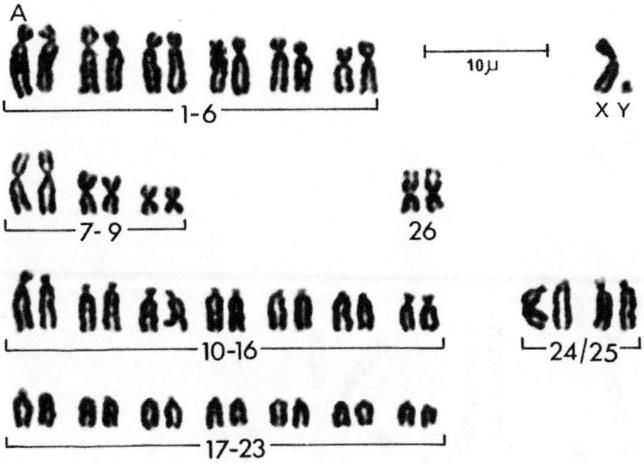

De Boer, L.E.M. (1974). *Genen en Phaenen* 17(1–2): 39.

Chiarelli, B. (1961). Some chromosome numbers in primates. *Mammalian Chromosomes Newsletter* 6: 3.
Chiarelli, B. (1963). Primi risultati di ricerche di genetica e cariologia comparata in primati e loro interesse evolutivo. *Rivista di Antropologia* 50: 87–124.

2.3.2.1. *Callicebus cupreus* 2n = 46–50?

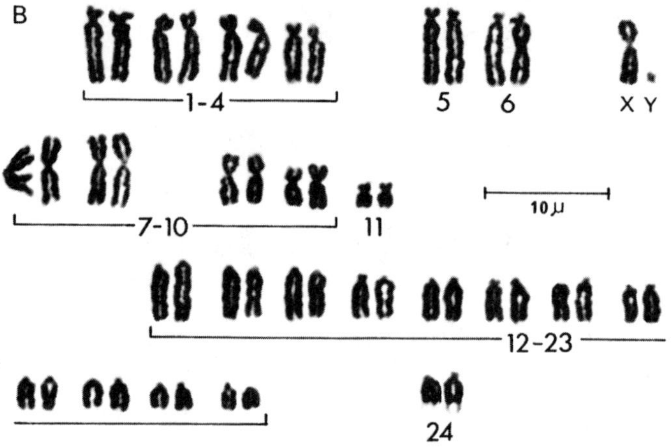

De Boer, L.E.M. (1974). *Genen en Phaenen* 17(1–2): 29.

Bender, M.A., and Mettler, L.E. (1958). Chromosome studies of primates. *Science* 128: 186–190.

2.3.2.2. *Callicebus torquatus* 2n = 20

Egozcue, J., et al. (1969). *Folia Primatologica* 11: 20.

Egozcue, J., Perkins, E.M., Hagemenas, F., and Ford, D.M. (1969). The chromosomes of some Platyrrhini (*Callicebus, Ateles* and *Saimiri*). *Folia Primatologica* 11: 17–27.

2.3.3.1. *Pithecia pithecia* 2n = 46–48

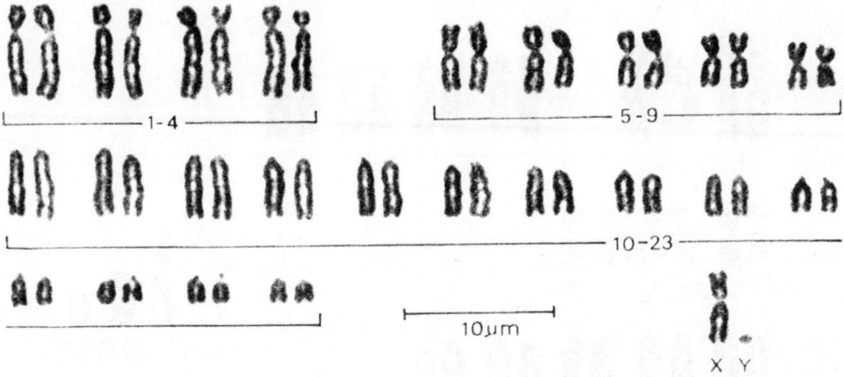

De Boer, L.E.M. (1975). *Folia Primatologica* 23: 152.

Bender, M.A., and Mettler, L.E. Quoted by
Bender, M.A., and Chu, E.H.Y. (1963). "The chromosomes of primates," in *Evolutionary and genetic biology of the primates*. Edited by J. Buettner-Janusch, 261–310. London, New York: Academic Press.

De Boer, L.E.M. (1975). The somatic chromosome complement and the idiogram of *Pithecia pithecia pithecia* (Linnaeus 1766). *Folia Primatologica* 23: 149–157.

2.3.4.1. *Chiropotes satanas* 2n = 54

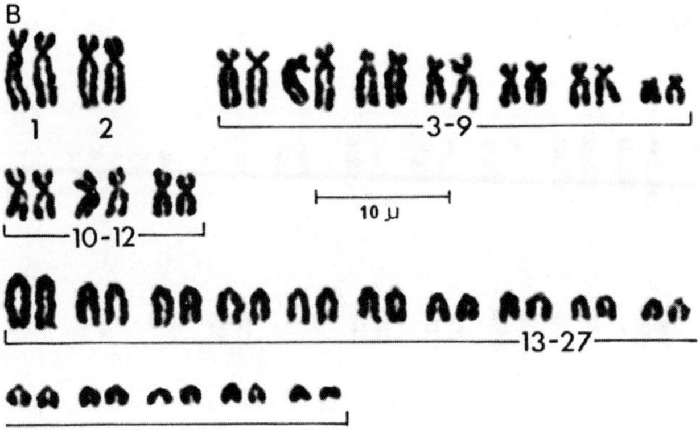

De Boer, L.E.M. (1974). *Genen en Phaenen* 17(1–2): 50.

De Boer, L.E.M. (1974).Cytotaxonomy of the Platyrrhini (primates). *Genen en Phaenen* 17(1–2): 1–115.

2.3.5.2. *Cacajao rubicundus* 2n = 45–46

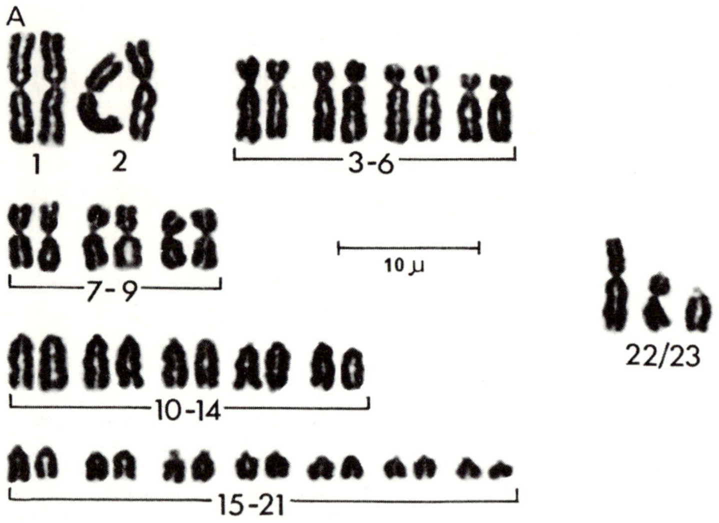

De Boer, L.E.M. (1974). *Genen en Phaenen* 17(1–2): 50.

Bender, M.A., and Mettler, L.E. Quoted by
Chu, E.H.Y., and Bender, L.E. (1961). Chromosome cytology and evolution in primates. *Science* 133: 1,399–1,405.
De Boer, L.E.M. (1974). Cytotaxonomy of the Platyrrhini (primates). *Genen en Phaenen* 17(1–2): 1–115.

2.3.6.1. *Alouatta villosa* 2n = 53–54

Ma, N.S.F., et al. (1975). *Journal of Medical Primatology* 4: 303.

Hsu, T.C. (1965). Foreword. *Mammalian Chromosomes Newsletter* 15: 98–99.
Ma, N.S.F., Jones, T.C., Thorington, R.W.,
Miller, A., and Morgan, L. (1975). Y-autosome translocation in the howler monkey, *Alouatta palliata. Journal of Medical Primatology* 4: 299–307.

2.3.6.2. *Alouatta fusca* 2n = 50

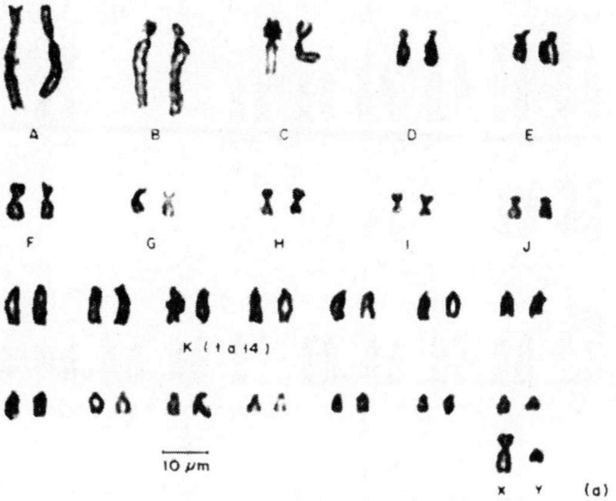

Koiffmann, C.P., and Saldanha, P.H. (1974). *Journal of Human Evolution* 3: 276.

Koiffmann, C.P., and Saldanha, P.H. (1974). Cytogenetics of Brazilian monkeys. *Journal of Human Evolution* 3: 275–282.

2.3.6.3. *Alouatta seniculus* 2n = 44

Yunis, E.J., Torres de Caballero, O.M., Ramirez, C., and Ramirez, Z.E. (1976). *Folia Primatologica* 25: 219.

Bender, M.A., and Mettler, L.E. Quoted by
Bender, M.A., and Chu, E.H.Y. (1963). "The chromosomes of primates," in *Evolutionary and genetic biology of the primates*. Edited by J. Buettner-Janusch, 261–310. London, New York: Academic Press.

2.3.6.5. *Alouatta caraya* 2n = 52

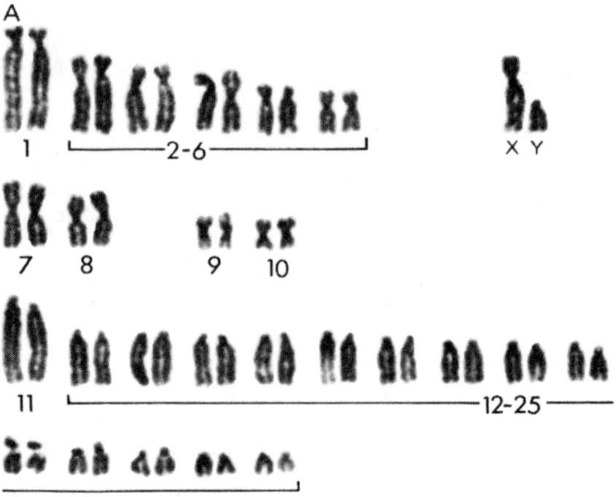

Egozcue, J., cited by De Boer, L.E.M. (1974). *Genen en Phaenen* 17(1–2): 68.

Egozcue, J., and Vilarasau de Egozcue, M. (1965). The chromosomes of the howler monkey (*Alouatta caraya*). *Mammalian Chromosomes Newsletter* 17: 84.

2.3.7.1. *Saimiri sciurea* 2n = 44

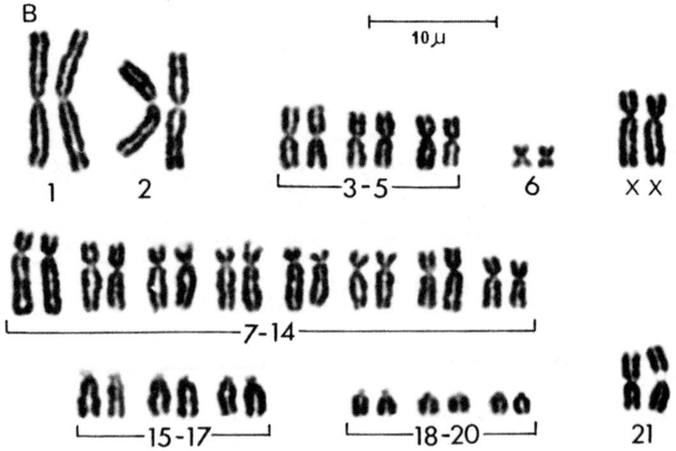

De Boer, L.E.M., (1974). *Genen en Phaenen* 17(1–2): 56.

Bender, M.A., and Mettler, L.E. (1958). Chromosome studies of primates. *Science* 128: 186–190.

2.3.8.1. *Cebus apella* 2n = 54

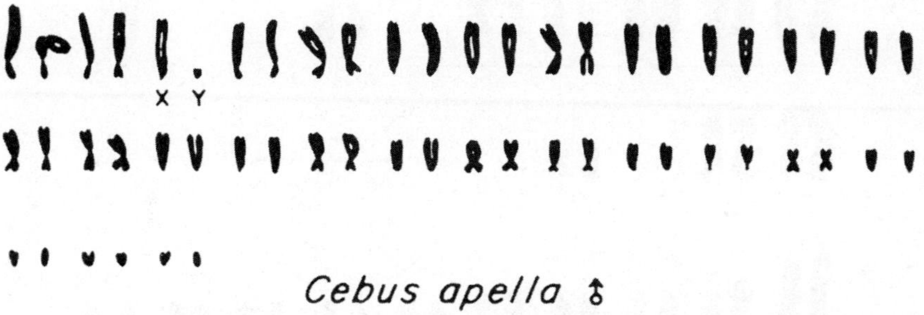

Bender, M.A., and Chu, E.H.Y. (1963). *Evolutionary and genetic biology of the primates*, volume one. Edited by J. Buettner-Janusch, 286. London, New York: Academic.

Bender, M.A., and Mettler, L.E. (1958). Chromosome studies of primates. *Science* 128: 186–190.

2.3.8.2. *Cebus capucinus* 2n = 52–54

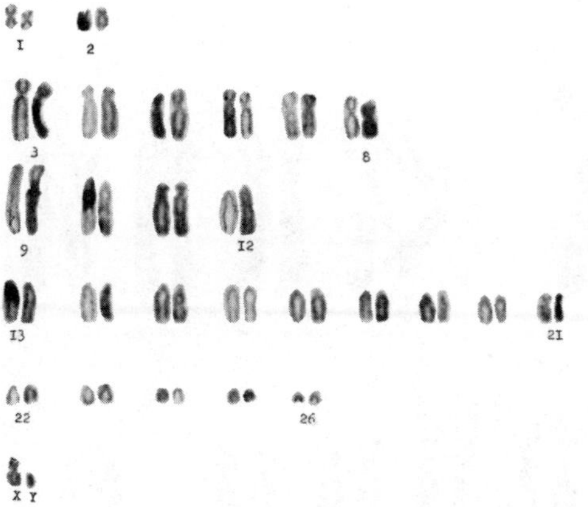

Chiarelli, B. (1963). *Rivista di Antropologia* 50: 96.

Bender, M.A., and Mettler, L.E. (1958). Chromosome studies of primates. *Science* 128: 186–190.
Chiarelli, B. (1963). Primi risultati di ricerche di genetica e cariologia comparata in primati e loro interesse evolutivo. *Rivista di Antropologia* 50: 87–124.

2.3.8.3. *Cebus albifrons* 2n = 52–54

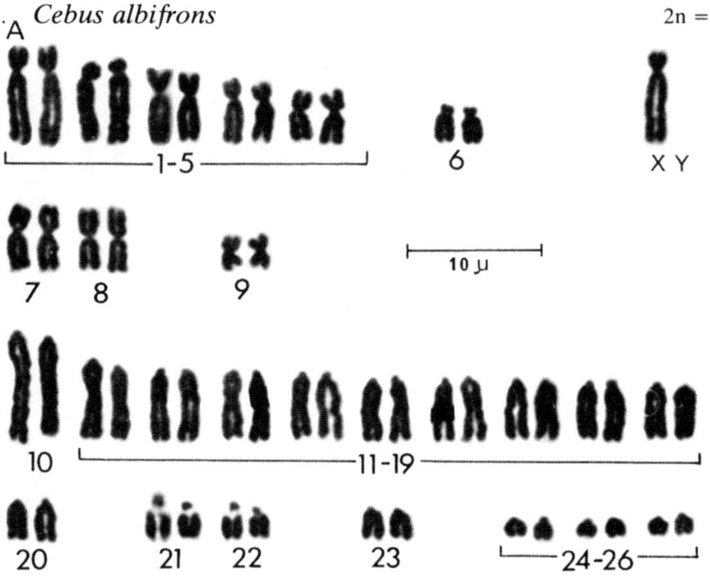

De Boer, L.E.M. (1974). *Genen en Phaenen* 17(1–2): 63.

Egozcue, J., and Vilarasau de Egozcue, M. (1967). The chromosome complement of *Cebus albifrons* (Erxleben, 1777). *Folia Primatologica* 5: 285–294.

2.3.9.1. *Ateles paniscus* 2n = 34

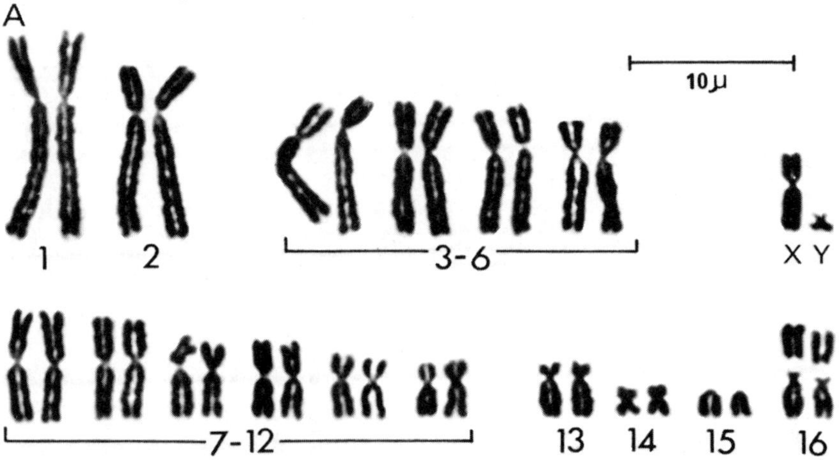

De Boer, L.E.M. (1974). *Genen en Phaenen* 17(1–2): 73.

Bender, M.A., and Mettler, L.E. (1958). Chromosome studies of primates. *Science* 128: 186–190.
Koiffmann, C.P., and Saldanha, P.H. (1974). Cytogenetics of Brazilian monkeys. *Journal of Human Evolution* 3: 275–282.

2.3.9.2. *Ateles fusciceps* 2n = 34

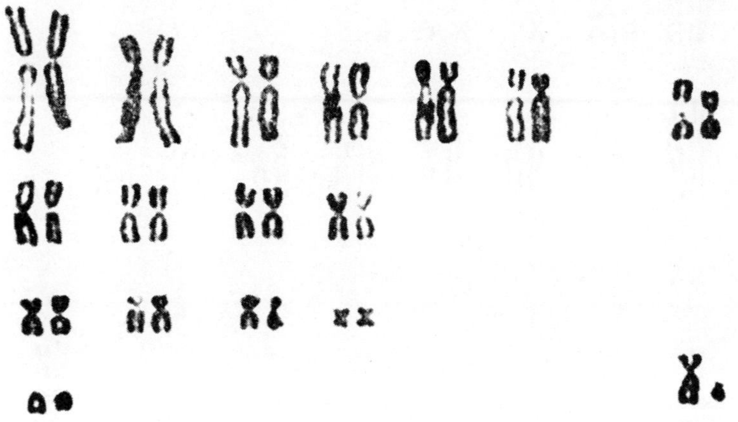

Wurster, D.H., and Benirschke, K. (1969). *Mammalian Chromosomes Newsletter* 10(1): 4.

Wurster, D.H., and Benirschke, K. (1969). Chromosomes of some primates. *Mammalian Chromosomes Newsletter* 10(1): 3–5.

2.3.9.3. *Ateles belzebuth* 2n = 34

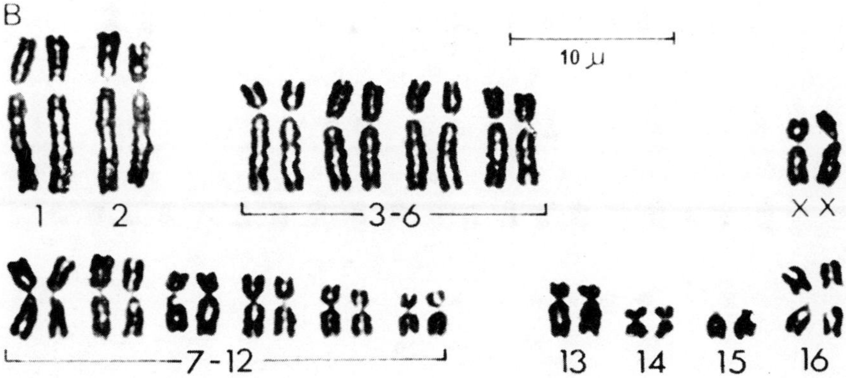

De Boer, L.E.M. (1974). *Genen en Phaenen* 17(1–2): 68.

Chu, E.H.Y., and Bender, L.E. (1961). Chromosome cytology and evolution in primates. *Science* 133: 1,399–1,405.
De Boer L.E.M. (1974). Cytotaxonomy of the Platyrrhini (primates). *Genen en Phaenen* 17(1–2): 1–115.

2.3.9.4. *Ateles geoffroyi* 2n = 34

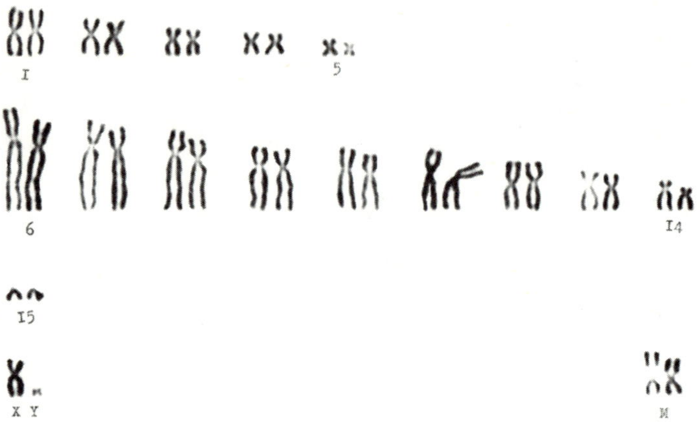

Egozcue, J., et al. (1969). *Folia Primatologica* 11: 24.

Bender, M.A., and Mettler, L.E. (1958). Chromosome studies of primates. *Science* 128: 186–190.

2.3.10.1 *Brachyteles arachnoides* 2n = 62

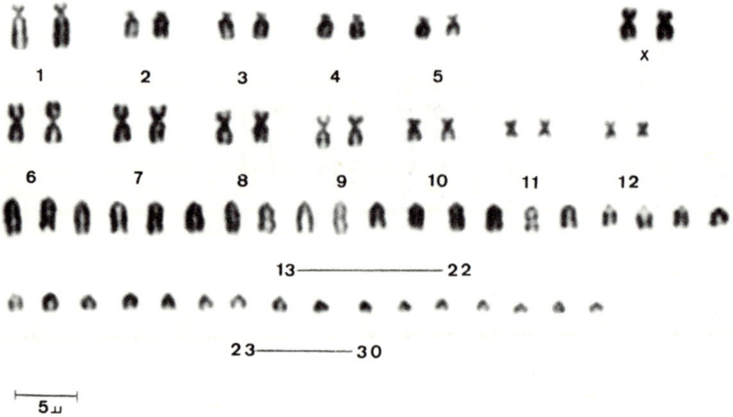

Koiffmann, C.P. – Original.

Koiffmann, C.P., and Saldanha, P.H. (1978). The karyotype of *Brachyteles arachnoides* (E. Geoffroy, 1806) (primates: Platyrrhini). *Genetica* 48(2): 129–130.

2.3.11.1. *Lagothrix lagothricha* 2n = 62

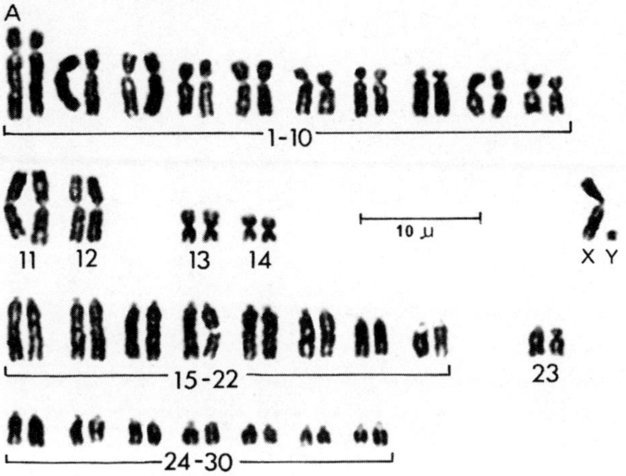

De Boer, L.E.M. (1974). *Genen en Phaenen* 17(1–2): 76.

Bender, M.A., and Chu, E.H.Y. (1963). "The chromosomes of primates," in *Evolutionary and genetic biology of the primates*. Edited by J. Buettner-Janusch. London, New York: Academic Press.

Egozcue, J., and Perkins, E.M. (1970). The chromosomes of Humboldt's woolly monkey (*Lagothrix lagothricha*, Humboldt 1812). *Folia Primatologica* 12: 77–80.

Table 3. Simiae catarrhinae

	2n	S–M	A	X	Y
Macaca					
3.1.1.1. *Macaca silenus*	42	40	–	S	S
3.1.1.2. *Macaca nigra*	42	40	–	S	S
3.1.1.3. *Macaca sylvana*	42	40	–	S	S
3.1.1.4. *Macaca arctoides*	42	40	–	S	S
3.1.1.5. *Macaca maura*	42	40	–	S	S
3.1.1.6. *Macaca sinica*	42	40	–	S	S
3.1.1.7. *Macaca radiata*	42	40	–	S	S
3.1.1.8. *Macaca cyclopis**	42	40	–	S	S
3.1.1.9. *Macaca mulatta*	42	40	–	S	S
3.1.1.10. *Macaca fuscata*	42	40	–	S	S
3.1.1.11. *Macaca nemestrina*	42	40	–	S	S
3.1.1.12. *Macaca fascicularis*	42	40	–	S	S
3.1.1.13. *Macaca assamensis*	42	40	–	S	S
Papio					
3.1.2.1. *Papio hamadryas*	42	40	–	S	S
3.1.2.2. *Papio ursinus*	42	40	–	S	S
3.1.2.3. *Papio anubis**	42	40	–	S	S
3.1.2.4. *Papio cynocephalus*	42	40	–	S	S
3.1.2.5. *Papio papio*	42	40	–	S	S
3.1.2.6. *Papio sphinx*	42	40	–	S	S
3.1.2.7. *Papio leucophaeus*	42	40	–	S	S
Theropithecus					
3.1.3.1. *Theropithecus gelada*	42	40	–	S	S
Cercocebus					
3.1.4.1. *Cercocebus galeritus*	42	40	–	S	S
3.1.4.3. *Cercocebus torquatus*	42	40	–	S	S
3.1.4.4. *Cercocebus aterrimus*	42	40	–	S	S
3.1.4.5. *Cercocebus albigena**	42	40	–	S	S
Cercopithecus					
3.1.5.1. *Cercopithecus aethiops*	60	34	24	S	A
3.1.5.2. *Cercopithecus cynosurus*	60	36	22	S	A
3.1.5.3. *Cercopithecus sabaeus*	60	36	22	S	A
3.1.5.4. *Cercopithecus cephus*	66	46	18	S	A
3.1.5.5. *Cercopithecus diana*	58	42	14	S	A
3.1.5.6. *Cercopithecus l'hoesti*	58–60–72	48	14–16	S	A
3.1.5.7. *Cercopithecus preussi**	66	–	–	–	–
3.1.5.8. *Cercopithecus hamlyni*	64	50	12	S	?
3.1.5.9. *Cercopithecus mona*	66–68	46	18–20	S	S–A
3.1.5.10. *Cercopithecus campbelli**	66–68	46	18–20	S	S–A
3.1.5.13. *Cercopithecus denti**	66	–	–	–	–
3.1.5.14. *Cercopithecus petaurista**	66	–	–	–	–
3.1.5.15. *Cercopithecus neglectus*	58–62	44–46	12–14	S	S
3.1.5.16. *Cercopithecus nictitans*	66–70	46	18–22	S	A
3.1.5.17. *Cercopithecus ascanius*	66	46	18	S	A
3.1.5.20. *Cercopithecus mitis*	72	44–52	22	S	A–S
3.1.5.21. *Cercopithecus nigroviridis*	48	46	–	S	A
3.1.5.22. *Cercopithecus talapoin*	54	38	14	S	A
Erythrocebus					
3.1.6.1. *Erythrocebus patas*	54	36	16	S	A–S

Table 3. Simiae catarrhinae (*continued*)

	2n	S–M	A	X	Y
Presbytis					
3.2.1.1. *Presbytis entellus*	44	40	2	S	A
3.2.1.2. *Presbytis senex*	44	40	2	S	A
3.2.1.8. *Presbytis cristatus*	44	40	2	S	A
3.2.1.10. *Presbytis obscurus*	44	40	2	S	A
Pygathrix					
3.2.2.1. *Pygathrix nemaeus*	44	42	–	S	A
Rhinopithecus					
3.2.3.1. *Rhinopithecus roxellanae*	44	40	2	S	A
Nasalis					
3.2.5.1. *Nasalis larvatus*	48	46	–	S	?
Colobus					
3.2.6.1. *Colobus polykomos*	44	42	–	S	?
3.2.6.4. *Colobus badius**	44	42	–	S	?
3.2.6.5. *Colobus kirkii**	44	42	–	S	A
Hylobates					
3.3.1.1. *Hylobates lar*	44	42	–	S	S
3.3.1.2. *Hylobates agilis*	44	42	–	S	S
3.3.1.3. *Hylobates moloch*	44	42	–	S	S
3.3.1.4. *Hylobates concolor*	52	44	6	S	A
3.3.1.5. *Hylobates hoolock*	44	42	–	S	S
3.3.1.6. *Hylobates klossii*	44	42	–	S	–
Symphalangus					
3.3.2.1. *Symphalangus syndactylus*	50	46	2	S	S
Pongo					
3.4.1.1. *Pongo pygmaeus*	48	26	20	S	S
Pan					
3.4.2.1. *Pan troglodytes*	48	34	12	S	A
3.4.2.2. *Pan paniscus*	48	34	12	S	A
Gorilla					
3.4.3.1. *Gorilla gorilla*	48	30	16	S	S
Homo					
3.5.1.1. *Homo sapiens*	46	34	10	S	A

The code numbers for the species are the ones used in Chiarelli, B. (1972). *Taxonomic atlas of living primates.* London: Academic Press.

Key:

* No published karyotypes for this species.

2n = Diploid number.

S–M: Number of submetacentric and metacentric chromosomes.

A: Number of acrocentric chromosomes.

X and Y: Morphology of X and Y chromosomes (M = metacentric; S = submetacentric; and A = acrocentric).

3.1.1.1. *Macaca silenus* 2n = 42

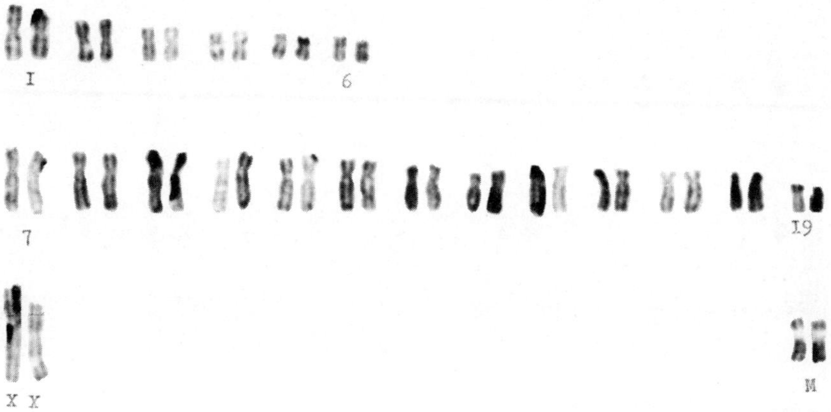

Chiarelli, B. (1960) – Original.

Chiarelli, B. (1962). Comparative morphometric analysis of primate chromosomes. II. The chromosomes of genera *Macaca*, *Papio*, *Theropithecus* and *Cercocebus*. *Caryologia* 15: 401–420.
Chiarelli, B. This volume.

3.1.1.2. *Macaca nigra* 2n = 42

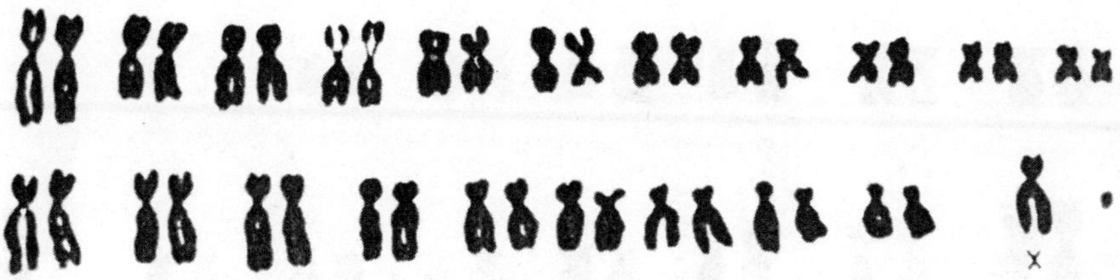

Ushijima, R.N., et al. (1964). *Science* 146: 78.

Chiarelli, B. (1962). Comparative morphometric analysis of primate chromosomes. II. The chromosomes of genera *Macaca*, *Papio*, *Theropithecus* and *Cercocebus*. *Caryologia* 15: 401–420.
Ushijima, R.N., Shininger, F.S., and Grand, T. (1964). Chromosome complements of two species of primates: *Cynopithecus niger* and *Presbytis entellus*. *Science* 146: 78–79.

3.1.1.3. *Macaca sylvana* — 2n = 42

Chiarelli, B. (1960) – Original.

Chiarelli, B. (1962). Comparative morphometric analysis of primate chromosomes. II. The chromosomes of genera *Macaca*, *Papio*, *Theropithecus* and *Cercocebus*. *Caryologia* 15: 401–420.
Chiarelli, B. This volume.

3.1.1.4. *Macaca arctoides* — 2n = 42

Huang, C.C., et al. (1969). *Folia Primatologica* 11: 30.

Chiarelli, B. (1962). Comparative morphometric analysis of primate chromosomes. II. The chromosomes of genera *Macaca*, *Papio*, *Theropithecus* and *Cercocebus*. *Caryologia* 15: 401–420.
Huang, C.C., Habbit, H., and Ambrus, J.L. (1969). Chromosomes and DNA synthesis in the stumptail monkey (*Macaca speciosa*), with special regard to marker and sex chromosomes. *Folia Primatologica* 11: 28–34.

3.1.1.5. *Macaca maura* 2n = 42

Chiarelli, B. (1960) – Original.

Chiarelli, B. (1962). Comparative morphometric analysis of primate chromosomes. II. The chromosomes of genera *Macaca*, *Papio*, *Theropithecus* and *Cercocebus*. *Caryologia* 15: 401–420.

Chiarelli, B. This volume.

3.1.1.6. *Macaca sinica* 2n = 42

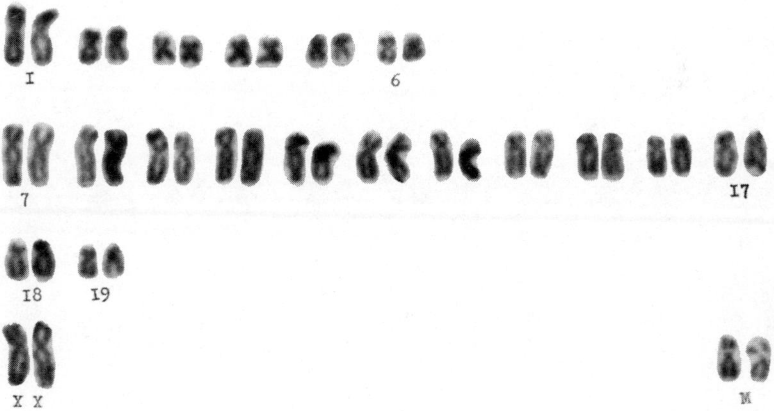

Chiarelli, B. (1960) – Original.

Chiarelli, B. (1962). Comparative morphometric analysis of primate chromosomes. II. The chromosomes of genera *Macaca*, *Papio*, *Theropithecus* and *Cercocebus*. *Caryologia* 15: 401–420.

Chiarelli, B. This volume.

3.1.1.7. *Macaca radiata* 2n = 42

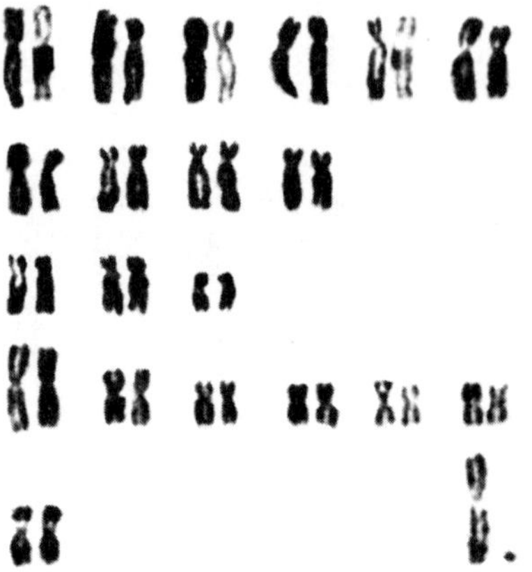

Chiarelli, B. (1962). *Caryologia* 15: 404.

Chiarelli, B. (1962). Comparative morphometric analysis of primate chromosomes. II. The chromosomes of genera *Macaca*, *Papio*, *Theropithecus* and *Cercocebus*. *Caryologia* 15: 401–420.

3.1.1.9. *Macaca mulatta* 2n = 42

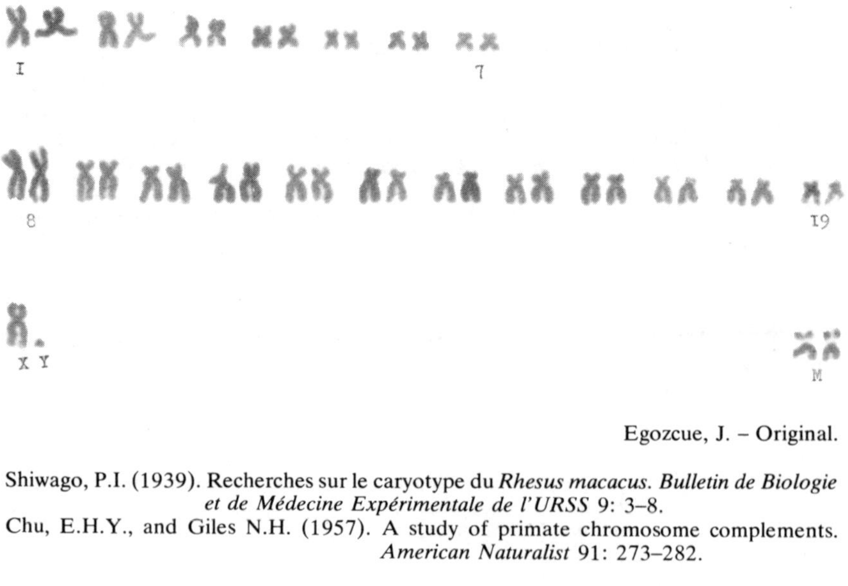

Egozcue, J. – Original.

Shiwago, P.I. (1939). Recherches sur le caryotype du *Rhesus macacus*. *Bulletin de Biologie et de Médecine Expérimentale de l'URSS* 9: 3–8.
Chu, E.H.Y., and Giles N.H. (1957). A study of primate chromosome complements. *American Naturalist* 91: 273–282.

3.1.1.10. *Macaca fuscata* 2n = 42

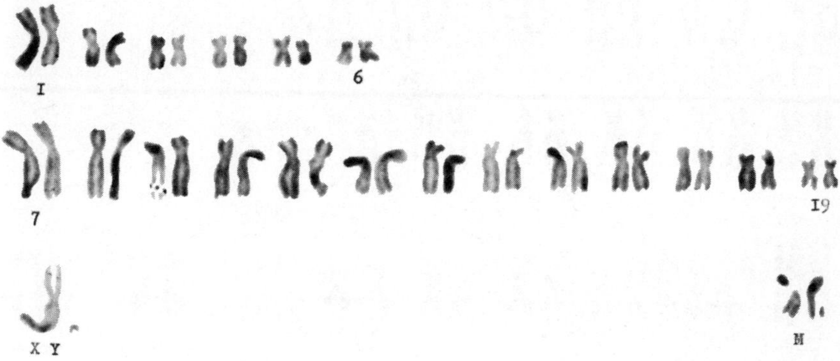

Chiarelli, B. (1960) – Original.

Chiarelli, B. (1962). Comparative morphometric analysis of primate chromosomes. II. The chromosomes of genera *Macaca*, *Papio*, *Theropithecus* and *Cercocebus*. *Caryologia* 15: 401–420.
Chiarelli, B. This volume.

3.1.1.11. *Macaca nemestrina* 2n = 42

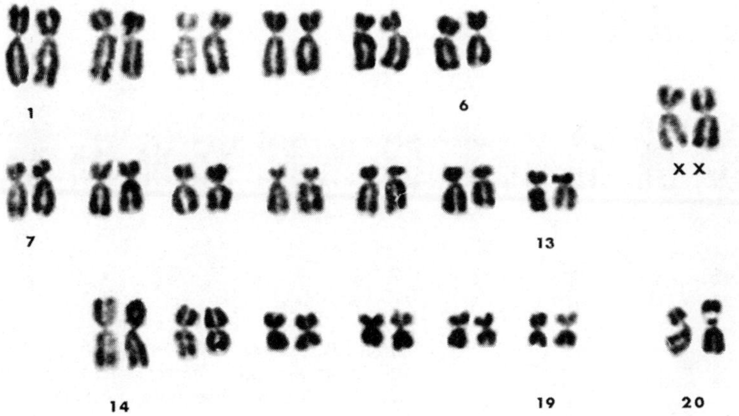

De Boer, L.E.M. – Original.

Darlington, C.D., and Hague, A. (1955). Chromosomes of monkeys and man. *Nature* 175: 32.

3.1.1.12. *Macaca fascicularis* 2n = 42

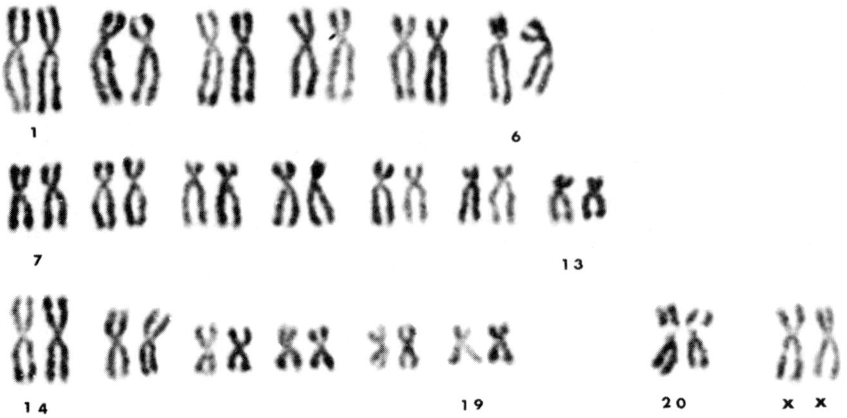

De Boer, L.E.M. – Original.

Chu, E.H.Y., and Bender, M.A. (1961). Chromosome cytology and evolution in primates. *Science* 133: 1,399–1,405

Fernandez-Donoso, R., Lindsten, J., and Norby, E. (1970). The chromosomes of the cynomolgus macaque (*Macaca fascicularis*). *Hereditas* 65: 269–276.

3.1.1.13. *Macaca assamensis* 2n = 42

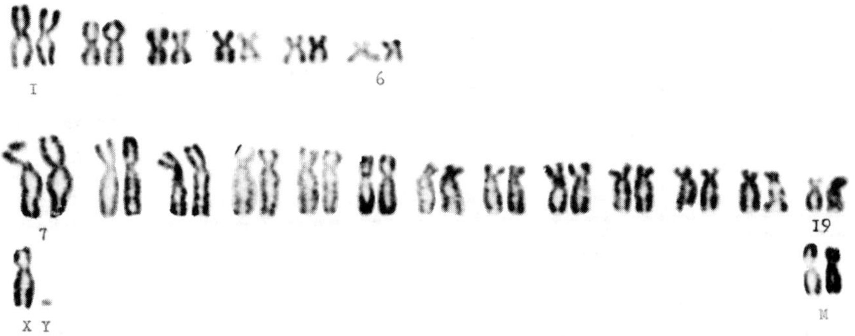

Chiarelli, B. (1968) in *Taxonomy and phylogeny of Old World primates with references to the origin of man*, 157. Turin: Rosenberg and Sellier.

Chiarelli, B. (1962). Comparative morphometric analysis of primate chromosomes. II. The chromosomes of genera *Macaca*, *Papio*, *Theropithecus* and *Cercocebus*. *Caryologia* 15: 401–420.

Chiarelli, B. (1968). *Taxonomy and phylogeny of Old World primates with references to the origin of man*, 151–186. Turin: Rosenberg and Sellier.

3.1.2.1. *Papio hamadryas* 2n = 42

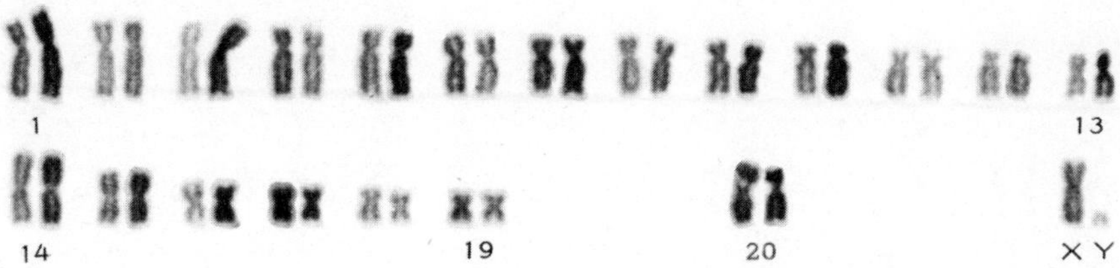

Chiarelli, B. (1962). *Caryologia* 15: 404.

Chiarelli, B. (1962). Comparative morphometric analysis of primate chromosomes. II. The chromosomes of genera *Macaca*, *Papio*, *Theropithecus* and *Cercocebus*. *Caryologia* 15: 401–420.

3.1.2.2. *Papio ursinus* 2n = 42

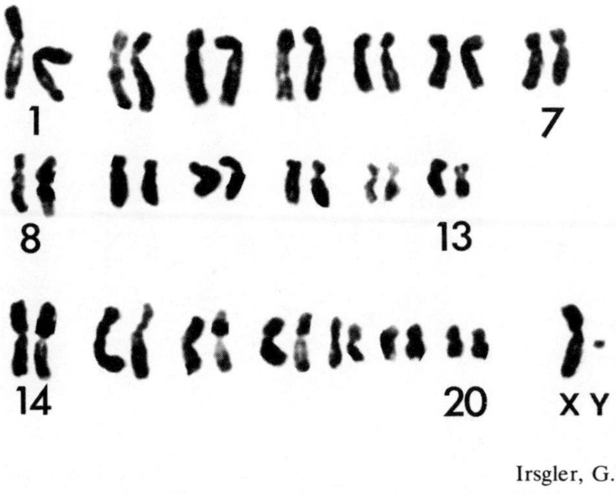

Irsgler, G.B. – Original.

Irsgler, G.B. (1967). Studies on the chromosomes of African mammals. IV. The karyotype of female chacma baboon (*Papio ursinus*). *South African Journal of Science* 63: 527–531.

3.1.2.4. *Papio cynocephalus* 2n = 42

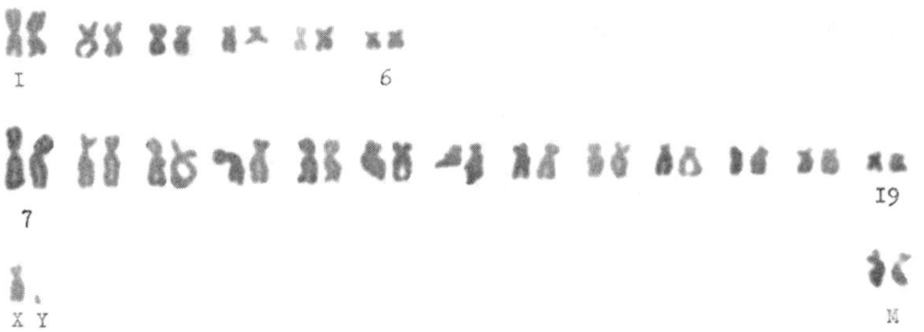

Ushijima, R.N. – Original.

Chiarelli, B. (1962). Comparative morphometric analysis of primate chromosomes. II. The chromosomes of genera *Macaca*, *Papio*, *Theropithecus* and *Cercocebus*. *Caryologia* 15: 401–420.

Chiarelli, B. (1963). Primi risultati di ricerche di genetica e cariologia comparata in primati e loro interesse evolutivo. *Rivista di Antropologia* 50: 87–124.

3.1.2.5. *Papio papio* 2n = 42

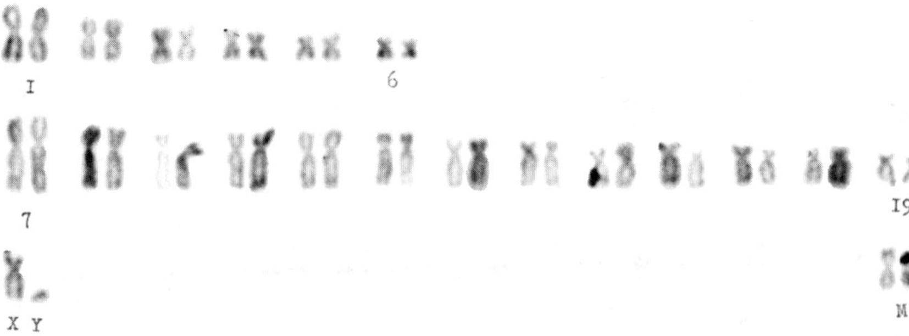

Chiarelli, B. (1960) – Original.

Darlington, C.D., and Hague, A. (1955). Chromosomes of monkeys and man. *Nature* 175: 32.

Chiarelli, B. This volume.

3.1.2.6. *Papio sphinx* 2n = 42

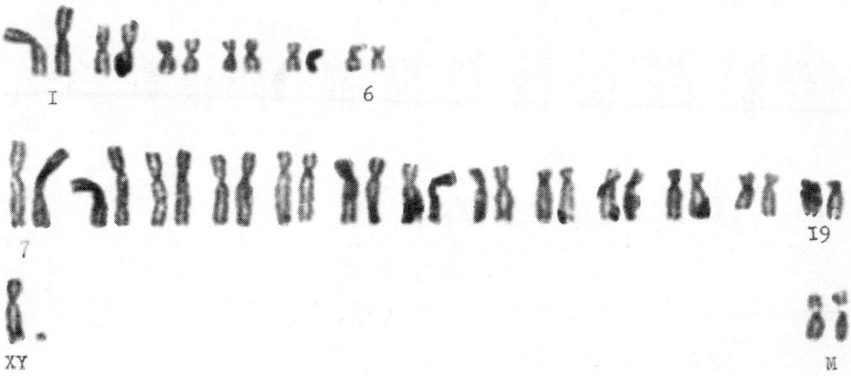

Chiarelli, B. (1966). *American Journal of Physical Anthropology* 24(2): 158.

Bender, M.A., and Mettler, L.E. (1958). Chromosome studies of primates. *Science* 128: 186–190.

Chiarelli, B. (1966). Caryology and taxonomy of the catarrhine monkeys. *American Journal of Physical Anthropology* 24(2): 155–169.

3.1.2.7. *Papio leucophaeus* 2n = 42

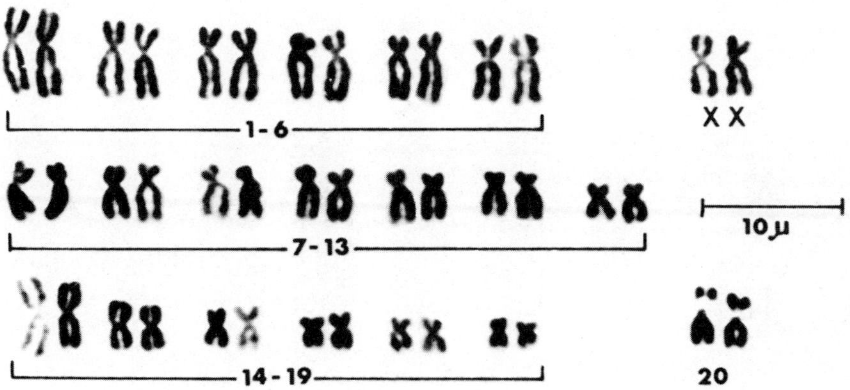

Chiarelli, B. (1960) – Original.

Chiarelli, B. (1962). Comparative morphometric analysis of primate chromosomes. II. The chromosomes of genera *Macaca*, *Papio*, *Theropithecus* and *Cercocebus*. *Caryologia* 15:401–420.

Chiarelli, B. This volume.

3.1.3.1. *Theropithecus gelada* 2n = 42

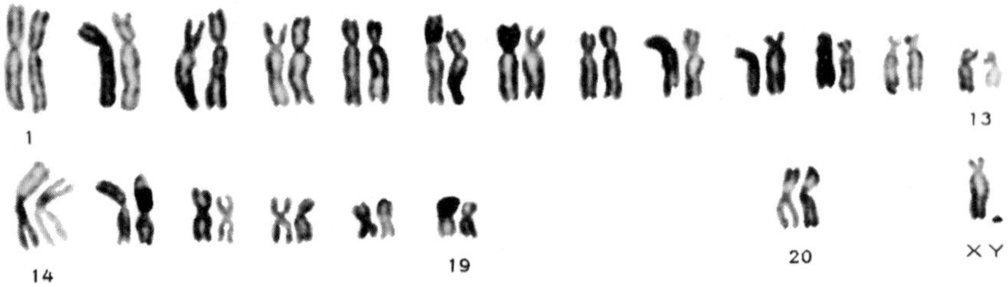

Chiarelli, B. (1962). *Caryologia* 15: 406.

Chiarelli, B. (1962). Comparative morphometric analysis of primate chromosomes. II. The chromosomes of genera *Macaca*, *Papio*, *Theropithecus* and *Cercocebus*. *Caryologia* 15: 401–420.

3.1.4.1. *Cercocebus galeritus* 2n = 42

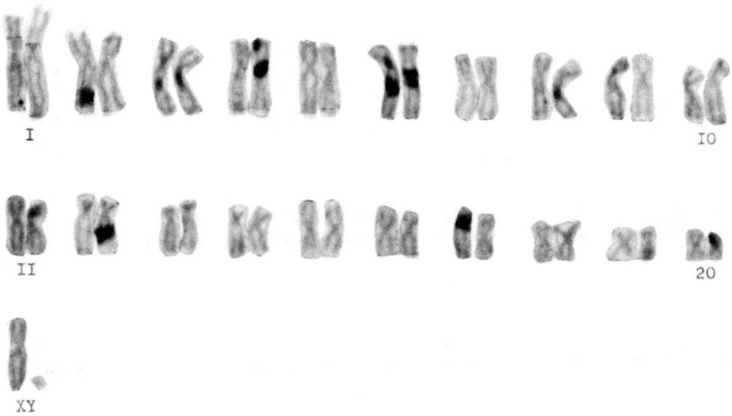

Chiarelli, B. (1960) – Original.

Tappen, N.C. (1960). Quoted by Bender, M.A., and Chu, E.H.Y. (1963). *Evolutionary and genetic biology of the primates*, volume one: chapter 7. Edited by J. Buettner-Janusch. London, New York: Academic Press.
Chiarelli, B. This volume.

3.1.4.3. *Cercocebus torquatus* 2n = 42

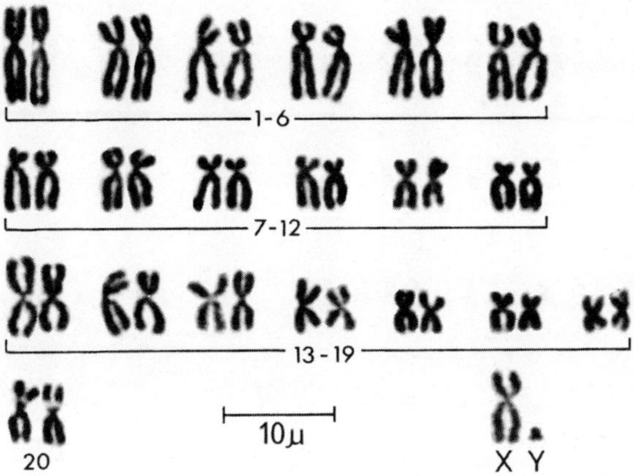

De Boer, L.E.M. (1971). *Genen en Phaenen* 14(4)

Chu, E.H.Y., and Giles, N.H. (1957). A study of primate chromosome complements. *American Naturalist* 91: 273–282.
Bender, M.A., and Mettler, L.E. (1958). Chromosome studies of primates. *Science* 128: 186–190.

3.1.4.4. *Cercocebus aterrimus* 2n = 42

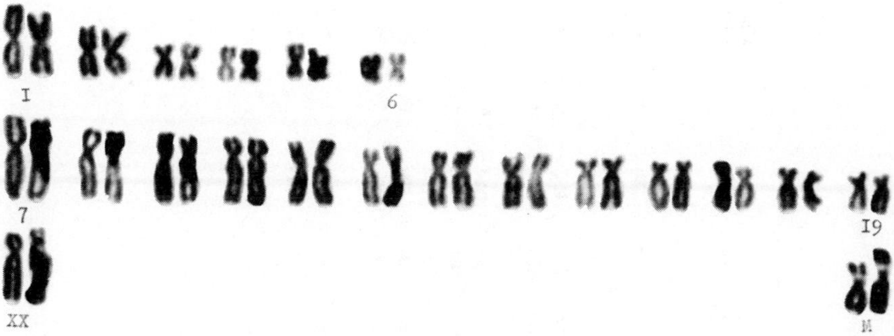

Chiarelli, B. (1963). *Rivista di Antropologia* 50: 103.

Chiarelli, B. (1962). Comparative morphometric analysis of primate chromosomes. II. The chromosomes of genera *Macaca*, *Papio*, *Theropithecus* and *Cercocebus*. *Caryologia* 15: 401–420.
Klinger, H.P. (1963). The somatic chromosomes of some primates (*Tupaia glis*, *Nycticebus coucang, Tarsius bancanus, Cercocebus aterrimus, Symphalangus syndactylus*). *Cytogenetics* 2: 140–151.

3.1.5.1. *Cercopithecus aethiops* 2n = 60

Ardito, G. – Original.

Chu, E.H.Y., and Giles, N.H. (1957). A study of primate chromosome complements. *American Naturalist* 91: 273–282.

3.1.5.2. *Cercopithecus cynosurus* 2n = 60

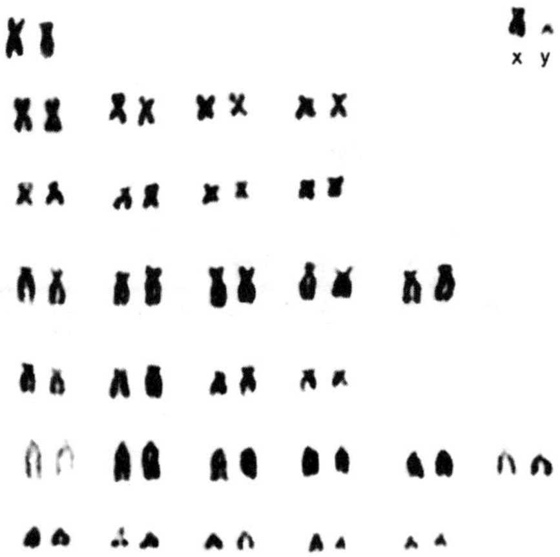

Farber, P.A. (1966). *Mammalian Chromosomes Newsletter* 22: 205.

Farber, P.A. (1966). The karyotype of *Cercopithecus pygerythrus*. *Mammalian Chromosomes Newsletter* 22: 205.

3.1.5.3. *Cercopithecus sabaeus* 2n = 60

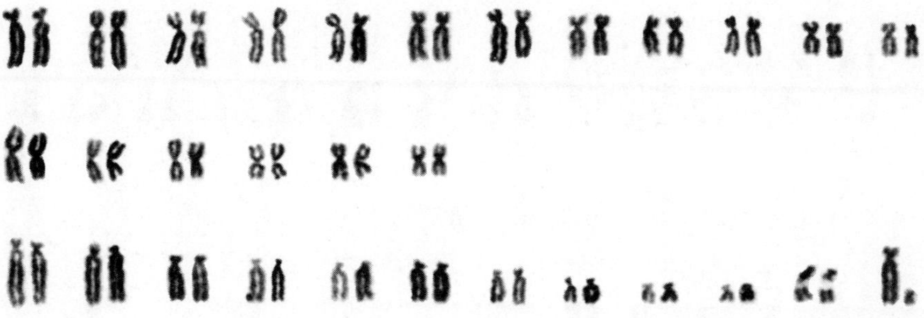

Ardito, G. – Original.

Chu, E.H.Y., and Giles, N.H. (1957). A study of primate chromosome complements. *American Naturalist* 91: 273–282.

3.1.5.4. *Cercopithecus cephus* 2n = 66

Ardito, G. – Original.

Chiarelli, B. (1963). Primi risultati di ricerche di genetica e cariologia comparata in primati e loro interesse evolutivo. *Rivista di Antropologia* 50: 87–124.

3.1.5.5. *Cercopithecus diana* 2n = 58

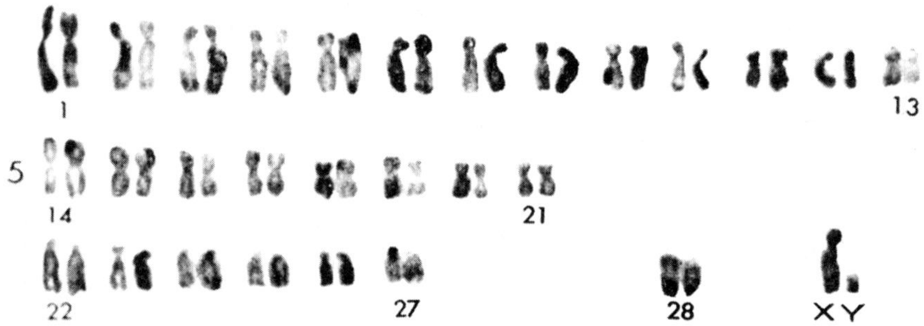

Chiarelli, B. (1968). *Cytologia* 33: 6.

Chu, E.H.Y., and Giles, N.H. (1957). A study of primate chromosome complements. *American Naturalist* 91: 273–282.
Chiarelli, B. (1968). Chromosome polymorphism in the species of the genus *Cercopithecus*. *Cytologia* 33: 1–16.

3.1.5.6. *Cercopithecus l'hoesti* 2n = 58–60–72

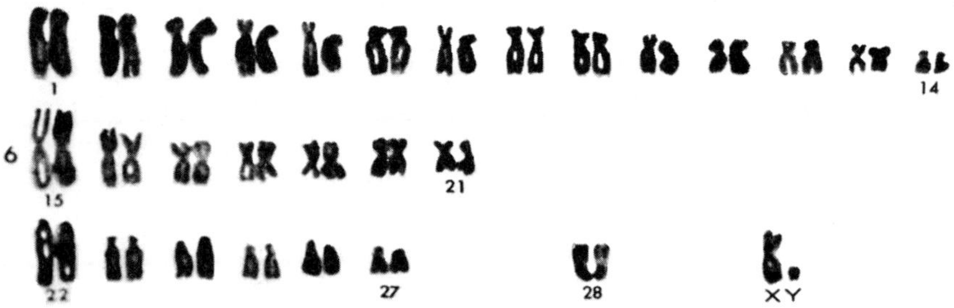

Chiarelli, B. (1968). *Cytologia* 33: 6.

Chu, E.H.Y., and Bender, M.A. (1961). Chromosome cytology and evolution in primates. *Science* 133: 1,399–1,405.
Chiarelli, B. (1968). Chromosome polymorphism in the species of the genus *Cercopithecus*. *Cytologia* 33: 1–16.

3.1.5.8. *Cercopithecus hamlyni* 2n = 64

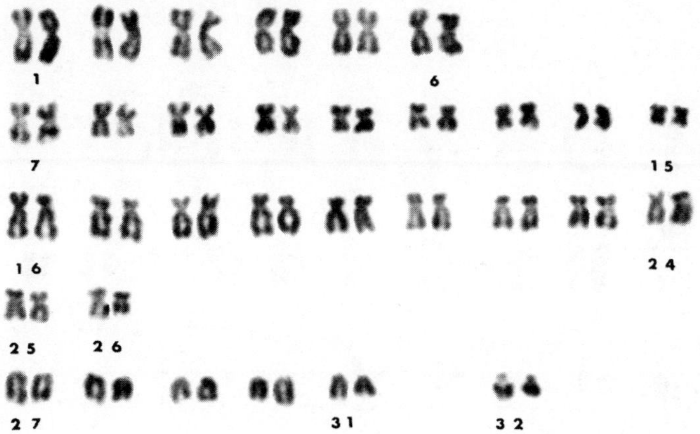

De Boer, L.E.M. – Original.

De Boer, L.E.M. (1971). Cytology of the Cercopithecidae with special reference to the karyotype of *Cercopithecus hamlyni* Pocock 1907. *Proceedings of the Third International Congress of Primatology, Zurich, 1970,* volume two: 121–126.

De Boer L.E.M. This volume.

3.1.5.9. *Cercopithecus mona* 2n = 66–68

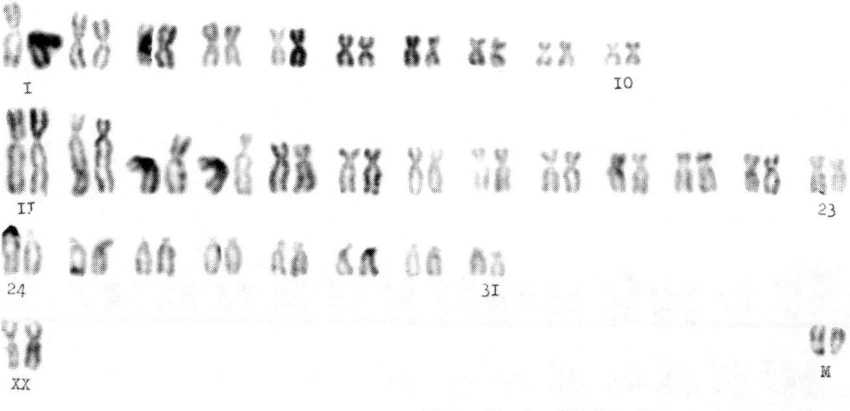

Chiarelli, B. (1968). *Taxonomy and phylogeny of Old World primates with references to the origin of man.* Edited by B. Chiarelli. Turin: Rosenberg and Sellier: 159.

Bender, M.A., and Mettler, L.E. (1958). Chromosome studies of primates. *Science* 128: 186–190.

Chiarelli, B. (1968). "Caryological and hybridological data for the taxonomy and phylogeny of the Old World primates," in *Taxonomy and phylogeny of Old World primates with references to the origin of man.* Edited by B. Chiarelli, 151–186. Turin: Rosenberg and Sellier.

3.1.5.15. *Cercopithecus neglectus* 2n = 58–62

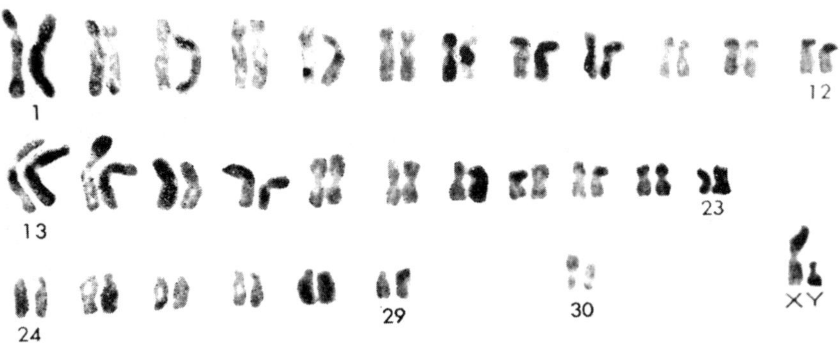

Chiarelli, B. (1968). *Cytologia* 33: 9.

Chiarelli, B. (1962). Comparative morphometric analysis of primate chromosomes. II. The chromosomes of genera *Macaca*, *Papio*, *Theropithecus* and *Cercocebus*. *Caryologia* 15: 401–420.

Chiarelli, B. (1968). Chromosome polymorphism in the species of the genus *Cercopithecus*. *Cytologia* 33: 1–16.

3.1.5.16. *Cercopithecus nictitans* 2n = 66–70

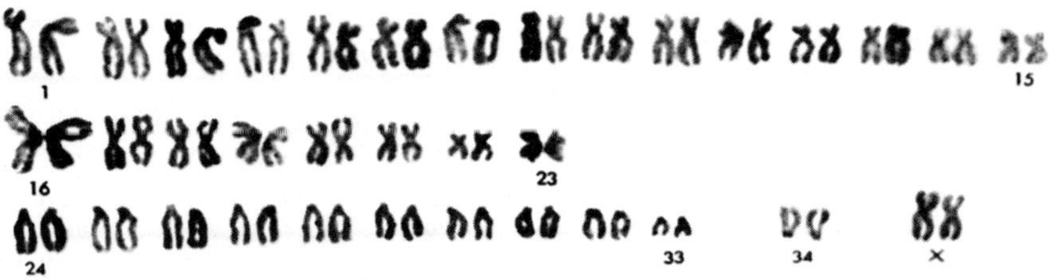

Chiarelli, B. (1968). *Cytologia* 33: 11.

Chu, E.H.Y., and Giles, N.H. (1957). A study of primate chromosome complements. *American Naturalist* 91: 273–282.

Chiarelli, B. (1968). Chromosome polymorphism in the species of the genus *Cercopithecus*. *Cytologia* 33: 1–16.

3.1.5.17. *Cercopithecus ascanius* 2n = 66

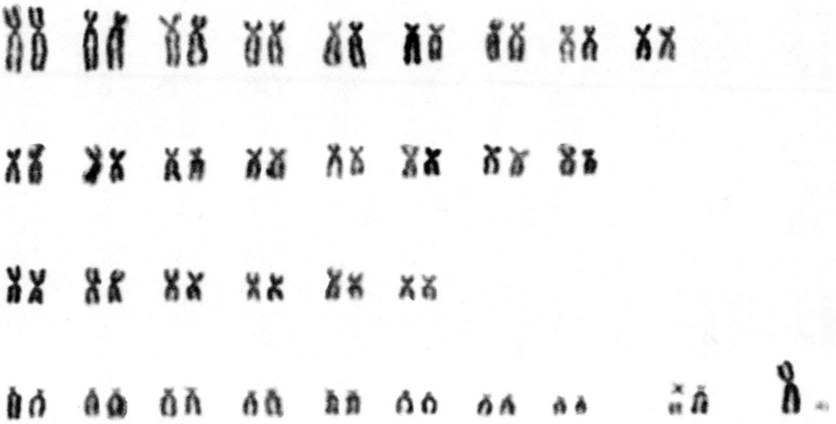

Ardito, G. – Original.

Hsu, T.C. and Benirschke, K. (1971). *An atlas of mammalian chromosomes*, volume six: 300.

3.3.5.20. *Cercopithecus mitis* 2n = 72

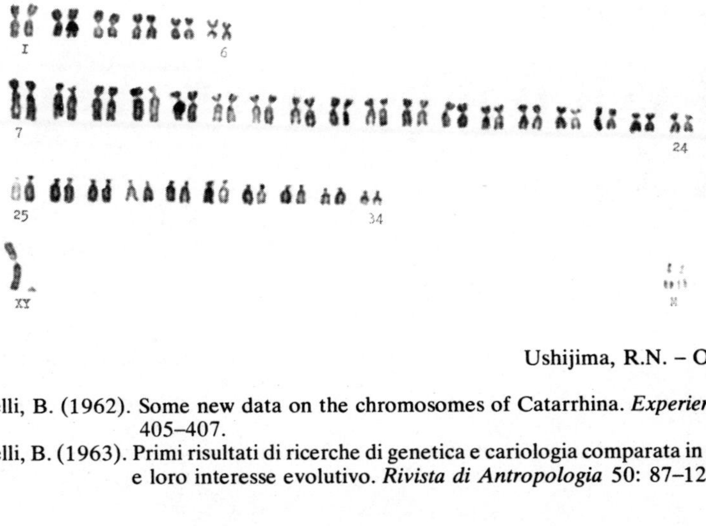

Ushijima, R.N. – Original.

Chiarelli, B. (1962). Some new data on the chromosomes of Catarrhina. *Experientia* 18: 405–407.

Chiarelli, B. (1963). Primi risultati di ricerche di genetica e cariologia comparata in primati e loro interesse evolutivo. *Rivista di Antropologia* 50: 87–124.

3.1.5.21. *Cercopithecus nigroviridis* 2n = 48

Ardito, G. – Original.

Chiarelli, B. (1962). Some new data on the chromosomes of Catarrhina. *Experientia* 18: 405–407.
Wurster, D.H., and Benirschke, K. (1969). Chromosomes of some primates. *Mammalian Chromosomes Newsletter* 10(1): 3–5.

3.1.5.22. *Cercopithecus talapoin* 2n = 54

Chiarelli, B. (1963). *Rivista di Antropologia* 50: 106.

Chiarelli, B. (1962). Some new data on the chromosomes of Catarrhina. *Experientia* 18: 405–407.
Chiarelli, B. (1963). Primi risultati di ricerche di genetica e cariologia comparata in primati e loro interesse evolutivo. *Rivista di Antropologia* 50: 87–124.

3.1.6.1. *Erythrocebus patas* 2n = 54

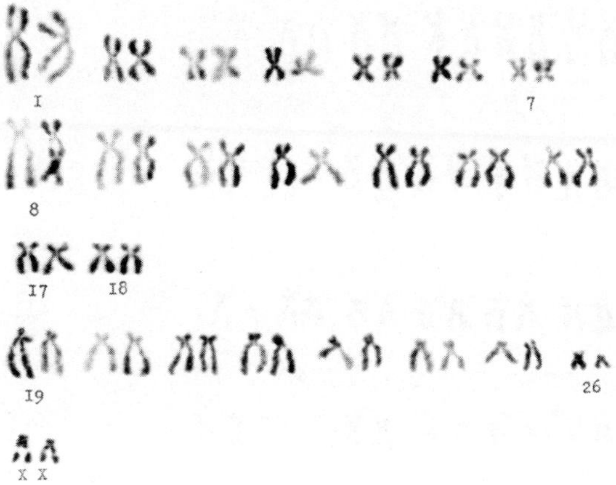

De Boer, L.E.M. – Original.

Chu, E.H.Y., and Giles, N.H. (1957). A study of primate chromosome complements. *American Naturalist* 91: 273–282.

3.2.1.1. *Presbytis entellus* 2n = 44

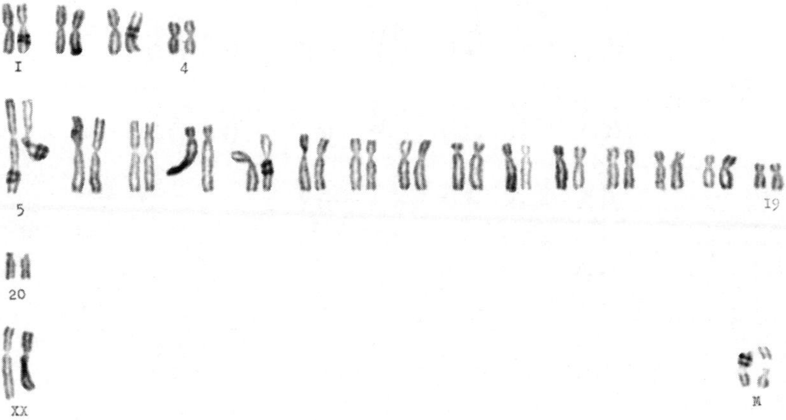

Ushijima, R.N. – Original.

Ushijima, R.N., Shininger, F.S., and Grand, T. (1964). Chromosome complements of two species of primates: *Cynopithecus niger* and *Presbytis entellus*. *Science* 146: 78–79.

3.2.1.2. *Presbytis senex* 2n = 44

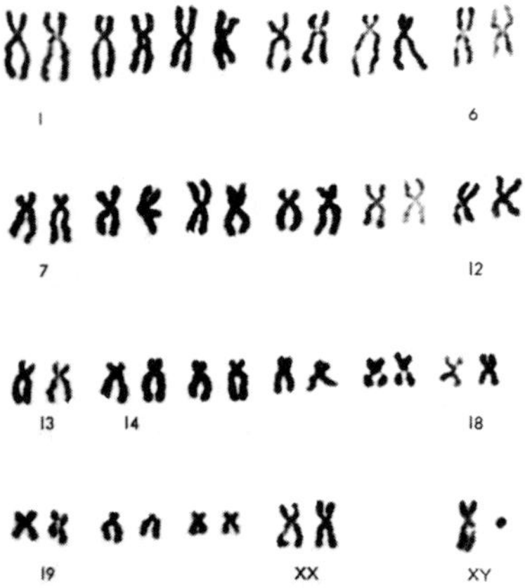

Goldstein, P.R., and Birdwell, R.T. (1967). *Mammalian Chromosomes Newsletter* 8: 197.

Goldstein, P.R., and Birdwell, R.T. (1967). Cytogenetic studies in the purple-faced langur. *Mammalian Chromosomes Newsletter* 8: 197.

3.2.1.8. *Prebytis cristatus* 2n = 44

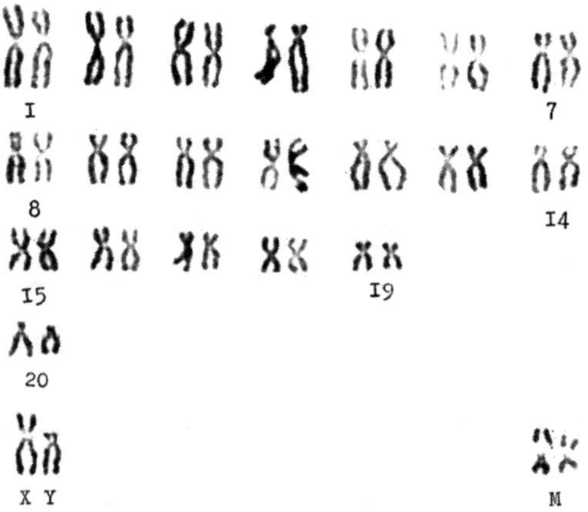

Wurster, D.H., and Benirschke, K. (1969). *Mammalian Chromosomes Newsletter* 10(1): 4.

Wurster, D.H., and Benirschke, K. (1969). Chromosomes of some primates. *Mammalian Chromosomes Newsletter* 10(1): 3–5.

3.2.1.10. *Presbytis obscurus* 2n = 44

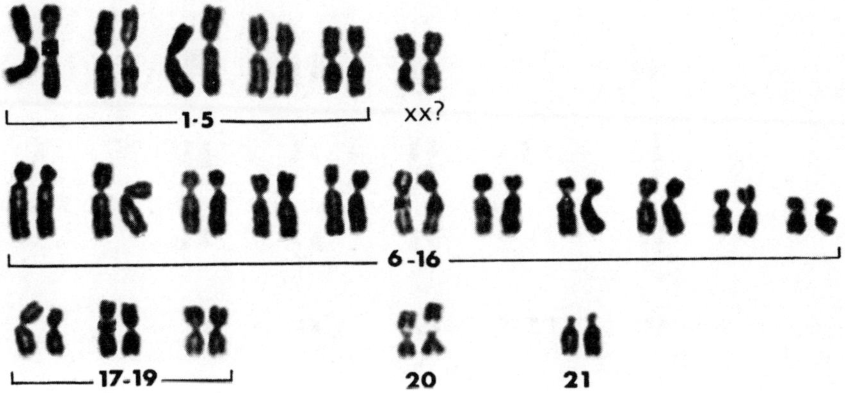

De Boer, L.E.M. – Original.

Chiarelli, B. (1962). Some new data on the chromosomes of Catarrhina. *Experientia* 18: 405–407.

Chiarelli, B. (1963). Primi risultati di ricerche di genetica e cariologia comparata in primati e loro interesse evolutivo. *Rivista di Antropologia* 50: 87–124.

3.2.2.1. *Pygathrix nemaeus* 2n = 44

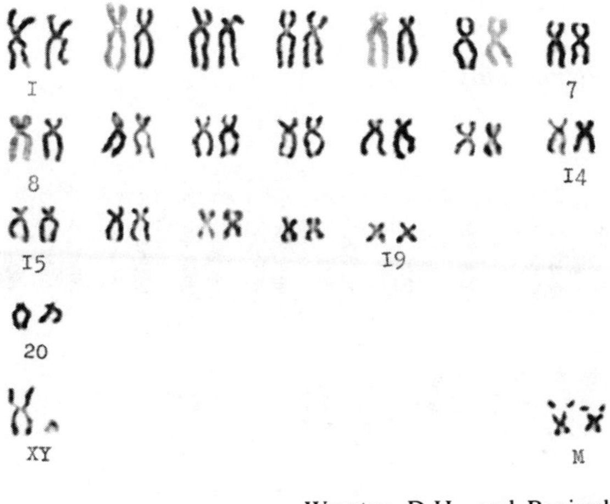

Wurster, D.H., and Benirschke, K. (1969). *Mammalian Chromosomes Newsletter* 10(1): 4.

Wurster, D.H., and Benirschke, K. (1969). Chromosomes of some primates. *Mammalian Chromosomes Newsletter* 10(1): 3–5.

3.2.3.1. *Rhinopithecus roxellanae* 2n = 44

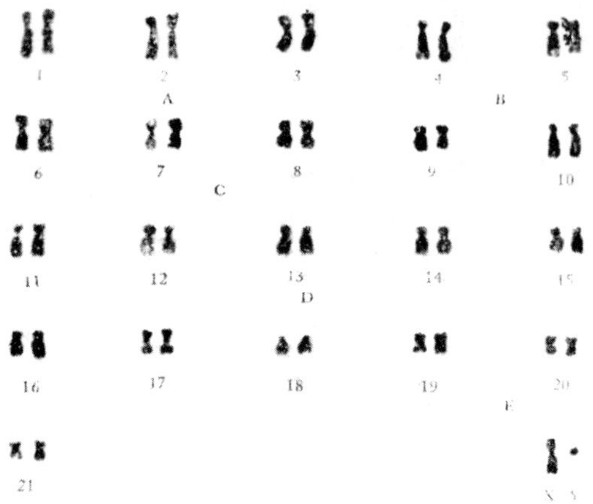

Chen Yifeng et al. (1978). *Acta Zoologica Sinica* 24(2): 130.

Chen Yifeng, Luo Lihua, Shan Xiangnian and Cao Xiaomei (1978). The karyotype of the golden monkey (*Rhinopithecus r. roxellanae*). *Acta Zoologica Sinica* 24(2): 127–135.

3.2.5.1. *Nasalis larvatus* 2n = 48

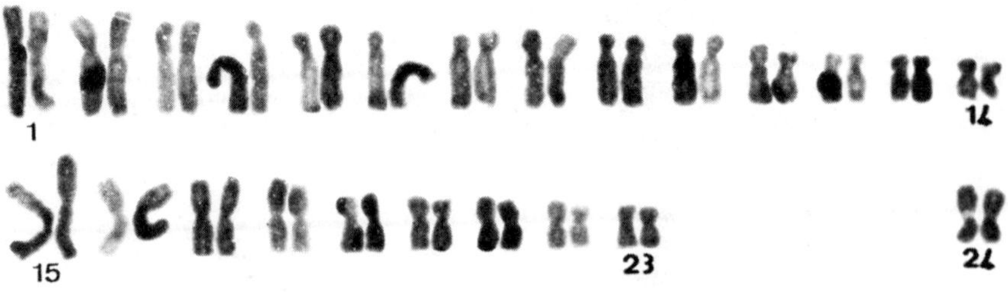

Chiarelli, B. (1965). *Rivista di Antropologia* 52: 160.

Chiarelli, B. (1965). Il cariotipo di *Nasalis larvatus* e la sua importanza nella citotassonomia delle scimmie del vecchio mondo. Nota preliminare. *Rivista di Antropologia* 52: 159–164.

3.2.6.1. *Colobus polykomos* 2n = 44

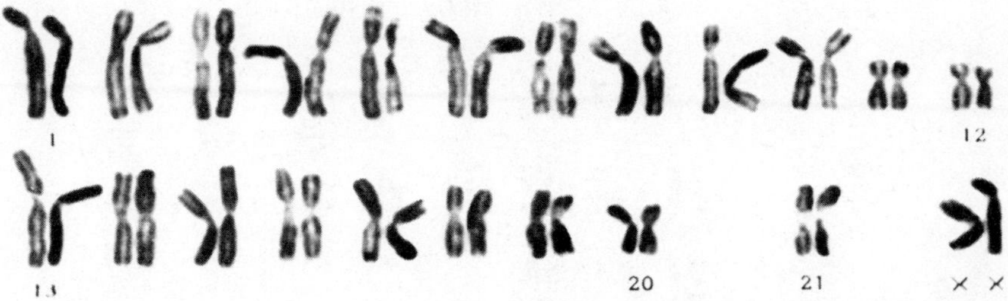

Chiarelli, B. (1966). *American Journal of Physical Anthropology* 24(2): 157.

Chiarelli, B. (1962). Some new data on the chromosomes of Catarrhina. *Experientia* 18: 405–407.

Bender, M.A., and Chu, E.H.Y. (1963). "The chromosomes of primates," in *Evolutionary and genetic biology of the primates*, volume one. Edited by J. Buettner-Janusch, 261–301. London: Academic Press.

3.3.1.1. *Hylobates lar*

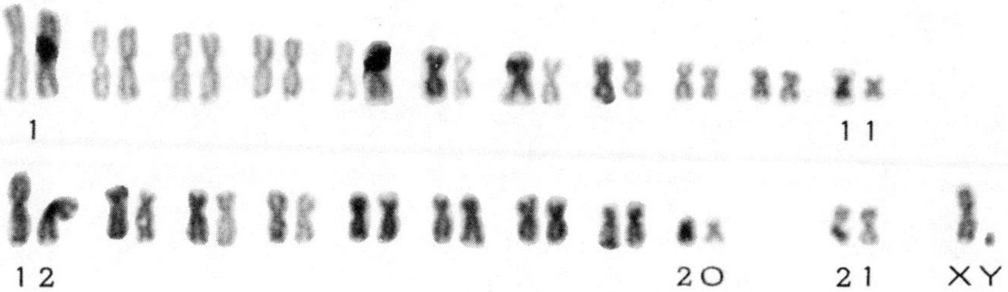

Chiarelli, B. (1972). *Gibbon and siamang*, volume one. Edited by D.M. Rumbaugh, 93. Basel: S. Karger.

Chiarelli, B. (1962). Some new data on the chromosomes of Catarrhina. *Experientia* 18: 405–407.

Hamerton, J.L. (1963). Primate chromosomes. *Symposia of the Zoological Society of London* 10: 211–219.

3.3.1.2. *Hylobates agilis* 2n = 44

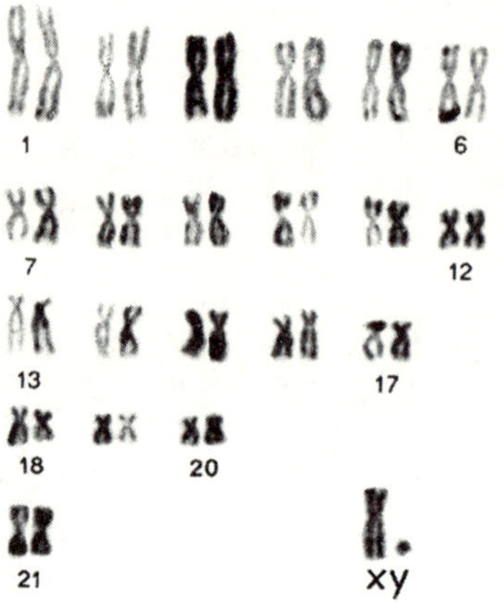

Chiarelli, B. (1963). *Caryologia* 16: 639.

Chiarelli, B. (1962). Some new data on the chromosomes of Catarrhina. *Experientia* 18: 405–407.

Chiarelli, B. (1963). Comparative morphometric analysis of primate chromosomes. III. The chromosomes of the genera *Hylobates*, *Colobus* and *Presbytis*. *Caryologia* 16: 637–648.

3.3.1.3. *Hylobates moloch* 2n = 44

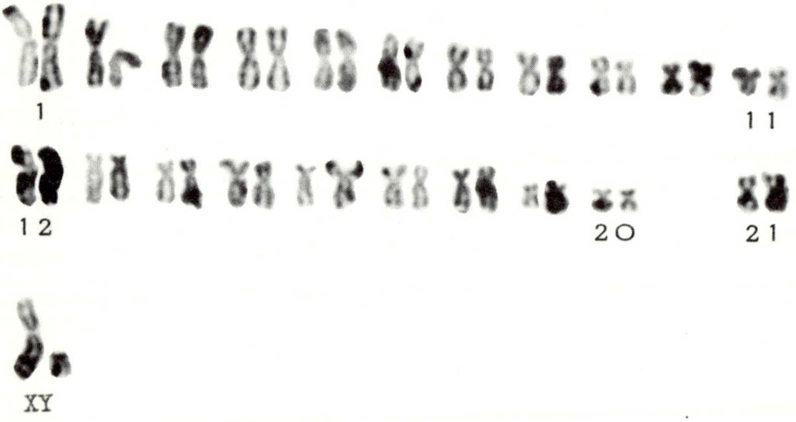

Chiarelli, B. (1972). *Gibbon and siamang*, volume one. Edited by D.M. Rumbaugh, 93. Basel: S. Karger.

Chiarelli, B. (1962). Some new data on the chromosomes of Catarrhina. *Experientia* 18: 405–407.

Hamerton, J.L., Klinger, H.P., Mutton, D.E., and Lang, E.M. (1963). The somatic chromosomes of the Hominoidea. *Cytogenetics* 2: 240–263.

3.3.1.4. *Hylobates concolor* 2n = 52

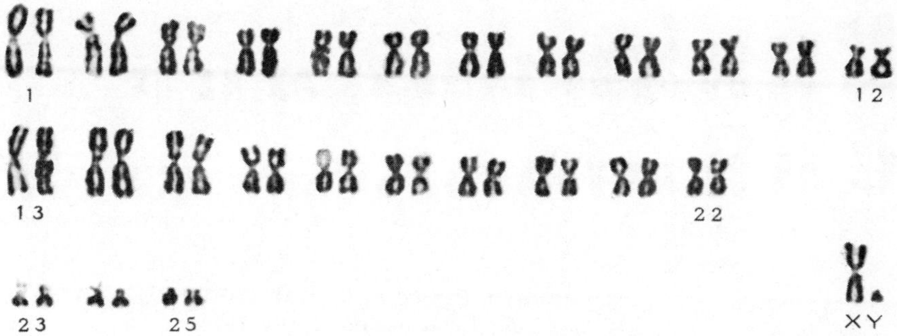

Wurster, D. Quoted by Chiarelli, B. (1972) *L'Anthropologie* 76: 529.

Wurster, D.H., and Benirschke, K. (1969). Chromosomes of some primates. *Mammalian Chromosomes Newsletter* 10(1): 3–5.

3.3.1.5. *Hylobates hoolock* 2n = 44

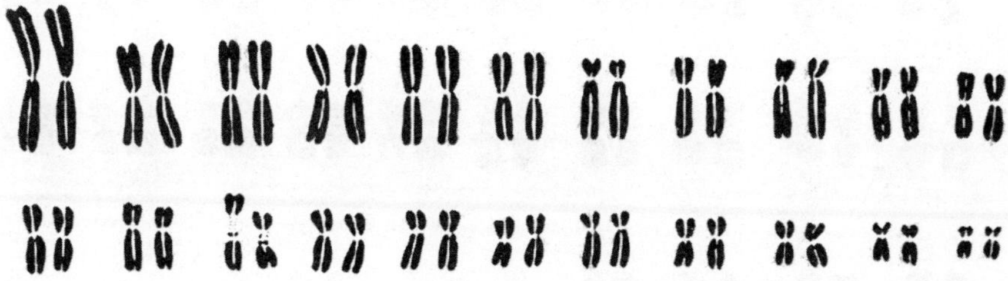

Bender, M.A., and Chu, E.H.Y. (1963). *Evolutionary and genetic biology of the primates*, volume one. Edited by J. Buettner-Janusch, 297. London: Academic Press.

Chu, E.H.Y., and Bender, M.A. (1961). Chromosome cytology and evolution in primates. *Science* 133: 1,399–1,405.

Bender, M.A., and Chu, E.H.Y. (1963). "The chromosomes of primates," in *Evolutionary and genetic biology of the primates*, volume one. Edited by J. Buettner-Janusch, 261–310. London: Academic Press.

3.3.1.6. *Hylobates klossii* 2n = 44

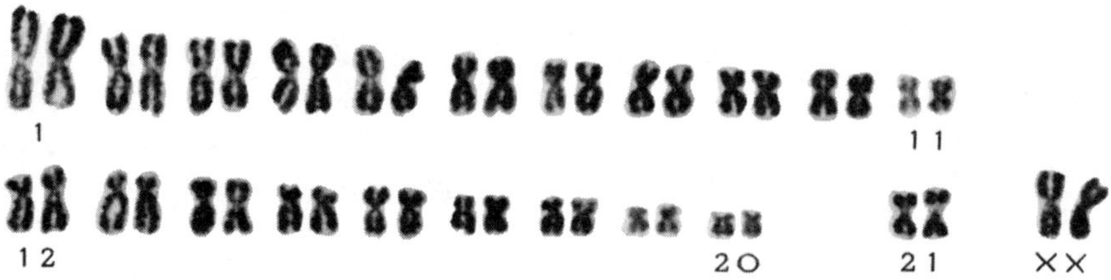

Hölsi, P., and Lang, E.M. (1970). *Schweizer Archiv für Tierheilkunde* 112(6): 297.

Hölsi, P., and Lang, E.M. (1970). Die Chromosomen des Zwergsiamang (*Symphalangus klossii*). *Schweizer Archiv für Tierheilkunde* 112(6): 296–297.

3.3.2.1. *Symphalangus syndactylus* 2n = 50

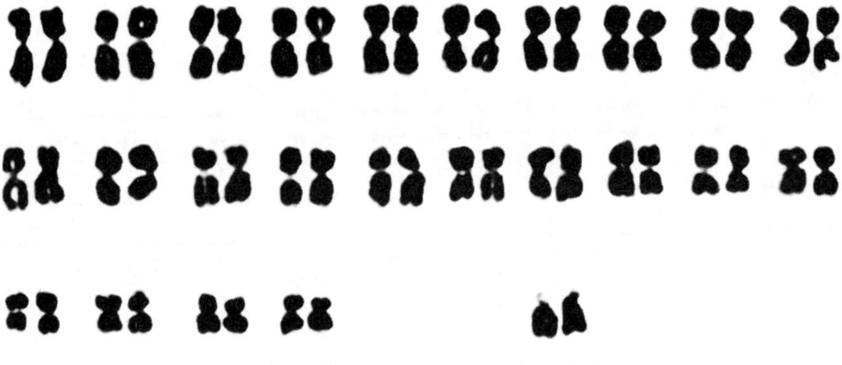

De Boer, L.E.M. – Original.

Bender, M.A., and Chu, E.H.Y. (1963). "The chromosomes of primates," in *Evolutionary and genetic biology of the primates*, volume one. Edited by J. Buettner-Janusch, 261–310. London: Academic Press.
Klinger, H.P. (1963). The somatic chromosomes of some primates (*Tupaia glis*, *Nycticebus coucang, Tarsius bancanus, Cercocebus atterrimus, Symphalangus syndactylus*). *Cytogenetics* 2: 140–151.

3.4.1.1. *Pongo pygmaeus* 2n = 48

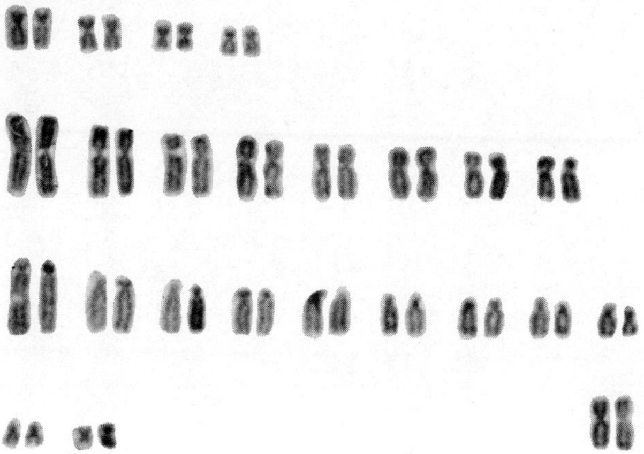

Chiarelli, B. (1962). *Caryologia* 15(1): 106.

Chiarelli, B. (1961). Chromosomes of the orangutan (*Pongo pygmaeus*). *Nature* 192: 285.

3.4.2.1. *Pan troglodytes* 2n = 48

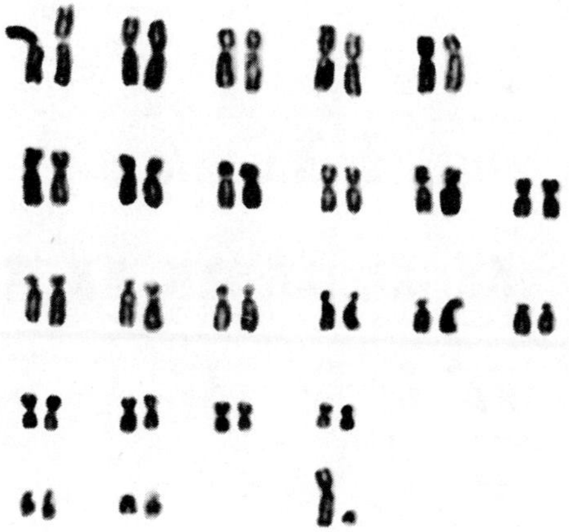

Chiarelli, B. (1962) – Original.

Yeager, C.H., Painter, T.S., and Yerkes, R.M. (1940). The chromosomes of the chimpanzee. *Science* 91: 74–75.
Young, W.J., Merz, T., Ferguson-Smith, M.A., and Johnston, A.W. (1960). Chromosome number of the chimpanzee, *Pan troglodytes*. *Science* 131: 1,672–1,673.

3.4.2.2. *Pan paniscus* 2n = 48

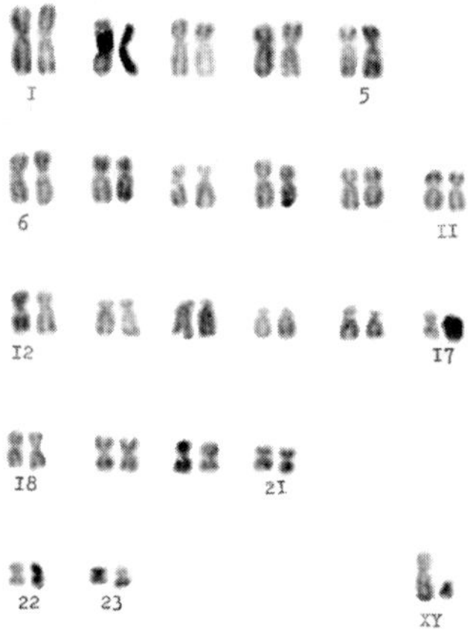

Chiarelli, B. (1962). *Caryologia* 15(1): 105.

Chiarelli, B. (1962). Comparative morphometric analysis of primate chromosomes. I. The chromosomes of anthropoid apes and of man. *Caryologia* 15: 99–121.

3.4.3.1. *Gorilla gorilla* 2n = 48

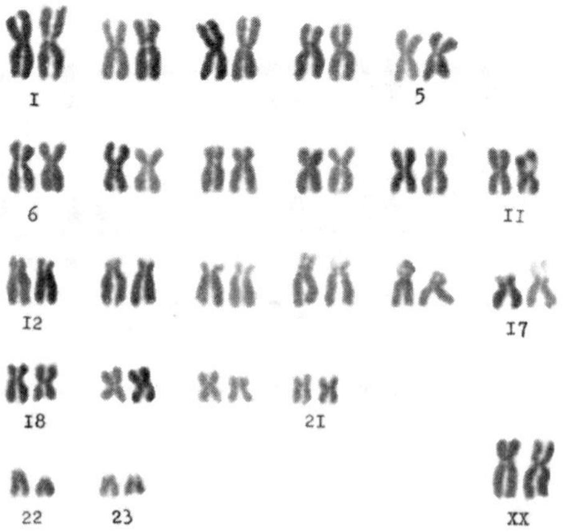

Egozcue, J. and B. Chiarelli (1967). *Folia Primatologica* 5: 239.

Hamerton. J.L., Fraccaro, M., de Carli, L., Nuzzo, F., Klinger, H.P., et al. (1961). Somatic chromosomes of the gorilla. *Nature* 192: 225–228.

3.5.1.1. *Homo sapiens* 2n = 46

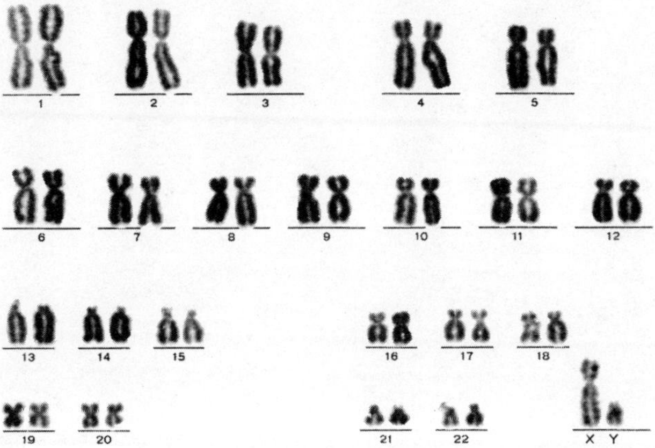

Ardito, G. – Original.

Tijo, J.H., and Levan, A. (1956). The chromosome number of man. *Hereditas* 42: 1–6.

APPENDICES

Proposed Cooperation on Tissue and Information Storage

PRIMATE CELL REPOSITORY

The increasing use of biological samples from nonhuman primate species and the continuously decreasing availability of these animals for biomedical research induced the primatologists present at the Detroit Primate Conference (August, 1973) to suggest the possibility of storing in liquid nitrogen biological samples of tissue from different primate species for further studies.

The feasibility of this project has been thoroughly discussed, and an international committee composed of Professors T. Caspersson, B. Chiarelli, A. Falek, A. Koen, and R. Eckhardt has been established for further development of the project.

PRIMATE CHROMOSOME ATLAS

Following the recommendation of the participants in the Primate Chromosome Conference (held in Detroit in August, 1973) and in anticipation of the publication of an atlas on primate chromosomes, the *Journal of Human Evolution* will devote a certain number of pages in each issue to the publication of banded karyotypes of different primates.

Cytogeneticists interested in publishing karyotypes in this atlas should submit to the editor at least three copies of a banded metaphase, along with their proposed karyotype arrangement of the chromosomes. The material must be accompanied by detailed information on the technique used for banding and the source of the animal. The date on which the pictures and the information are received will be quoted in the publication.

Statement on the Definition of Primate Species

Members of different primate populations have usually been classified as different species or genera on the basis of gross morphological characteristics. The present state of knowledge, derived from studies in captivity and under natural conditions, indicates that at least some of the species belonging to *Lemur, Macaca, Papio*, and others can hybridize, producing viable, fertile offspring.

The widely accepted biological definition of what constitutes a species ("species are groups of interfertile populations reproductively isolated from other similar groups": Mayr 1963) would suggest that in cases in which hybrids occur, taxonomic subdivision at the species level would not be justified. Data derived from biochemical, cytogenetic, and ethological studies must be considered in species determination. It was suggested by the physical anthropologists and primatologists at the IXth International Congress of Anthropological and Ethnological Sciences, meeting in Detroit, that this requirement is necessary in order to ensure that current taxonomic practice conforms with present biological concepts.

Resolution Concerning the Conservation of Primates, Proposed and Approved by the IXth International Congress of Anthropological and Ethnological Sciences

Primates are being used, and their continued use is contemplated, in research on population growth and change.

Therefore, being aware of the unique value of nonhuman primates as man's closest relatives and as models serving the biological and medical sciences in the advancement of human health, welfare, and knowledge;

being aware of their responsibility to preserve the existence of the whole spectrum of contemporary primate species;

being aware that expanding human populations and the growing use of nonhuman primates have threatened some species with extinction, while others have become drastically reduced;

anthropologists at the ICAES urgently request that the International Union of Anthropological and Ethnological Sciences encourage governments to assure the following:

1. That scientists be selective in the use of nonhuman primates and employ other animal models when appropriate, and that they require that use of endangered and rare species be limited to investigations in which other species of nonhuman primates are unsuitable.

2. That scientists contribute to the conservation needs of nonhuman primates:

by introducing and insisting upon humanitarian and efficient procedures for their capture, translocation, and maintenance prior to and during their use;

by insisting upon methods of acquisition which ensure the enforcement of national laws covering capture and transportation and upon the international support of these laws;

by promoting the development of knowledge of the distribution and status of nonhuman primate populations in the areas where they occur, and of management and husbandry methods which ensure their survival in natural ecosystems and as economic and scientific resources;

by promoting the development of permanent breeding programs according to a long-term requirement for different species;

by encouraging urgent cooperation, financial support, and investments from research institutions and pharmaceutical and other industries to accomplish these aims for ensuring the continued existence and development of nonhuman primate populations in their natural ecosystems and/or in especially designed environments, so that their place in the ecosystem be preserved.

Complete Bibliography on Primate Karyology

CHRONOLOGICAL BIBLIOGRAPHY

1922

PAINTER, T. S.
The sex chromosomes of the monkeys. *Science* 56:286–287.

1924

PAINTER, T. S.
Studies in mammalian spermatogenesis. IV. The sex chromosomes of monkeys. *Journal of Experimental Zoology* 39:433–462.

1939

SHIWAGO, P. I.
Recherches sur le caryotype du *Rhesus macacus. Bulletin de Biologie et de Médecine Expérimentale* (U.S.S.R.) 9:3–8.

1940

YEAGER, C. H., T. S. PAINTER, R. M. YERKES
The chromosomes of the chimpanzee. *Science* 91:74–75.

1952

MAKINO, S.
A contribution to the study of the chromosomes in some Asiatic mammals. *Cytologia* 16:288–301.

1955

DARLINGTON, C. D., A. HAGUE
Chromosomes of monkeys and man. *Nature* 175:32.

MATTHEY, R.
Les chromosomes de *Galago senegalensis* Geoffroy (Prosimii-Lorisidae-Galaginae). *Revue Suisse de Zoologie* (supplement) 62:190–197.

PRINCE, R. H., M. A. GRAHAM, M. L. BARR
Nuclear morphology according to sex in *Macacus rhesus. Anatomical Record* 122:153–171.

1956

CHU, E. H. Y., N. H. GILES
Primate chromosome complements determined by tissue culture techniques. *Records of the Genetic Society of America* 25:637.

FORD, C. E., J. L. HAMERTON
The chromosomes of man. *Nature* 178:1,020–1,023.

ISMAIL, H., P. V. TOBIAS
In *Chromosomes, sex-cells and evolution in a mammal.* Edited by P. V. Tobias. London: Lund.

TIJO, H., A. LEVAN
The chromosome number of man. *Hereditas* 42:1–6.

1957

CHU, E. H. Y., N. H. GILES
A study of primate chromosome complements. *American Naturalist* 91:273–282.

PARK, W. W.
The occurrence of sex chromatin in early human and macaque embryos. *Journal of Anatomy* 91:369–373.

1958

BENDER, M. A., L. E. METTLER
Chromosome studies of primates. *Science* 128:186–190.

CHIARELLI, B.
Tavole cromosomiche dei primati. *Caryologia* 11:99–104.

KLEINFELD, R., E. H. Y. CHU
DNA determination of kidney cell cultures of three species of monkeys. *Cytologia* 23:452–459.

METTLER, L., M. A. BENDER
Chromosome studies and their evolutionary significance in New World primates. *Proceedings of the Tenth International Congress of Genetics*, volume two:186.

ROTHFELS, K. H., L. SIMINOVITCH
The chromosome complement of the rhesus monkey (*Macaca mulatta*) determined in kidney cells cultivated "in vitro." *Chromosoma* 9:163–175.

1959

CHIARELLI, B., L. DE CARLI
Evoluzione delle conoscenze sui cromosomi umani e loro significato in antropologia. *Archivio per l'Antropologia e la Etnologia* 89:149–167

1960

BENDER, M. A., L. E. METTLER
Chromosome studies of primates. II. *Callithrix, Leontocebus* and *Callimico. Cytologia* 25:400–404.

CHIARELLI, B., F. NUZZO, L. DE CARLI, B. AULISA
Morphometric analysis of human chromosomes. *Atti Associazione Genetica Italiana* 5:265–273.

YOUNG, W. J., T. MERZ, M. A. FERGUSON-SMITH, A. W. JOHNSTON
Chromosome number of the chimpanzee, *Pan troglodytes. Science* 131:1,672–1,673.

1961

CHIARELLI, B.

Ibridologia e sistematica in primati. I. Raccolta di dati. *Atti Associazione Genetica Italiana* 6:213–220.

Cariologia e sistematica dei primati con cenni sull'origine del cariotipo umano. Comunicazione solo in riassunto pubblicata alla Società Italiana di Antropologia ed Etnologia nella riunione del 19 dicembre 1961.

Some chromosome numbers in primates. *Mammalian Chromosomes Newsletter* 6:3–4.

Chromosomes of the orangutan (*Pongo pygmaeus*). *Nature* 192:285.

Comparative morphometric analysis of the primate chromosomes. *Mammalian Chromosomes Newsletter* 6:4–5.

CHU, E. H. Y., M. A. BENDER

Chromosome cytology and evolution in primates. *Science* 133:1,399–1,405.

Cytogenetics and evolution of primates. *Annals of the New York Academy of Sciences* 102:253–266.

CHU, E. H. Y., B. A. SWOMLEY

Chromosomes of lemurine lemurs. *Science* 133:1,925–1,926.

HAMERTON, J. L., M. FRACCARO, L. DE CARLI, F. NUZZO, H. P. KLINGER, *et al.*

Somatic chromosomes of the gorilla. *Nature* 192:225–228.

1962

BEATH, M. N., K. BENIRSCHKE

The sex chromatin of interphase nuclei in platyrrhine monkeys. *Cytologia* 27:1–10.

BENDER, M. A., P. E. EIDE

Separation of monkey (*Ateles*) leucocytes for culture. *Mammalian Chromosomes Newsletter* 8:3.

BENIRSCHKE, K., J. M. ANDERSON, L. E. BROWNHILL

Marrow chimerism in marmosets. *Science* 138:513–515.

BENIRSCHKE, K., L. E. BROWNHILL

Further observations on marrow chimerism in marmosets. *Cytogenetics* 1:245–257.

CHIARELLI, B.

Comparative morphometric analysis of primate chromosomes. I. The chromosomes of anthropoid apes and of man. *Caryologia* 15(1):99–121.

Comparative morphometric analysis of primate chromosomes. II. The chromosomes of genera *Macaca, Papio, Theropithecus* and *Cercocebus*. *Caryologia* 15:401–420.

Some new data of the chromosomes of Catarrhina. *Experientia* 18:405.

Karyological evolution in primates and origin of the human karyotype. *Atti Associazione Genetica Italiana* 7:284–285.

MATTHEY, R.

Les nombres diploides des mammifères euthériens. Liste critique. *Mammalian Chromosomes Newsletter* 8:17–25.

SRIVASTAVA, P. K., A. K. SRIVASTAVA, F. V. LUCAS

Somatic chromosomes of the squirrel monkey (*Saimiri sciureus*). *Primates* 10:171–180.

1963

BENDER, M. A., E. H. Y. CHU

"The chromosomes of primates", in *Evolutionary and genetic biology of the*

BENDER, M.A., E.H.Y. CHU — *Cont.*
primates, volume one. Edited by J. Buettner-Janusch, 261–310. London, New York: Academic Press.

BENDER, M. A., P. E. EIDE
The karyotype of the woolly monkey (*Lagothrix*). *Mammalian Chromosomes Newsletter* 9:40.

BENIRSCHKE, K., L. E. BROWNHILL
Heterosexual cells in testes of chimeric marmoset monkeys. *Cytogenetics* 2:231–240.
Mitotic chromosomes in some species of marmosets. *Mammalian Chromosomes Newsletter* 9:6.

BENIRSCHKE, K., L. E. BROWNHILL, R. LOW
Chromosome complement of two further species of marmoset monkeys, *Tamarinus leucopus* and *Tamarinus mystax*. *Mammalian Chromosomes Newsletter* 10:83.

CHIARELLI, B.
Primi risultati di ricerche di genetica e cariologia comparata in primati e loro interesse evolutivo. *Rivista di Antropologia* 50:87–124.
Comparative morphometric analysis of primate chromosomes. III. The chromosomes of the genera *Hylobates, Colobus* and *Presbytis*. *Caryologia* 16:637–648.

HAMERTON, J. L.
Primate chromosomes. *Symposia of the Zoological Society of London* 10:211–219.

HAMERTON, J. L., H. P. KLINGER, D. E. MUTTON, E. M. LANG
The somatic chromosomes of the Hominoidea. *Cytogenetics* 2:240–263.

HSU, T. C., M. L. JOHNSON
Karyotypes of two mammals from Malaya. *American Naturalist* 97:127–129.

KLINGER, H. P.
The somatic chromosomes of some primates (*Tupaia glis, Nycticebus coucang, Tarsius bancanus, Cercocebus aterrimus, Symphalangus syndactylus*). *Cytogenetics* 2:140–151.

KLINGER, H. P., J. L. HAMERTON, D. E. MUTTON, E. M. LANG
The chromosomes of the Hominoidea. *Burg Wartenstein Symposium for Anthropological Research*. New York: Wenner-Gren Foundation.

1964

BAYLET, R., H. GRATTEPANCHE
Sur les chromosomes des Cercopithecidae *Papio papio, Macaca mulatta, Cercopithecus aethiops, Erythrocebus patas*. *Comptes Rendus des Séances de la Société de Biologie et de ses Filiales* 158:1,382–1,385.
Chromosomes de *Cercopithecus aethiops*. *Comptes Rendus Hebdomadaires des Séances de l'Académie des Sciences* 259:3,096–3,099.

CHIARELLI, B.
Genetica comparata in primati e genetica umana. *Atti XI Convegno della Salute. Ferrara, 1964*:53–79.
Incremento genomico nel genere *Cercopithecus*. *Archivio per l'Antropologia e la Etnologia* 94:27–29.
A pair of marked chromosomes in Catarrhina monkeys. *Mammalian Chromosomes Newsletter* 15:3–4.

CHIARELLI, B., L. BARBERIS
Drumsticks in the leucocytes of primates. *Experientia* 20:679–681.

Il drumstick nei leucociti dei primati. *Caryologia* 17:567–573.
CHIARELLI, B., C. VACCARINO
Cariologia ed evoluzione nel genere *Cercopithecus. Atti Associazione Genetica Italiana* 9:328–339.
GENGOZIAN, N., J. S. BATSON, P. EIDE
Hematologic and cytogenetic evidence for hematopoietic chimerism in the marmoset *Tamarinus nigricollis. Cytogenetics* 3:384–393.
USHIJIMA, R. N., F. S. SHININGER, T. GRAND
Chromosome complements of two species of primates: *Cynopithecus niger* and *Presbytis entellus. Science* 146:78–79.

1965

BORGAONKAR, D. S.
Primate chromosome bibliography. *Mammalian Chromosomes Newsletter* 17:135–137.
Primate chromosome bibliography (supplement). *Mammalian Chromosomes Newsletter* 18:184–185.
CHIARELLI, B.
Il cariotipo di *Nasalis larvatus* e la sua importanza nella citotassonomia delle scimmie del vecchio mondo. Nota preliminare. *Rivista di Antropologia* 52:159–164.
EGOZCUE, J., M. VILARASAU DE EGOZCUE
The chromosomes of the howler monkey (*Alouatta caraya*). *Mammalian Chromosomes Newsletter* 17:84.
MANNA, G. K., M. TALUKDAR
Somatic chromosome number in twenty species of mammals from India. *Mammalian Chromosomes Newsletter* 17:77–78.

1966

BORGAONKAR, D. S.
A list of chromosome number in primates. *Mammalian Chromosomes Newsletter* 20:55–65.
A list of chromosome numbers in primates. *Journal of Heredity* 57:60–64.
CHIARELLI, B.
Karyological and hybridological data available for a taxonomic revision of the Old-World monkeys at a supergenetic level. *Mammalian Chromosomes Newsletter* 22:212–215.
Caryology and taxonomy of the catarrhine monkeys. *American Journal of Physical Anthropology* 24(2):155–169.
The chromosome complement of *Nasalis larvatus* (Wurm 1781). *Experientia* 22:797.
Marked chromosome in catarrhine monkeys. *Folia Primatologica* 4:74–80.
The chromosome complement of *Nasalis larvatus. Mammalian Chromosomes Newsletter* 20:68.
Survey on the caryology of the genus *Cercopithecus. Mammalian Chromosomes Newsletter* 22:209–210.
Summary of the karyology of Old-World primates. *Laboratory Primate Newsletter* 5:1–4.
Interesse tassonomico e filogenetico dei dati cariologici ed ibridologici nei primati del vecchio mondo. *Bollettino di Zoologia* 33:381–387.
Ibridologia e sistematica in primati. Deduzioni sul cariotipo degli ibridi. *Rivista di Antropologia* 53:113–117.

CHIARELLI, B. — *Cont.*
Dati cariologici e ibridologici per una revisione, al livello sopragenerico, della tassonomia delle scimmie del vecchio mondo. *Archivio per l'Antropologia e la Etnologia* 46:105–109.

CHIARELLI, B., L. BARBERIS
Some data on the chromosomes of *Prosimiae* and of New-World monkeys. *Mammalian Chromosomes Newsletter* 22:216.

CHIARELLI, B., M. SARTI
Nuclear protrusion in the polymorphonucleated leucocytes of the anthropoid apes. *Experientia* 22:652–653.

EGOZCUE, J.
Primate chromosome bibliography. *Mammalian Chromosomes Newsletter* 21:173.

EGOZCUE, J., R. N. USHIJIMA, M. VILARASAU DE EGOZCUE
The chromosomes of mongoose lemur (*Lemur mongoz*). *Mammalian Chromosomes Newsletter* 22:197–198.
The chromosomes of the slender loris (*Loris tardigradus*). *Mammalian Chromosomes Newsletter* 22:204.

EGOZCUE, J., M. VILARASAU DE EGOZCUE
The chromosomes of the slow loris (*Nycticebus coucang*). *Mammalian Chromosomes Newsletter* 20:50.
The chromosomes of the angwantibo (*Arctocebus calabarensis*). *Mammalian Chromosomes Newsletter* 20:53–54.
The chromosomes of *Cebus albifrons. Mammalian Chromosomes Newsletter* 20:65–66.
The chromosomes of *Cebus capucinus. Mammalian Chromosomes Newsletter* 20:71–72.
The chromosome complement of the howler monkey (*Alouatta caraya* Humboldt 1812). *Cytogenetics* 5:20–27.

FARBER, P. A.
Chromosomes of *Macaca mulatta* leukocytes. *Folia Primatologica* 4:409–415.
The karyotype of *Cercopithecus pygerythrus. Mammalian Chromosomes Newsletter* 22:205.

HABERLANDT, W. F.
Chromosomenforschung bei Mensch und Primaten. *Arzneimittel Forschung* 11:573–578.

SHARMA, T., S. KAKATI
Karyotype of *Presbytis entellus entellus* (primate). *Cytologia* 32:111–116.
Chromosomes of Indian common langur. *Mammalian Chromosomes Newsletter* 20:70.

WOHNUS, J. F., K. BENIRSCHKE
Chromosome analysis of four species of marmosets (*Callithrix jacchus, Tamarinus mystax, Tamarinus nigricollis, Cebuella pygmaea*). *Cytogenetics* 5:94–105.

1967

BARBERIS, L.
"The idiograms of three species of the genus *Cercopithecus: C. talapoin, C. aethiops* and *C. mona,*" in *Neue Ergebnisse der Primatologie. Progress in primatology*. Edited by D. Stark, R. Schneider, and H. J. Kuhn, 155–159. Stuttgart: G. Fischer.

BORGAONKAR, D. S.
Addition to the lists of chromosome numbers in the insectivores and primates. *Journal of Heredity* 58:211–213.
CHIARELLI, B.
"Somatic and meiotic chromosomes in the different species of Papio," in *The baboon in medical research*, volume two. Edited by H. Vagtborg, 271–281. Austin: University of Texas Press.
La morfologia del cromosoma "Y" nelle differenti specie di primati. *Rivista di Antropologia* 54:137–140.
"Caryological and hybridological data available for a taxonomic revision of the Old World primates at a supergenetic level," in *Neue Ergebnisse der Primatologie. Progress in primatology*. Edited by D. Stark, R. Schneider, and H. J. Kuhn, 160–163. Stuttgart: G. Fischer.
Chromosomy a taksonomia i filogeneza naczelnych. *Przeglad Antropologiczny* 33:115–127.
Statistical analysis of the relation between the length of the genoma and the diploid number of the chromosomes in *Cercopithecus*. *Experientia* 23:672.
CHIARELLI, B., J. EGOZCUE
The meiotic chromosomes of two *Macaca. Caryologia* 20:339–346.
Observations on the nuclear morphology of the spermatogenetic cells of *Macaca. Fertility and Sterility* 19:812–818.
CHIARELLI, B., M. SARTI
"Les protrusions nucléaires des leucocytes polymorphonuclées dans les singes anthropomorphes (note preliminaire)," in *Neue Ergebnisse der Primatologie. Progress in primatology.* Edited by D. Stark, R. Schneider, and H. J. Kuhn, 171–176. Stuttgart: G. Fischer.
EGOZCUE, J.
New data on the chromosomes of Lemuridae. *Mammalian Chromosomes Newsletter* 8:3–4.
Chromosome variability in the Lemuridae. *American Journal of Physical Anthropology* 26(3):341–348.
Chromosome dimorphism in primates. *Folia Primatologica* 7:231–237.
Position of the centromere in the marked acrocentric chromosomes of primates. *Folia Primatologica* 7:238–242.
EGOZCUE, J., B. CHIARELLI
The idiogram of the lowland gorilla (*Gorilla gorilla gorilla*). *Folia Primatologica* 5:237–240.
EGOZCUE, J., F. HAGEMENAS
The chromosomes of the hooded spider monkey (*Ateles geoffroyi cucullatus*). *Mammalian Chromosomes Newsletter* 8:12–13.
EGOZCUE, J., M. VILARASAU DE EGOZCUE, F. HAGEMENAS
The chromosomes of two species of *Saimiri: Saimiri madeirae juruanus* and *Saimiri boliviensis nigriceps. Mammalian Chromosomes Newsletter* 8:14.
GOLDSTEIN, P. R., R. T. BIRDWELL
Cytogenetic studies in the purple-faced langur. *Mammalian Chromosomes Newsletter* 8:197–201.
IRSGLER, G.B.
Studies on the chromosomes of African mammals. IV. The karyotype of female chacma baboon. *South African Journal of Science* 63:527–531.
MANFREDI ROMANINI, M. G.
"Quantitative relative determination of the nuclear DNA in the Old World primates," in *Taxonomy and phylogeny of the Old World primates with refer-*

MANFREDI ROMANINI, M. G.—*Cont.*
ences to the origin of man. Edited by B. Chiarelli, 141–150. Turin: Rosenberg and Sellier.

MATTHEY, R.
Introduction to the fifth supplement of the list of chromosome numbers by eutherian mammals. *Mammalian Chromosomes Newsletter* 8:82–93.

RUZICSKA, P., T. C. HSU
The karyotype of the African green monkey (*Cercopithecus aethiops sabaceus*). *Texas Reports on Biology and Medicine* (supplement) 25:151–155.

STATON, D.
Karyotypes and the phylogeny of New World primates. *Mammalian Chromosomes Newsletter* 8:203–219.
A preliminary analysis of karyotypic evolution in catarrhines. *Mammalian Chromosomes Newsletter* 8:220–224.

STATON, D., E. M. ANDREWS
The chromosomes of the owl monkey (*Aotus trivirgatus*). *Mammalian Chromosomes Newsletter* 8.

TODD, N. B.
A theory of karyotypic fissioning, genetic potentiation and eutherian evolution. *Mammalian Chromosomes Newsletter* 8:268–279.

1968

CHIARELLI, B.
"Caryological and hybridological data for the taxonomy and phylogeny of the Old World primates," in *Taxonomy and phylogeny of Old World primates with references to the origin of man*. Edited by B. Chiarelli, 151–186. Turin: Rosenberg and Sellier.
Chromosome polymorphism in the species of the genus *Cercopithecus. Cytologia* 33:1–16.
From the karyotype of the apes to the human karyotype. *South African Journal of Science* 64:72–80.
"Chromosome changes in nonhuman primates," in *Use of nonhuman primates in drug evaluation*. Edited by H. Vagtborg, 172–183. Austin: University of Texas Press.
"Caryologie comparée et évolution chez les singes Catarrhina," in *VIIième Congrès International des Sciences Anthropologiques et Ethnologiques, Moscou, 1964,* volume three:441–449. Moscow: Izdatelstvo "Nauka".
Dati comparativi preliminari sui cromosomi meiotici in diacinesi di alcuni primati. *Atti Associazione Genetica Italiana* 13:183–192.
Dane kariologiczne i hybrydologiczne umozliwia jace rewizje taksonomii małp zwierzokstaltnych starego swiata na poziomie supergenetvcznym. *Przeglad Antropologiczny* 34:341–343.

CHIARELLI, B., J. EGOZCUE
The meiotic chromosomes of some primates. *Mammalian Chromosomes Newsletter* 9:85–86.

CHIARELLI, B., A. FALEK, K. J. BACK, G. T. COWART
Spermatogenesis in man. I. Nuclear morphology during spermatogenesis in man. *Fertility and Sterility* 19:758–765.

EGOZCUE, J.
The meiotic chromosomes of the lesser bushbaby (*Galago senegalensis*). *Mammalian Chromosomes Newsletter* 9:92–93.

EGOZCUE, J., B. CHIARELLI, M. SARTI-CHIARELLI
The somatic and meiotic chromosomes of *Cebuella Pygmaea* (Spix 1823) with special reference to the behavior of the sex chromosomes during spermatogenesis. *Folia Primatologica* 8:50–57.

EGOZCUE, J., B. CHIARELLI, M. SARTI-CHIARELLI, F. HAGEMENAS
Chromosome polymorphism in the tree shrew (*Tupaia glis*). *Folia Primatologica* 8:150–158.

EGOZCUE, J., F. HAGEMENAS, M. VILARASAU DE EGOZCUE
Chromosome studies in Catarrhinae monkeys. *Mammalian Chromosomes Newsletter* 9:65–68.

EGOZCUE, J., E. M. PERKINS, F. HAGEMENAS
Chromosomal evolution in marmosets, tamarins and pinchés. *Folia Primatologica* 9:81–94.

FALEK, A., B. CHIARELLI
Meiotic chromosomes of man. *American Journal of Physical Anthropology* 28:351–354.

LOW, R. J., K. BENIRSCHKE
Chromosome study of a marmoset hybrid. *Folia Primatologica* 8:180–191.

MANNA, G. K., M. TALUKDAR
Analysis of somatic chromosome complements of both sexes of two primates, the slender loris, *Loris tardigradus* and rhesus monkey, *Macaca mulatta*. *Mammalia* 32:118.

MATTHEY, R.
Chromosome numbers of eutherian mammals. *Mammalian Chromosomes Newsletter* 9:72–77.

1969

ARRIGHI, F. E.
Chromosomes of tree shrews (Tupaiidae). *Mammalian Chromosomes Newsletter* 10(1):27.

ARRIGHI, F. E., M. W. SORENSON, L. R. SHIRLEY
Chromosomes of the tree shrews (Tupaiidae). *Cytogenetics* 8:199–208.

BENIRSCHKE, K.
Discussion paper: cytogenetics contribution to primatology. *Annals of the New York Academy of Sciences* 162:217–224.

CHIARELLI, B.
The phylogeny of primates from a karyological point of view. *Acta Zoologica Pathologica Antwerpiensia* 48:11–20.

ECKHARDT, R. B.
A chromosome arm number index and its applications to the phylogeny and classification of lemurs. *American Journal of Physical Anthropology* 31:85–88.

EGOZCUE, J.
"Primates," in *Comparative mammalian cytogenetics*. Edited by K. Benirschke, 357–389. New York, Berlin, Heidelberg: Springer.
Meiosis in five *Macaca* species. *Folia Primatologica* 11:1–16.

EGOZCUE, J., E. M. PERKINS, F. HAGEMENAS
The chromosomes of *Saguinus fuscicollis illigeri* (Pucheran 1845) and *Aotus trivirgatus* (Humboldt 1811). *Folia Primatologica* 10:154–159.

EGOZCUE, J., E. M. PERKINS, F. HAGEMENAS, D. M. FORD
The chromosomes of some Platyrrhini *Callicebus, Ateles* and *Saimiri*). *Folia Primatologica* 11:17–27.

ELLIOT, O. S., M. WONG, D. S. BORGAONKAR
Karyological study of Tupaia from Thailand. *Journal of Heredity* 60:153–157.
HUANG, C. C., H. HABBIT, J. L. AMBRUS
Chromosomes and DNA synthesis in the stumptail monkey (*Macaca speciosa*), with special regard to marker and sex chromosomes. *Folia Primatologica* 11:28–34.
LOW, R. J., K. BENIRSCHKE
"The replicating pattern of the chromosomes of *Pan troglodytes*," in *Proceedings of the Second International Congress of Primatology, Atlanta, 1968*, volume two: 95–102. Basel: S. Karger.
MARKARYAN, D. S., M. G. MACHAVARIANI, M. V. AVJIAN
The normal karyotype of the green monkey (*Cercopithecus aethiops*). *Genetika* 5:143–148.
McCLURE, H. M., K. H. BELDEN, W. A. PIEPER
Autosomal trisomy in a chimpanzee: resemblance to Down's syndrome. *Science* 165:1,010–1,012.
RUMPLER, Y., R. ALBIGNAC
Existence d'une variabilité chromosomique intraspécifique chez certains lémuriens. *Comptes Rendus des Séances de la Société de Biologie et de ses Filiales* 163:1,989–1,992.
Etude cytogénétique de deux lémuriens, *Lemur macaco macaco,* Linné 1766, et *Lemur fulvus rufus* (Audebert 1800) et d'un hybride *macaco macaco/fulvus rufus. Comptes Rendus Hebdomadaires des Séances et Mémoires de la Société de Biologie* 163:1,247–1,250.
Etude cytogénétique de quelques hybrides intraspécifiques et interspécifiques de lémuriens. *Annales Scientifiques de l'Université de Besançon* 6:1–3.
WURSTER, D. H., K. BENIRSCHKE
Chromosomes of some primates. *Mammalian Chromosomes Newsletter* 10(1):3–5.

1970

BRANDOM, W. F., Y. GAUDE, M. W. PARROTT
Chromosome preparations from short-term bone marrow cultures of *Macaca mulatta. Mammalian Chromosomes Newsletter* 11:113–116.
CHIARELLI, B.
"The chromosomes of the chimpanzee," in *The chimpanzee*, volume two. Edited by G. H. Bourne, 254–264. Basel: S. Karger.
DE BOER, L. E. M.
Het karyotype van *Cercopithecus hamlyni* (Pocock 1907). *Genen en Phaenen* 13:99–103.
EGOZCUE, J.
The chromosomes of the lesser bushbaby (*Galago senegalensis*) and the greater bushbaby (*Galago crassicaudatus*). *Folia Primatologica* 12:236–240.
EGOZCUE, J., E. M. PERKINS
The chromosomes of Humboldt's woolly monkey (*Lagothrix lagothricha,* Humboldt 1812). *Folia Primatologica* 12:77–80.
FERNANDEZ-DONOSO, R., J. LINDSTEN, E. NORBY
The chromosomes of the cynomolgus macaque (*Macaca fascicularis*). *Hereditas* 65:269–276.
HÖLSI, P., E. M. LANG
Die Chromosomen des Zwergsiamang (*Symphalangus klossii*). *Schweizer Archiv für Tierheilkunde* 112(6):296–297.

Die Chromosomen des Kleideraffen (*Pygathrix nemaeus*). *Schweizer Archiv für Tierheilkunde* 112:338–339.

HSU, T. C., S. H. HAMPTON

Chromosomes of Callithricidae with special reference to an XX/"XO" sex chromosome system in Goeldi's marmoset (*Callimico goeldii* Thomas 1904). *Folia Primatologica* 13:183–195.

PASZTOR, L.

Antenatal diagnosis of sex chromatin examination. *Mammalian Chromosomes Newsletter* 11:119.

RAO, S. R., V. C. SHAH, S. ANANDA RAO, V. K. SHARMA

Male meiotic chromosomes of the slow loris *Nycticebus coucang*. *Mammalian Chromosomes Newsletter* 11:96–97.

RUFFIE, J., P. COLOMBIES, CH. GINOUX-MOUNIE, E. CARLES-TROCHAIN

Etude cytogénétique de 4 spécies de primates. Comparaison avec le caryotype humain. *Annales de Genetique* 13:1–8.

RUMPLER, Y.

Etude cytogénétique de *Lemur catta*. *Cytogenetics* 9:239–244.

RUMPLER, Y., R. ALBIGNAC

Evolution chromosomique des lémuriens malgaches. *Annales de l'Université du Madagascar* 12(13):3–11.

1971

BRUMBACK, R. A., R. D. STATON, S. A. BENJAMIN, C. M. LANG

The chromosomes of *Aotus trivirgatus* Humboldt 1812. *Folia Primatologica* 15:264–273.

CHIARELLI, B.

"Comparative cytogenetics in primates and its relevance for human cytogenetics," in *Comparative genetics in monkeys, apes and man*. Edited by B. Chiarelli, 273–308. London: Academic.

"New data for the comparison of the karyotype of the anthropoid apes with that of man," in *Proceedings of the Third International Congress of Primatology, Zurich, 1970*, volume two:104–109. Basel: S. Karger.

CHIARELLI, B., R. DOBLE

Revisione della posizione sistematica e filogenetica delle Hylobatinae su basi cariologiche. *Bollettino di Zoologia* 38:212.

DE BOER, L. E. M.

A case of abnormal chromosome number in a female *Aotes trivirgatus* (Humboldt 1811). (*Cebidae,* primates). *Genen en Phaenen* 14:51–55.

Structural heterozygosity of the marked chromosome in pair in a female *Cercocebus torquatus* Kerr 1792. *Genen en Phaenen* 14:91–96.

Cytology of the Cercopithecidae with special reference to the karyotype of *Cercopithecus hamlyni* Pocock 1907. *Proceedings of the Third International Congress of Primatology, Zurich, 1970,* volume two:121–126. Basel: S. Karger.

EGOZCUE, J.

A note on the chromosomes of *Aotus trivirgatus* Humboldt 1812. *Folia Primatologica* 15:274–276.

Evolution of marked chromosomes of primates. Experimental evidence for pericentric inversion. *Folia Primatologica* 16:276–281.

A possible case of chromosome fission in a primate. *Experientia* 27:969–970.

EGOZCUE, J., E. M. PERKINS

"Chromosomal evolution in the Platyrrhini," in *Proceedings of the Third*

EGOZCUE, J., E. M. PERKINS — *Cont.*
International Congress of Primatology, Zurich, 1970:131–134. Basel: S. Karger.

FONTANA, F., B. CHIARELLI
"The karyotypes of the baboon," in *Proceedings of the Third International Congress of Primatology, Zurich, 1970*, volume two:127–130. Basel: S. Karger.

KANOGAWA, H., E. S. E. HAFEZ, M. NAWAR
The chromosomes of the crab-eating macaque (*Macaca fascicularis*). *Folia Primatologica* 16:270–275.

LAMBERTI, L., L. BARBERIS
"The idiogram of different species of Cercopithecus," in *Proceedings of the Third International Congress of Primatology. Zurich, 1970,* volume two:116–120. Basel: S. Karger.

McCLURE, H. M., W. A. PIEPER, M. E. KEELING, C. B. JACOBSON
"Mongoloid-like condition in a chimpanzee," in *Proceedings of the Third International Congress of Primatology, Zurich, 1970:* 110–115. Basel: S. Karger.

RUMPLER, Y., R. ALBIGNAC
Etude cytogénétique de *Varecia variegata* et de *Lemur rubriventer*. *Comptes Rendus des Séances de la Société de Biologie et de ses Filiales* 165(3):741–745.

SCAPITTA, C., M. CHIARELLI-SARTI, W. C. OSMAN HILL
Investigations on the so-called Barr body in smears from the oral mucosa in *Pongo pygmaeus, Gorilla gorilla* and *Pan troglodytes. Bollettino di Zoologia* 38:171–175.

TOSCO, F., M. CHIARELLI-SARTI
Les protrusions nucléaires des leucocytes polymorphonuclées de l'homme, des singes anthropomorphes et de deux espèces de macaque. *Bollettino di Zoologia* 38:165–169.

YING, K. L., H. BUTLER
Chromosome polymorphism in the lesser bushbabies (*Galago senegalensis*). *Canadian Journal of Genetics and Cytology* 13:793–801.

1972

CHIARELLI, B.
New data on Prosimiae chromosomes. *Journal of Human Evolution* 1:61–63.
Comparative chromosome analysis between man and chimpanzee. *Journal of Human Evolution* 1:389–393.
"The karyotype of the gibbons," in *Gibbon and siamang,* volume one. Edited by D. M. Rumbaugh, 90–102. Basel: S. Karger.
Spéculations sur les chromosomes et la phylogenèse des singes catarrhiniens. *L'Anthropologie* 76:525–534.

CHIARELLI, B., C. C. LIN
Comparative fluorescent patterns in human and chimpanzee chromosomes. *Genen en Phaenen* 15(2–3):103–106.

CHU, E. W., A. S. RABSON
Chimerism in lymphoid cell culture line derived from lymph node of marmoset infected with *Herpes virus saimiri. Journal of the National Cancer Institute* 48:771–775.

DATTA, M.
Study of meiosis in the chimpanzee, *Pan troglodytes troglodytes* (Blumenbach 1779). *Cytologia* 37:207–216.

DE BOER, L. E. M.
Chromosome studies in some primates. *Mammalian Chromosomes Newsletter* 13:4–6.
The karyotype of *Galago alleni* Waterhouse 1873 (2n = 40) compared with that of *Galago senegalensis* Geoffroy 1796 (2n = 38) (primates, Prosimii: Galagidae). *Genetica* 43:183–189.
Marked chromosome associations in catarrhine monkeys, with a note on chromosome associations in other primate groups. *Journal of Human Evolution* 1:83–86.
The karyotype of *Galago demidovii* Fischer 1808 (2n = 58) (primates, Prosimii: Galagidae). *Genen en Phaenen* 15:19–22.

DE GROUCHY, J., C. TURLEAU, M. ROUBIN, M. KLEIN
Evolutions caryotypiques de l'homme et du chimpanzé. Etude comparative des topographies de bandes après dénaturation menagée. *Annales de Génétique* 15:79–84.

EBERLE, P.
Chromosomenbefunde bei Primaten und cytogenetische Gesichtspunkte in der Stammesgeschichte des Menschen. *Anthropologischer Anzeiger* 33:161–181.

EGOZCUE, J.
The chromosomes of lemurs. *Folia Primatologica* 17:171–176.
XX male *Presbytis entellus*. A retrospective study. *Folia Primatologica* 17:292–296.

MANFREDI ROMANINI, M. G.
Nuclear DNA content and area of primate lymphocytes as a cytotaxonomical tool. *Journal of Human Evolution* 1:23–40.

MANFREDI ROMANINI, M. G., L. E. M. DE BOER, B. CHIARELLI, S. TINOZZI MASSARI
DNA nuclear content in the cytotaxonomy of *Galago senegalensis* and *Galago crassicaudatus*. *Journal of Human Evolution* 1:473–476.

McCLURE, H. M.
Features of Down's-like syndrome in a chimpanzee. *American Journal of Pathology* 67:413–416.

SCHMAGER, J.
Cytotaxonomy and geographical distribution of the Papinae. *Journal of Human Evolution* 1:477–485.

TURLEAU, C., J. DE GROUCHY
Caryotypes de l'homme et du chimpanzé. Comparaison de la topographie des bandes. Mécanismes évolutifs possibles. *Comptes Rendus Hebdomadaires des Séances de l'Académie des Sciences* 274:2,355–2,357.
Entre le chimpanzé et l'homme: l'évolution des chromosomes. *La Recherche* 3:779–782.

TURLEAU, C., J. DE GROUCHY, M. KLEIN
Phylogénie chromosomique de l'homme et des primates hominiens (*Pan troglodytes, Gorilla gorilla* et *Pongo pygmaeus*). Essai de reconstitution du caryotype de l'ancêtre commun. *Annales de Génétique* 15:225–240.

CYTOTAXONOMY AND EVOLUTION

ARRIGHI, F. E., M. W. SORENSON, L. R. SHIRLEY
1969 Chromosomes of the tree shrews (Tupaiidae). *Cytogenetics* 8:199–208.

BARBERIS, L.
1967 "The idiograms of three species of the genus *Cercopithecus: C. talapoin, C. aethiops* and *C. mona,*" in *Neue Ergebnisse der Primatologie. Progress in primatology*. Edited by D. Stark, R. Schneider, and H. J. Kuhn, 155–159. Stuttgart: G. Fischer.

BAYLET, R., H. GRATTEPANCHE
1964a Sur les chromosomes des Cercopithecidae *Papio papio, Macaca mulatta, Cercopithecus aethiops, Erythrocebus patas. Comptes Rendus des Séances de la Société de Biologie et de ses Filiales* 158:1,382–1,385.
1964b Chromosomes de *Cercopithecus aethiops. Comptes Rendus Hebdomadaires des Séances de l'Académie des Sciences* 259:3,096–3,099.

BENDER, M. A., E. H. Y. CHU
1963 "The chromosomes of primates," in *Evolutionary and genetic biology of the primates,* volume one. Edited by J. Buettner-Janusch, 261–310. London, New York: Academic Press.

BENDER, M. A., L. E. METTLER
1958 Chromosome studies of primates. *Science* 128:186–190.
1960 Chromosome studies of primates. II. *Callithrix, Leontocebus* and *Callimico. Cytologia* 25:400–404.

BENIRSCHKE, K.
1969 Discussion paper: cytogenetics contribution to primatology. *Annals of the New York Academy of Sciences* 162:217–224.

BORGAONKAR, D. S.
1966 A list of chromosome numbers in primates. *Journal of Heredity* 57:60–64.
1967 Addition to the lists of chromosome numbers in the insectivores and primates. *Journal of Heredity* 58:211–213.

BRUMBACK, R. A., R. D. STATON, S. A. BENJAMIN, C. M. LANG
1971 The chromosomes of *Aotus trivirgatus* Humboldt 1812. *Folia Primatologica* 15:264–273.

CHIARELLI, B.
1958 Tavole cromosomiche dei primati. *Caryologia* 11:99–104.
1961 Chromosomes of the orangutan (*Pongo pygmaeus*). *Nature* 192:285.
1962a Some new data of the chromosomes of Catarrhina. *Experientia* 18:405.
1962b Comparative morphometric analysis of primate chromosomes. I. The chromosomes of anthropoid apes and of man. *Caryologia* 15(1):99–121.
1962c Karyological evolution in primates and origin of the human karyotype. *Atti Associazione Genetica Italiana* 7:284–285.
1962d Comparative morphometric analysis of primate chromosomes. II. The chromosomes of genera *Macaca, Papio, Theropithecus* and *Cercocebus*. *Caryologia* 15:401–420.
1963a Primi risultati di ricerche di genetica e cariologia comparata in primati e loro interesse evolutivo. *Rivista di Antropologia* 50:87–124.
1963b Comparative morphometric analysis of primate chromosomes. III. The chromosomes of the genera *Hylobates, Colobus* and *Presbytis. Caryologia* 16:637–648.
1964a Genetica comparata in primati e genetica umana. *Atti XI Convegno della Salute. Ferrara, 1964*:53–79.
1964b Incremento genomico nel genere *Cercopithecus. Archivio per l'Antropologia e la Etnologia* 94:27–29.
1965 Il cariotipo di *Nasalis larvatus* e la sua importanza nella citotassonomia

delle scimmie del vecchio mondo. Nota preliminare. *Rivista di Antropologia* 52:159–164.
1966a Caryology and taxonomy of the catarrhine monkeys. *American Journal of Physical Anthropology* 24(2):155–169.
1966b The chromosome complement of *Nasalis larvatus* (Wurm 1781). *Experientia* 22:797.
1966c Marked chromosome in catarrhine monkeys. *Folia Primatologica* 4:74–80.
1967a Chromosomy a taksonomia i filogeneza naczelnych. *Przeglad Antropologiczny* 33:115–127.
1967b Statistical analysis of the relation between the length of the genoma and the diploid number of the chromosomes in *Cercopithecus. Experientia* 23:672.
1967c La morfologia del cromosoma "Y" nelle differenti specie di primati. *Rivista di Antropologia* 54:137–140.
1967d "Somatic and meiotic chromosomes in the different species of Papio," in *The baboon in medical research,* volume two. Edited by H. Vagtborg, 271–281. Austin:University of Texas Press.
1967e "Caryological and hybridological data available for a taxonomic revision of the Old World primates at a supergenetic level," in *Neue Ergebnisse der Primatologie. Progress in primatology*. Edited by D. Stark, R. Schneider, and H. J. Kuhn, 160–163. Stuttgart: G. Fischer.
1968a "Caryological and hybridological data for the taxonomy and phylogeny of the Old World primates," in *Taxonomy and phylogeny of Old World primates with references to the origin of man.* Edited by B. Chiarelli, 151–186. Turin: Rosenberg and Sellier.
1968b From the karyotype of the apes to the human karyotype. *South African Journal of Science* 64:72–80.
1968c Chromosome polymorphism in the species of the genus *Cercopithecus. Cytologia* 33:1–16.
1968d "Caryologie comparée et évolution chez les singes Catarrhina," in *VIIième Congrès International des Sciences Anthropologiques et Ethnologiques, Moscou, 1964*, volume three:441–449. Moscow: Izdatelstvo "Nauka".
1968e "Chromosome changes in nonhuman primates," in *Use of nonhuman primates in drug evaluation*. Edited by H. Vagtborg, 172–183. Austin: University of Texas Press.
1968f Dane kariologiczne i hybrydologiczne umozliwia jace rewizje taksonomii malp zwierzokstaltnych starego swiata na poziomie supergenetvcznym. *Przeglad Antropologiczny* 34:341–343.
1969 The phylogeny of primates from a karyological point of view. *Acta Zoologica Pathologica Antwerpiensia* 48:11–20.
1970 "The chromosomes of the chimpanzee," in *The chimpanzee,* volume two. Edited by G. H. Bourne, 254–264. Basel: S. Karger.
1971a "Comparative cytogenetics in primates and its relevance for human cytogenetics", in *Comparative genetics in monkeys, apes and man.* Edited by B. Chiarelli, 273–308. London: Academic Press.
1971b "New data for the comparison of the karyotype of the anthropoid apes with that of man," in *Proceedings of the Third International Congress of Primatology, Zurich, 1970*, volume two:104–109. Basel: S. Karger.
1972a New data on Prosimiae chromosomes. *Journal of Human Evolution* 1:61–63.

CHIARELLI, B., — *Cont.*
1972b Comparative chromosome analysis between man and chimpanzee. *Journal of Human Evolution* 1:389–393.
1972c "The karyotype of the gibbons," in *Gibbon and siamang*, volume one. Edited by D. M. Rumbaugh, 90–102. Basel: S. Karger.
1972d Spéculations sur les chromosomes et la phylogenèse des singes catarrhiniens. *L'Anthropologie* 76:525–534.

CHIARELLI, B., L. DE CARLI
1959 Evoluzione delle conoscenze sui cromosomi umani e loro signifcato in antropologia. *Archivio per l'Antropologia e la Etnologia* 89:149–167.

CHIARELLI, B., C. C. LIN
1972 Comparative fluorescent patterns in human and chimpanzee chromosomes. *Genen en Phaenen* 15(2–3):103–106.

CHIARELLI, B., F. NUZZO, L. DE CARLI, B. AULISA
1960 Morphometric analysis of human chromosomes. *Atti Associazione Genetica Italiana* 5:265–273.

CHIARELLI, B., C. VACCARINO
1964 Cariologia ed evoluzione nel genere *Cercopithecus*. *Atti Associazione Genetica Italiana* 9:328–339.

CHU, E. H. Y., M. A. BENDER
1961a Chromosome cytology and evolution in primates. *Science* 133:1,399–1,405.
1961b Cytogenetics and evolution of primates. *Annals of the New York Academy of Sciences* 102:253–266.

CHU, E. H. Y., N. H. GILES
1957 A study of primate chromosome complements. *American Naturalist* 91:273–282.

CHU, E. H. Y., B. A. SWOMLEY
1961 Chromosomes of lemurine lemurs. *Science* 133:1,925–1,926.

DARLINGTON, C. D., A. HAGUE
1955 Chromosomes of monkeys and man. *Nature* 175:32.

DATTA, M.
1972 Study of meiosis in the chimpanzee, *Pan troglodytes troglodytes* (Blumenbach 1779). *Cytologia* 37:207–216.

DE BOER, L. E. M.
1970 Het karyotype van *Cercopithecus hamlyni* (Pocock 1907). *Genen en Phaenen* 13:99–103.
1971a Cytology of the Cercopithecidae with special reference to the karyotype of *Cercopithecus hamlyni* Pocock 1907. *Proceedings of the Third International Congress of Primatology, Zurich, 1970,* volume two:121–126. Basel: S. Karger.
1971b A case of abnormal chromosome number in a female *Aotes trivirgatus* (Humboldt 1811). (*Cebidae*, primates). *Genen en Phaenen* 14:51–55.
1971c Structural heterozygosity of the marked chromosome in pair in a female *Cercocebus torquatus* Kerr 1792. *Genen en Phaenen* 14:91–96.
1972a Marked chromosome associations in catarrhine monkeys, with a note on chromosome associations in other primate groups. *Journal of Human Evolution* 1:83–86.
1972b The karyotype of *Galago alleni* Waterhouse 1873 (2n = 40) compared

with that of *Galago senegalensis* Geoffroy 1796 (2n = 38) (primates, Prosimii: Galagidae). *Genetica* 43:183–189.

1972c The karyotype of *Galago demidovii* Fischer 1808 (2n = 58) (primates, Prosimii: Galagidae). *Genen en Phaenen* 15:19–22.

DE GROUCHY, J., C. TURLEAU, M. ROUBIN, M. KLEIN

1972 Evolutions caryotypiques de l'homme et du chimpanzé. Etude comparative des topographies de bandes après dénaturation menagée. *Annales de Génétique* 15:79–84.

EBERLE, P.

1972 Chromosomenbefunde bei Primaten und cytogenetische Gesichtspunkte in der Stammesgeschichte des Menschen. *Anthropologischer Anzeiger* 33:161–181.

ECKHARDT, R. B.

1969 A chromosome arm number index and its applications to the phylogeny and classification of lemurs. *American Journal of Physical Anthropology* 31:85–88.

EGOZCUE, J.

1967a Chromosome dimorphism in primates. *Folia Primatologica* 7:231–237.

1967b Chromosome variability in the Lemuridae. *American Journal of Physical Anthropology* 26(3):341–348.

1967c Position of the centromere in the marked acrocentric chromosomes of primates. *Folia Primatologica* 7:238–242.

1969 "Primates," in *Comparative mammalian cytogenetics*. Edited by K. Benirschke, 357–389. New York, Berlin, Heidelberg: Springer.

1970 The chromosomes of the lesser bushbaby (*Galago senegalensis*) and the greater bushbaby (*Galago crassicaudatus*). *Folia Primatologica* 12:236–240.

1971a A note on the chromosomes of *Aotus trivirgatus* Humboldt 1812. *Folia Primatologica* 15:274–276.

1971b Evolution of marked chromosomes of primates. Experimental evidence for pericentric inversion. *Folia Primatologica* 16:276–281.

1971c A possible case of chromosome fission in a primate. *Experientia* 27:969–970.

1972a The chromosomes of lemurs. *Folia Primatologica* 17:171–176.

1972b XX male *Presbytis entellus.* A retrospective study. *Folia Primatologica* 17:292–296.

EGOZCUE, J., B. CHIARELLI

1967 The idiogram of the lowland gorilla (*Gorilla gorilla gorilla*). *Folia Primatologica* 5:237–240.

EGOZCUE, J., B. CHIARELLI, M. SARTI-CHIARELLI

1968 The somatic and meiotic chromosomes of *Cebuella Pygmaea* (Spix 1823) with special reference to the behavior of the sex chromosomes during spermatogenesis. *Folia Primatologica* 8:50–57.

EGOZCUE, J., B. CHIARELLI, M. SARTI-CHIARELLI, F. HAGEMENAS

1968 Chromosome polymorphism in the tree shrew (*Tupaia glis*). *Folia Primatologica* 8:150–158.

EGOZCUE, J., E. M. PERKINS

1970 The chromosomes of Humboldt's woolly monkey (*Lagothrix lagothricha*, Humboldt 1812). *Folia Primatologica* 12:77–80.

1971 "Chromosomal evolution in the Platyrrhini," in *Proceedings of the Third International Congress of Primatology, Zurich, 1970*:131–134. Basel: S. Karger.

EGOZCUE, J., E. M. PERKINS, F. HAGEMENAS

1968 Chromosomal evolution in marmosets, tamarins and pinchés. *Folia Primatologica* 9:81–94.

EGOZCUE, J., E. M. PERKINS, F. HAGEMENAS

1969 The chromosomes of *Sanguinas fuscicollis illigeri* (Pucheran 1845) and *Aotus trivirgatus* (Humboldt 1811). *Folia Primatologica* 10:154–159.

EGOZCUE, J., E. M. PERKINS, F. HAGEMENAS, D. M. FORD

1969 The chromosomes of some Platyrrhini (*Callicebus, Ateles* and *Saimiri*). *Folia Primatologica* 11:17–27.

EGOZCUE, J., M. VILARASAU DE EGOZCUE

1966 The chromosome complement of the howler monkey (*Alouatta caraya* Humboldt 1812). *Cytogenetics* 5:20–27.

ELLIOT, O. S., M. WONG, D. S. BORGAONKAR

1969 Karyological study of Tupaia from Thailand. *Journal of Heredity* 60:153–157.

FARBER, P. A.

1966 Chromosomes of *Macaca mulatta* leukocytes. *Folia Primatologica* 4:409–415.

FERNANDEZ-DONOSO, R., J. LINDSTEN, E. NORBY

1970 The chromosomes of the cynomolgus macaque (*Macaca fascicularis*). *Hereditas* 65:269–276.

FONTANA, F., B. CHIARELLI

1971 "The karyotypes of the baboon," in *Proceedings of the Third International Congress of Primatology, Zurich, 1970*, volume two:127–130. Basel: S. Karger.

FORD, C. E., J. L. HAMERTON

1956 The chromosomes of man. *Nature* 178:1,020–1,023.

HABERLANDT, W. F.

1966 Chromosomenforschung bei Mensch und Primaten. *Arzneimittel Forschung* 11:573–578.

HAMERTON, J. L.

1963 Primate chromosomes. *Symposia of the Zoological Society of London* 10:211–219.

HAMERTON, J. L., M. FRACCARO, L. DE CARLI, F. NUZZO, H. P. KLINGER, *et al.*

1961 Somatic chromosomes of the gorilla. *Nature* 192:225–228.

HAMERTON, J. L., H. P. KLINGER, D. E. MUTTON, E. M. LANG

1963 The somatic chromosomes of the Hominoidea. *Cytogenetics* 2:240–263.

HÖLSI, P., E. M. LANG

1970a Die Chromosomen des Kleideraffen (*Pygathrix nemaeus*). *Schweizer Archiv für Tierheilkunde* 112(6):338–339.

1970b Die Chromosomen des Zwergsiamang (*Symphalangus klossii*). *Schweizer Archiv für Tierheilkunde* 112:296–297.

HSU, T. C., M. L. JOHNSON

1963 Karyotypes of two mammals from Malaya. *American Naturalist* 97:127–129.

HUANG, C. C., H. HABBIT, J. L. AMBRUS

1969 Chromosomes and DNA synthesis in the stumptail monkey (*Macaca speciosa*), with special regard to marker and sex chromosomes. *Folia Primatologica* 11:28–34.

IRSGLER, G. B.

1967 Studies on the chromosomes of African mammals. IV. The karyotype of female chacma baboon. *South African Journal of Science* 63:527–531.

ISMAIL, H., P. V. TOBIAS
1956 In *Chromosomes, sex-cells and evolution in a mammal*. Edited by P. V. Tobias. London: Lund.

KANOGAWA, H., E. S. E. HAFEZ, M. NAWAR
1971 The chromosomes of the crab-eating macaque (*Macaca fascicularis*). *Folia Primatologica* 16:270–275.

KLINGER, H. P.
1963 The somatic chromosomes of some primates (*Tupaia glis, Nycticebus coucang, Tarsius bancanus, Cercocebus aterrimus, Symphalangus syndactylus*). *Cytogenetics* 2:140–151.

KLINGER, H. P., J. L. HAMERTON, D. E. MUTTON, E. M. LANG
1963 The chromosomes of the Hominoidea. *Burg Wartenstein Symposium for Anthropological Research*. New York: Wenner-Gren Foundation.

LAMBERTI, L., L. BARBERIS
1971 "The idiogram of different species of Cercopithecus," in *Proceedings of the Third International Congress of Primatology, Zurich, 1970*, volume two:116–120. Basel: S. Karger.

LOW, R. J., K. BENIRSCHKE
1968 Chromosome study of a marmoset hybrid. *Folia Primatologica* 8:180–191.
1969 "The replicating pattern of the chromosomes of *Pan troglodytes*," in *Proceedings of the Second International Congress of Primatology, Atlanta, 1968*, volume two:95–102. Basel: S. Karger.

MAKINO, S.
1952 A contribution to the study of the chromosomes in some Asiatic mammals. *Cytologia* 16:288–301.

MANNA, G. K., M. TALUKDAR
1968 Analysis of somatic chromosome complements of both sexes of two primates, the slender loris, *Loris tardigradus* and rhesus monkey, *Macaca mulatta*. *Mammalia* 32:118.

MARKARYAN, D. S., M. G. MACHAVARIANI, M. V. AVJIAN
1969 The normal karyotype of the green monkey (*Cercopithecus aethiops*). *Genetika* 5:143–148.

MATTHEY, R.
1955 Les chromosomes de *Galago senegalensis* Geoffroy (Prosimii-Lorisidae-Galaginae). *Revue Suisse de Zoologie* (supplement) 62:190–197.

McCLURE, H. M.
1972 Features of Down's-like syndrome in a chimpanzee. *American Journal of Pathology* 67:413–416.

McCLURE, H. M., K. H. BELDEN, W. A. PIEPER
1969 Autosomal trisomy in a chimpanzee: resemblance to Down's syndrome. *Science* 165:1,010–1,012.

McCLURE, H. M., W. A. PIEPER, M. E. KEELING, C. B. JACOBSON
1971 "Mongoloid-like condition in a chimpanzee," in *Proceedings of the Third International Congress of Primatology, Zurich, 1970*:110–115. Basel: S. Karger.

PAINTER, T. S.
1922 The sex chromosomes of the monkeys. *Science* 56:286–287.
1924 Studies in mammalian spermatogenesis. IV. The sex chromosomes of monkeys. *Journal of Experimental Zoology* 39:433–462.

ROTHFELS, K. H., L. SIMINOVITCH
1958 The chromosome complement of the rhesus monkey (*Macaca mulatta*)

ROTHFELS, K. H., L. SIMINOVITCH — *Cont.*
determined in kidney cells cultivated "in vitro." *Chromosoma* 9:163–175.

RUFFIE, J., P. COLOMBIES, CH. GINOUX-MOUNIE, E. CARLES-TROCHAIN
1970 Etude cytogénétique de 4 spécies de primates. Comparaison avec le caryotype humain. *Annales de Genetique* 13:1–8.

RUMPLER, Y.
1970 Etude cytogénétique de *Lemur catta. Cytogenetics* 9:239–244.

RUMPLER, Y., R. ALBIGNAC
1969a Etude cytogénétique de quelques hybrides intraspécifiques et interspécifiques de lémuriens. *Annales Scientifiques de l'Université de Besançon* 6:1–3.
1969b Existence d'une variabilité chromosomique intraspécifique chez certains lémuriens. *Comptes Rendus des Séances de la Société de Biologie et de ses Filiales* 163:1,989–1,992.
1969c Etude cytogénétique de deux lémuriens, *Lemur macaco macaco*, Linné 1766, et *Lemur fulvus rufus* (Audebert 1800) et d'un hybride *macaco macaco/fulvus rufus. Comptes Rendus Hebdomadaires des Séances et Mémoires de la Société de Biologie* 163:1,247–1,250.

RUMPLER, Y., R. ALBIGNAC
1970 Evolution chromosomique des Lémuriens malgaches. *Annales de l'Université du Madagascar* 12(13):3–11.

RUMPLER, Y., R. ALBIGNAC
1971 Etude cytogénétique de *Varecia variegata* et de *Lemur rubriventer. Comptes Rendus des Séances de la Société de Biologie et de ses Filiales* 165(3):741–745.

RUZICSKA, P., T. C. HSU
1967 The karyotype of the African green monkey (*Cercopithecus aethiops sabaceus*). *Texas Reports on Biology and Medicine* (supplement) 25:151–155.

SCHMAGER, J.
1972 Cytotaxonomy and geographical distribution of the Papinae. *Journal of Human Evolution* 1:477–485.

SHARMA, T., S. KAKATI
1966 Karyotype of *Presbytis entellus entellus* (primate). *Cytologia* 32:111–116.

SHIWAGO, P. I.
1939 Recherches sur le caryotype du *Rhesus macacus. Bulletin de Biologie et de Médecine Expérimentale* (U.S.S.R.) 9:3–8.

SRIVASTAVA, P. K., A. K. SRIVASTAVA, F. V. LUCAS
1962 Somatic chromosomes of the squirrel monkey (*Saimiri sciureus*). *Primates* 10:171–180.

TIJO, H., A. LEVAN
1956 The chromosome number of man. *Hereditas* 42:1–6.

TURLEAU, C., J. DE GROUCHY
1972a Entre le chimpanzé et l'homme: l'évolution des chromosomes. *La Recherche* 3:779–782.
1972b Caryotypes de l'homme et du chimpanzé. Comparaison de la topographie des bandes. Mécanismes évolutifs possibles. *Comptes Rendus Hebdomadaires des Séances de l'Académie des Sciences* 274:2,355–2,357.

TURLEAU, C., J. DE GROUCHY, M. KLEIN
1972 Phylogénie chromosomique de l'homme et des primates hominiens

(*Pan troglodytes, Gorilla gorilla* et *Pongo pygmaeus*). Essai de reconstitution du caryotype de l'ancêtre commun. *Annales de Génétique* 15:225–240.

USHIJIMA, R. N., F. S. SHININGER, T. GRAND

1964 Chromosome complements of two species of primates: *Cynopithecus niger* and *Presbytis entellus. Science* 146:78–79.

WOHNUS, J. F., K. BENIRSCHKE

1966 Chromosome analysis of four species of marmosets (*Callithrix jacchus, Tamarinus mystax, Tamarinus nigricollis, Cebuella pygmaea*). *Cytogenetics* 5:94–105.

YEAGER, C. H., T. S. PAINTER, R. M. YERKES

1940 The chromosomes of the chimpanzee. *Science* 91:74–75.

YING, K. L., H. BUTLER

1971 Chromosome polymorphism in the lesser bushbabies (*Galago senegalensis*). *Canadian Journal of Genetics and Cytology* 13:793–801.

YOUNG, W. J., T. MERZ, M. A. FERGUSON-SMITH, A. W. JOHNSTON

1960 Chromosome number of the chimpanzee, *Pan troglodytes. Science* 131:1,672–1,673.

DNA NUCLEAR CONTENT

KLEINFELD, R., E. H. Y. CHU

1958 DNA determination of kidney cell cultures of three species of monkeys. *Cytologia* 23:452–459.

MANFREDI ROMANINI, M. G.

1967 "Quantitative relative determination of the nuclear DNA in the Old World primates," in *Taxonomy and phylogeny of the Old World primates with references to the origin of man*. Edited by B. Chiarelli, 141–150. Turin: Rosenberg and Sellier.

1972 Nuclear DNA content and area of primate lymphocytes as a cytotaxonomical tool. *Journal of Human Evolution* 1:23–40.

MANFREDI ROMANINI, M. G., L. E. M. DE BOER, B. CHIARELLI, S. TINOZZI MASSARI

1972 DNA nuclear content in the cytotaxonomy of *Galago senegalensis* and *Galago crassicaudatus. Journal of Human Evolution* 1:473–476.

HYBRIDS

CHIARELLI, B.

1961 Ibridologia e sistematica in primati. I. Raccolta di dati. *Atti Associazione Genetica Italiana* 6:213–220.

1966a Interesse tassonomico e filogenetico dei dati cariologici ed ibridologici nei primati del vecchio mondo. *Bollettino di Zoologia* 33:381–387.

1966b Ibridologia e sistematica in primati. Deduzioni sul cariotipo degli ibridi. *Rivista di Antropologia* 53:113–117.

1966c Dati cariologici e ibridologici per una revisione, al livello sopragenerico, della tassonomia delle scimmie del vecchio mondo. *Archivio per l'Antropologia e la Etnologia* 46:105–109.

MEIOTIC CHROMOSOMES

CHIARELLI, B.

1968 Dati comparativi preliminari sui cromosomi meiotici in diacinesi di alcuni primati. *Atti Associazione Genetica Italiana* 13:183–192.

CHIARELLI, B., J. EGOZCUE

1967a The meiotic chromosomes of two *Macaca. Caryologia* 20:339–346.

1967b Observations on the nuclear morphology of the spermatogenetic cells of *Macaca. Fertility and Sterility* 19:812–818.

CHIARELLI, B., A. FALEK, K. J. BACK, G. T. COWART

1968 Spermatogenesis in man. I. Nuclear morphology during spermatogenesis in man. *Fertility and Sterility* 19:758–765.

DATTA, M.

1972 Study of meiosis in the chimpanzee, *Pan troglodytes troglodytes* (Blumenbach 1779). *Cytologia* 37:207–216.

EGOZCUE, J.

1969 Meiosis in five *Macaca* species. *Folia Primatologica* 11:1–16.

FALEK, A., B. CHIARELLI

1968 Meiotic chromosomes of man. *American Journal of Physical Anthropology* 28:351–354.

CHIMERISM AND SEX CHROMOSOME ABNORMALITIES

BENIRSCHKE, K., J. M. ANDERSON, L. E. BROWNHILL

1962 Marrow chimerism in marmosets. *Science* 138:513–515.

BENIRSCHKE, K., L. E. BROWNHILL

1962 Further observations on marrow chimerism in marmosets. *Cytogenetics* 1:245–257.

BENIRSCHKE, K., L. E. BROWNHILL

1963 Heterosexual cells in testes of chimeric marmoset monkeys. *Cytogenetics* 2:231–240.

CHU, E. W., A. S. RABSON

1972 Chimerism in lymphoid cell culture line derived from lymph node of marmoset infected with *Herpes virus saimiri. Journal of the National Cancer Institute* 48:771–775.

GENGOZIAN, N., J. S. BATSON, P. EIDE

1964 Hematologic and cytogenetic evidence for hematopoietic chimerism in the marmoset *Tamarinus nigricollis. Cytogenetics* 3:384–393.

HSU, T. C., S. H. HAMPTON

1970 Chromosomes of Callithricidae with special reference to an XX/"XO" sex chromosome system in Goeldi's marmoset (*Callimico goeldii* Thomas 1904). *Folia Primatologica* 13:183–195.

SEX CHROMATIN AND DRUMSTICKS

BEATH, M. N., K. BENIRSCHKE

1962 The sex chromatin of interphase nuclei in platyrrhine monkeys. *Cytologia* 27:1–10.

CHIARELLI, B., L. BARBERIS
1964a Drumsticks in the leucocytes of primates. *Experientia* 20:679–681
1964b Il drumstick nei leucociti dei primati. *Caryologia* 17:567–573.

CHIARELLI, B., M. SARTI
1966 Nuclear protrusion in the polymorphonucleated leucocytes of the anthropoid apes. *Experientia* 22:652–653.
1967 "Les protrusions nucléaires des leucocytes polymorphonucléees dans les singes anthropomorphes (note preliminaire)," in *Neue Ergebnisse der Primatologie. Progress in primatology*. Edited by D. Stark, R. Schneider, and H. J. Kuhn, 171–176. Stuttgart: G. Fischer.

PARK, W. W.
1957 The occurrence of sex chromatin in early human and macaque embryos. *Journal of Anatomy* 91:369–373.

PRINCE, R. H., M. A. GRAHAM, M. L. BARR
1955 Nuclear morphology according to sex in *Macacus rhesus*. *Anatomical Record* 122:153–171.

SCAPITTA, C., M. CHIARELLI-SARTI, W. C. OSMAN HILL
1971 Investigations on the so-called Barr body in smears from the oral mucosa in *Pongo pygmaeus, Gorilla gorilla* and *Pan troglodytes*. *Bollettino di Zoologia* 38:171–175.

TOSCO, F., M. CHIARELLI-SARTI
1971 Les protrusions nucléaires des leucocytes polymorphonuclées de l'homme, des singes anthropomorphes et de deux espèces de macaque. *Bollettino di Zoologia* 38:165–169.

ABSTRACTS AND PRELIMINARY PUBLICATION IN *MAMMALIAN CHROMOSOMES NEWSLETTER*

ARRIGHI, F. E.
1969 Chromosomes of tree shrews (Tupaiidae). *Mammalian Chromosomes Newsletter* 10(1):27.

BENDER, M. A., P. E. EIDE
1962 Separation of monkey (*Ateles*) leucocytes for culture. *Mammalian Chromosomes Newsletter* 8:3.
1963 The karyotype of the woolly monkey (*Lagothrix*). *Mammalian Chromosomes Newsletter* 9:40.

BENIRSCHKE, K., L. E. BROWNHILL
1963 Mitotic chromosomes in some species of marmosets. *Mammalian Chromosomes Newsletter* 9:6.

BENIRSCHKE, K., L. E. BROWNHILL, R. LOW
1963 Chromosome complement of two further species of marmoset monkeys, *Tamarinus leucopus* and *Tamarinus mystax*. *Mammalian Chromosomes Newsletter* 10:83.

BORGAONKAR, D. S.
1965a Primate chromosome bibliography. *Mammalian Chromosomes Newsletter* 17:135–137.
1965b Primate chromosome bibliography (supplement). *Mammalian Chromosomes Newsletter* 18:184–185.
1966 A list of chromosome number in primates. *Mammalian Chromosomes Newsletter* 20:55–65.

BRANDOM, W. F., Y. GAUDE, M. W. PARROTT
1970 Chromosome preparations from short-term marrow cultures of *Macaca mulatta*. *Mammalian Chromosomes Newsletter* 11:113–116.
CHIARELLI, B.
1961a Cariologia e sistematica dei primati con cenni sull'origine del cariotipo umano. Comunicazione solo in riassunto pubblicata alla Società Italiana di Antropologia ed Etnologia nella riunione del 19 dicembre 1961.
1961b Some chromosome numbers in primates. *Mammalian Chromosomes Newsletter* 6:3–4.
1961c Comparative morphometric analysis of the primate chromosomes. *Mammalian Chromosomes Newsletter* 6:4–5.
1964 A pair of marked chromosomes in Catarrhina monkeys. *Mammalian Chromosomes Newsletter* 15:3–4.
1966a The chromosome complement of *Nasalis larvatus*. *Mammalian Chromosomes Newsletter* 20:68.
1966b Karyological and hybridological data available for a taxonomic revision of the Old-World monkeys at a supergenetic level. *Mammalian Chromosomes Newsletter* 22:212–215.
1966c Survey on the caryology of the genus *Cercopithecus*. *Mammalian Chromosomes Newsletter* 22:209–210.
1966d Summary of the karyology of Old-World primates. *Laboratory Primate Newsletter* 5:1–4.
CHIARELLI, B., L. BARBERIS
1966 Some data on the chromosomes of *Prosimiae* and of New-World monkeys. *Mammalian Chromosomes Newsletter* 22:216.
CHIARELLI, B., R. DOBLE
1971 Revisione della posizione sistematica e filogenetica delle Hylobatinae su basi cariologiche. *Bollettino di Zoologia* 38:212.
CHIARELLI, B., J. EGOZCUE
1968 The meiotic chromosomes of some primates. *Mammalian Chromosomes Newsletter* 9:85–86.
CHU, E. H. Y., N. H. GILES
1956 Primate chromosome complements determined by tissue culture techniques. *Records of the Genetic Society of America* 25:637.
DE BOER, L. E. M.
1972 Chromosome studies in some primates. *Mammalian Chromosomes Newsletter* 13:4–6.
EGOZCUE, J.
1966 Primate chromosome bibliography. *Mammalian Chromosomes Newsletter* 21:173.
1967 New data on the chromosomes of Lemuridae. *Mammalian Chromosomes Newsletter* 8:3–4.
1968 The meiotic chromosomes of the lesser bushbaby (*Galago senegalensis*). *Mammalian Chromosomes Newsletter* 9:92–93.
EGOZCUE, J., F. HAGEMENAS
1967 The chromosomes of the hooded spider monkey (*Ateles geoffroyi cucullatus*). *Mammalian Chromosomes Newsletter* 8:12–13.
EGOZCUE, J., F. HAGEMENAS, M. VILARASAU DE EGOZCUE
1968 Chromosome studies in Catarrhinae monkeys. *Mammalian Chromosomes Newsletter* 9:65–68.
EGOZCUE, J., R. N. USHIJIMA, M. VILARASAU DE EGOZCUE
1966a The chromosomes of the slender loris (*Loris tardigradus*). *Mammalian Chromosomes Newsletter* 22:204.

EGOZCUE, J., R. N. USHIJIMA, M. VILARASAU DE EGOZCUE
1966b The chromosomes of mongoose lemur (*Lemur mongoz*). *Mammalian Chromosomes Newsletter* 22:197–198.

EGOZCUE, J., M. VILARASAU DE EGOZCUE
1965 The chromosomes of the howler monkey (*Alouatta caraya*). *Mammalian Chromosomes Newsletter* 17:84.

EGOZCUE, J., M. VILARASAU DE EGOZCUE
1966a The chromosomes of the slow loris (*Nycticebus concang*). *Mammalian Chromosomes Newsletter* 20:50.
1966b The chromosomes of *Cebus albifrons*. *Mammalian Chromosomes Newsletter* 20:65–66.
1966c The chromosomes of the angwantibo (*Arctocebus calabarensis*). *Mammalian Chromosomes Newsletter* 20:53–54.
1966d The chromosomes of *Cebus capucinus*. *Mammalian Chromosomes Newsletter* 20:71–72.

EGOZCUE, J., M. VILARASAU DE EGOZCUE, F. HAGEMENAS
1967 The chromosomes of two species of *Saimiri: Saimiri madeirae juruanus* and *Saimiri boliviensis nigriceps*. *Mammalian Chromosomes Newsletter* 8:14.

FARBER, P. A.
1966 The karyotype of *Cercopithecus pygerythrus*. *Mammalian Chromosomes Newsletter* 22:205.

GOLDSTEIN, P. R., R. T. BIRDWELL
1967 Cytogenetic studies in the purple-faced langur. *Mammalian Chromosomes Newsletter* 8:197–201.

MANNA, G. K., M. TALUKDAR
1965 Somatic chromosome number in twenty species of mammals from India. *Mammalian Chromosomes Newsletter* 17:77–78.

MATTHEY, R.
1962 Les nombres diploides des mammifères euthériens. Liste critique. *Mammalian Chromosomes Newsletter* 8:17–25.
1967 Introduction to the fifth supplement of the list of chromosome numbers by eutherian mammals. *Mammalian Chromosomes Newsletter* 8:82–93.
1968 Chromosome numbers of eutherian mammals. *Mammalian Chromosomes Newsletter* 9:72–77.

METTLER, L., M. A. BENDER
1958 Chromosome studies and their evolutionary significance in new world primates. *Proceedings of the Tenth International Congress of Genetics*, volume two:186.

PASZTOR, L.
1970 Antenatal diagnosis of sex chromatin examination. *Mammalian Chromosomes Newsletter* 11:119.

RAO, S. R., V. C. SHAH, S. ANANDA RAO, V. K. SHARMA
1970 Male meiotic chromosomes of the slow loris *Nycticebus coucang*. *Mammalian Chromosomes Newsletter* 11:96–97.

SHARMA, T., S. KAKATI
1966 Chromosomes of Indian common langur. *Mammalian Chromosomes Newsletter* 20:70.

STATON, D. A.
1967a A preliminary analysis of karyotypic evolution in catarrhines. *Mammalian Chromosomes Newsletter* 8:220–224.
1967b Karyotypes and the phylogeny of New World primates. *Mammalian Chromosomes Newsletter* 8:203–219.

STATON, D., E. M. ANDREWS
1967 The chromosomes of the owl monkey (*Aotus trivirgatus*). *Mammalian Chromosomes Newsletter* 8.
TODD, N. B.
1967 A theory of karyotypic fissioning, genetic potentiation and eutherian evolution. *Mammalian Chromosomes Newsletter* 8:268–279.
WURSTER, D. H., K. BENIRSCHKE
1969 Chromosomes of some primates. *Mammalian Chromosomes Newsletter* 10(1):3–5.

Biographical Notes

M. Fathi Abdel-Hameed (1938–) was born in Tanta, Egypt. He studied agricultural sciences and poultry production at Alexandria University, where he received his B.S. and M.S. in 1958 and 1961 respectively. He was awarded a Fulbright Fellowship to study environmental horticulture and genetics at the University of California at Davis, where he received his M.S. and Ph.D. in 1963 and 1967 respectively. He has engaged in postdoctoral research on the ultrastructure and function of chromosomes at Argonne National Laboratory and in medical genetics and cancer cytogenetics at the University of Kentucky Medical Center. He is now an associate professor of biological sciences at Northern Illinois University and has published many articles in *Science* and other reputed journals.

R. Albignac. No biographical data available.

Giuseppe Ardito (1945–) took his degree in biological sciences at the University of Turin. After a fellowship assignment in anthropology he received postgraduate training in cytogenetics and cytotaxonomy. He also carried out several studies in the field of epigenetic characters in both humans and nonhuman primates. Since 1975 he has been associate professor of anthropology at the University of Turin.

Torbjörn Caspersson (1910–) completed his academic studies at the Karolinska Institutet (the medical faculty of the University of Stockholm). He received his M.D. in 1936 and the same year became docent in biophysics. He has lectured in biochemistry and was given the chair in medical cell research and genetics at Karolinska Institutet in 1944. The same year a special research institute was created (the Nobel Institute for

Medical Cell Research and Genetics), later supplemented with a department for experimental cytology (the Wallenberg Laboratory of Experimental Cell Research), of which institutions he is the director. He has since then received numerous professional honors and appointments.

BRUNETTO CHIARELLI (1934–) was born in Florence, Italy. He received degrees in natural sciences in 1957 and in biological sciences in 1960, and his Ph.D. in anthropology in 1964 and in general biology and zoology in 1965. Since 1969 he has been full professor of primatology at the University of Turin. There he has also, since 1962, been Director of the Institute and Museum of Anthropology. His publications include works on comparative genetics, cytogenetics, and taxonomy of primates. He has been editor in chief of the *Journal of Human Evolution*, published by Academic Press of London, since 1972.

P. COLOMBIES. No biographical data available.

JACQUES CONSTANS (1943–) received his doctorate in pharmaceutical sciences from the University of Toulouse in 1967 and presented his thesis for the Docteur de l'Université in 1969, with specializations in biochemistry and hematology. From 1968 to 1973 his research was devoted to human adaptation to high altitude (Peru and Bolivia) and to the genetics of Amerindian populations. In 1947, he began working for his Ph.D. in anthropology, studying genetic polymorphisms in the populations of the Pyrenees. He continues his laboratory investigations of serum polymorphisms (Gc, Hp, Transferrin, α_1 antitrypsin) and the epidemiological considerations in the geographical distribution of these factors at the Center of Hemotypology at Toulouse.

RORY DANIEL. No biographical data available.

MADHURI DATTA (1939–) was born in Calcutta, India. He received his M.S. in zoology from Calcutta University and his Ph.D. from Kalyani University. After lecturing for several years in zoology at Bethune and Lady Brabourne College in Calcutta, he went as a National Institutes of Health postdoctoral research assistant to Dartmouth Medical School. Since 1972 he has been an assistant professor of zoology at the University of Kalyani and since 1974 also at Bidhan Chandra Agriculture University. His fields of special interest are vertebrate and invertebrate karyology, cytotaxonomy, and evolution.

JEAN DE GROUCHY (1926–) received his medical degree in Paris and is now Directeur de Recherche at the Centre National de la Recherche Scientifique. He is head of the Department of Human and Compared

Cytogenetics at the Hospital of Enfants-Malades. He is author of some 300 papers on cytogenetics and has written many books, the latest of which is *Clinical atlas of human chromosomes*, coauthored by Catherine Turleau.

GIAN FRANCO DESTEFANO (1939–) was born in Turin, Italy. He studied at Pavia University where he received his degree in biological sciences in 1965 and his diploma of specialization in biophysics in 1973. Since 1971 he has been professor of human palaeontology and of human anatomy in the Faculty of Mathematical, Physical, and Natural Sciences of Pavia University and since 1976 professor of anthropometry at Rome University. His main fields of interest are the study of nuclear DNA quantitative variations in relation to primate evolution and the study of the mechanisms of somatic and genetic differentiation in isolated human populations.

ROBERT B. ECKHARDT (1942–) received his B.S. from Rutgers University in 1964. He then attended the University of Michigan, where he received an M.S. in human genetics (1966), an M.A. in anthropology (1967), and a Ph.D. jointly in anthropology and human genetics (1971). He was a lecturer in anthropology at the University of Michigan from 1968–1971, and at present is an assistant professor of anthropology at the Pennsylvania State University. His subject area is primate and human population biology, with a major research interest in rates of evolution, particularly of polygenic characteristics.

ARTHUR FALEK (1924–) is professor of psychiatry and associate professor of anatomy at Emory University and is Director of the Laboratory of Human and Behavioral Genetics at the Georgia Mental Health Institute in Atlanta, Georgia. He received his Ph.D. in human genetics from Columbia University in New York and continued to work there at the Department of Medical Genetics at the New York State Psychiatric Institute. His area of interest is the early detection of genetic disorders and, in this field, he has conducted studies at the morphological, biochemical, and cytogenetic levels. One aspect of his cytogenetic studies has been work in meiotic chromosomes of man and, in particular, the identification of the centromere position.

G. S. GARG. No biographical data available.

LINDA M. GLUESING. No biographical data available.

KENNETH W. JONES is a reader in genetics at the University of Edinburgh. He was educated in Wales at the University of Wales, Aberystwyth, where he took both his B.Sc. and Ph.D. Early work involved research on tumor biology from the cytogenetic standpoint and on human congenital karyotypic abnormalities with Charles Ford F.R.S. at Harwell. This stimulated an interest in normal developmental processes which was pursued under C. H. Waddington F.R.S. A technical development, that of *in situ* nucleic acid hybridization, which resulted from these studies in 1969, has shaped the direction of subsequent work, which has involved interests ranging from viral oncology and human genetic disease to primate evolution.

ANN L. KOEN (1922–) is associate professor of gynecology/obstetrics at Wayne State University in Detroit, Michigan. Born in Detroit, she received her education at the University of Detroit (B.S.) and Wayne University (M.S. and Ph.D.). Before coming to Wayne State, she served as Research Associate at Queen's University in Kingston, Ontario, Canada, and as Director of both the Research Laboratory of Hawthorn Center, Northville, Michigan and the Research Laboratory of the Hawthorn-Plymouth Research Unit, Northville, Michigan.

C. C. LIN (1933–) was born in China. He received his B.S. from the National Taiwan University. He came to Canada in 1962 and received his M.S. in cytogenetics and his Ph.D. in genetics from the University of Manitoba. After completing his postdoctoral training in the City of Hope National Medical Center in the United States he returned to Canada. At present he is an associate professor in the Division of Pediatrics and Medical Biochemistry, Faculty of Medicine, at the University of Calgary. His main research interest is in the nature and function of mammalian chromosomes.

BODIL ELIZABETH LOMHOLT (1932–) is an associate professor at the Institute of Medical Genetics, University of Copenhagen, Denmark. She received her Ph.D. in biochemistry in 1965 from the University of Copenhagen. From 1966 to 1969 she was supported by a three-year University of Copenhagen grant. During this period she worked with enzyme polymorphisms in relation to taxonomic studies at the Institute of Zoology, University of Copenhagen. This work was further supported by public University funds for twelve months. In 1970 she was employed at the Institute of Medical Genetics. Since 1971 she has occupied her present position. Her work and publications include enzyme polymorphisms and structural studies of mitotic chromosomes.

L. MALASPINA. No biographical data available.

MARIA GABRIELLA MANFREDI ROMANINI (1924–) was born in Piacenza and studied at the University of Pavia, where she received her degrees in biological sciences and in natural sciences. In 1954 she qualified for university teaching in comparative anatomy, and since 1956 she has been professor of anthropology. Since 1957 she has been Director of the Institute of Histology, Embryology and Anthropology at the University of Pavia. As professor of histology and embryology she is mainly concerned with the application of cytotopochemical methods to the study of nuclear chromatin; in particular she studies the quantitative variations of nuclear DNA and the respective philogenetic implications in mammals and particularly in primates.

J. C. QUILICI. No biographical data available.

EMMA RABINO MASSA (1939–) was born in Turin, Italy. She received her Ph.D. in natural science from the University of Turin. She is currently associate professor of general biology at the Institute of Anthropology of Turin. Her scientific publications include works on the biology of early populations, with particular regard to the histology of Egyptian mummies; cytogenetics, with particular regard to comparative kariology between primates and man; and the biology of human isolate populations.

YVES RUMPLER (1938–) studied medicine in Strasbourg, where he received his M.D. in 1963. From 1968 to 1978 he was professor in the Faculty of Medicine of Tananarive, Madagascar, where he worked especially on lemur cytogenetics and evolution, in which areas he received his Ph.D. in Paris in 1977. Now he is professor of embryology and cytogenetics in the Faculty of Medicine of the University of Strasbourg.

T. SHARMA. No biographical data available.

DUSHAN SOUDEK (1920–) was educated at the University of Brno, Czechoslovakia, where he received his M.D. and Ph.D. and became docent. He was originally interested in cell biology and later switched to medical cytogenetics. In the years 1962 through 1968, he was head of the Genetics Department of the Research Institute of Pediatrics in Brno. He has lived in Canada since 1968. He is the Ontario Mental Health Foundation Research Associate and an associate professor of Queen's University in Kingston, Ontario. His main interests are chromosomal errors causing mental retardation and abnormal behavior, chromosomal polymorphism, and chromosomal evolution in man.

CATHERINE TURLEAU (1944–) received her medical degree from the University of Paris and is currently Chargé de Recherche at the Centre

National de la Recherche Scientifique. A collaborator of Dr. Jean de Grouchy, she was the first to show the types of chromosomal changes which occurred in the evolution of the primates. She is coauthor of *Clinical atlas of human chromosomes.*

Index of Names

Index of Subjects